Sensing Techniques for Next Generation Cognitive Radio Networks

Ashish Bagwari
Uttrakhand Technical University, India

Jyotshana Bagwari
Uttrakhand Technical University, India

Geetam Singh Tomar
THDC Institute of Hydropower Engineering and Technology, India

A volume in the Advances in Wireless Technologies and Telecommunication (AWTT) Book Series

IGI Global
DISSEMINATOR OF KNOWLEDGE

Published in the United States of America by
IGI Global
Information Science Reference (an imprint of IGI Global)
701 E. Chocolate Avenue
Hershey PA, USA 17033
Tel: 717-533-8845
Fax: 717-533-8661
E-mail: cust@igi-global.com
Web site: http://www.igi-global.com

Library of Congress Cataloging-in-Publication Data

Names: Bagwari, Ashish, editor. | Bagwari, Jyotshana, editor. | Tomar, Geetam, editor.
Title: Sensing techniques for next generation cognitive radio networks / Ashish Bagwari, Jyotshana Bagwari, and Geetam Singh Tomar, editors.
Description: Hershey, Pa : Information Science Reference, [2018] | Includes bibliographical references.
Identifiers: LCCN 2017041201| ISBN 9781522553540 (hardcover) | ISBN 9781522553557 (ebook)
Subjects: LCSH: Cognitive radio networks.
Classification: LCC TK5103.4815 .S465 2018 | DDC 621.384--dc23 LC record available at https://lccn.loc.gov/2017041201

This book is published in the IGI Global book series Advances in Wireless Technologies and Telecommunication (AWTT) (ISSN: 2327-3305; eISSN: 2327-3313)

British Cataloguing in Publication Data
A Cataloguing in Publication record for this book is available from the British Library.

For electronic access to this publication, please contact: eresources@igi-global.com.

Advances in Wireless Technologies and Telecommunication (AWTT) Book Series

Xiaoge Xu
Xiamen University Malaysia, Malaysia

ISSN:2327-3305
EISSN:2327-3313

Mission

The wireless computing industry is constantly evolving, redesigning the ways in which individuals share information. Wireless technology and telecommunication remain one of the most important technologies in business organizations. The utilization of these technologies has enhanced business efficiency by enabling dynamic resources in all aspects of society.

The **Advances in Wireless Technologies and Telecommunication Book Series** aims to provide researchers and academic communities with quality research on the concepts and developments in the wireless technology fields. Developers, engineers, students, research strategists, and IT managers will find this series useful to gain insight into next generation wireless technologies and telecommunication.

Coverage

- Wireless Technologies
- Wireless Sensor Networks
- Mobile Technology
- Global Telecommunications
- Grid Communications
- Network Management
- Radio Communication
- Digital Communication
- Cellular Networks
- Mobile Communications

IGI Global is currently accepting manuscripts for publication within this series. To submit a proposal for a volume in this series, please contact our Acquisition Editors at Acquisitions@igi-global.com or visit: http://www.igi-global.com/publish/.

The Advances in Wireless Technologies and Telecommunication (AWTT) Book Series (ISSN 2327-3305) is published by IGI Global, 701 E. Chocolate Avenue, Hershey, PA 17033-1240, USA, www.igi-global.com. This series is composed of titles available for purchase individually; each title is edited to be contextually exclusive from any other title within the series. For pricing and ordering information please visit http://www.igi-global.com/book-series/advances-wireless-technologies-telecommunication/73684. Postmaster: Send all address changes to above address.

Titles in this Series

For a list of additional titles in this series, please visit: www.igi-global.com/book-series

Centrality Metrics for Complex Network Analysis Emerging Research and Opportunities
Natarajan Meghanathan (Jackson State University, USA)
Information Science Reference • copyright 2018 • 183pp • H/C (ISBN: 9781522538028) • US $175.00 (our price)

Mobile Technologies and Socio-Economic Development in Emerging Nations
Fredrick Japhet Mtenzi (Dublin Institute of Technology, Ireland & Aga Khan University, Tanzania) George S. Oreku (Open University of Tanzania (OUT), Tanzania) Dennis M. Lupiana (Institute of Finance Management, Tanzania) and Jim James Yonazi (Tanzania Standard (Newspapers) Limited, Tanzania)
Information Science Reference • copyright 2018 • 366pp • H/C (ISBN: 9781522540298) • US $195.00 (our price)

Affordability Issues Surrounding the Use of ICT for Development and Poverty Reduction
Sam Takavarasha Jr. (University of Fort Hare, South Africa & University of Zimbabwe, Zimbabwe) and Carl Adams (University of Portsmouth, UK)
Information Science Reference • copyright 2018 • 319pp • H/C (ISBN: 9781522531791) • US $195.00 (our price)

Handbook of Research on Environmental Policies for Emergency Management and Public Safety
Augustine Nduka Eneanya (University of Lagos, Nigeria)
Engineering Science Reference • copyright 2018 • 393pp • H/C (ISBN: 9781522531944) • US $295.00 (our price)

Positioning and Navigation in Complex Environments
Kegen Yu (Wuhan University, China)
Information Science Reference • copyright 2018 • 577pp • H/C (ISBN: 9781522535287) • US $195.00 (our price)

Examining Cloud Computing Technologies Through the Internet of Things
Pradeep Tomar (Gautam Buddha University, India) and Gurjit Kaur (Gautam Buddha University, India)
Information Science Reference • copyright 2018 • 311pp • H/C (ISBN: 9781522534457) • US $215.00 (our price)

Advanced Mobile Technologies for Secure Transaction Processing Emerging Research and Opportunities
Raghvendra Kumar (LNCT Group of Colleges, India) Preeta Sharan (The Oxford College of Engineering, India) and Aruna Devi (Surabhi Software, India)
Information Science Reference • copyright 2018 • 177pp • H/C (ISBN: 9781522527596) • US $130.00 (our price)

IGI Global
DISSEMINATOR OF KNOWLEDGE

701 East Chocolate Avenue, Hershey, PA 17033, USA
Tel: 717-533-8845 x100 • Fax: 717-533-8661
E-Mail: cust@igi-global.com • www.igi-global.com

Dedicated to our real God, our parents, Baba Kedarnath, and Anshul, Anushka, and Agrim.

Table of Contents

Detailed Table of Contents

The major hurdle that restricts the efficient utilization of radio spectrum is its inflexible and rigid licensing scheme. This constraint limits the emerging wireless systems to operate in static spectrum band. This deficiency compelled the innovators to introduce new paradigm (i.e., cognitive radio [CR] technique to boost productive spectrum utilization). In this chapter, the authors briefly introduce the CR to exploit wireless spectrum opportunistically. Cognitive radio network architecture is also discussed in detail. Authors also mentioned the software-defined radio (SDR) platform as a support to regulate spectrum reuse. IEEE 802.22, the first worldwide standard for wireless regional area network (WRAN), based of CR technology is presented in descriptive way along with its capacity, coverage, MAC layer frame structure. It has been observed that CR strikes its presence in today's modern wireless systems, applications, various projects, and trends, and become their desire.

Cognitive radio (CR) offers a novel way for effective usage of wireless spectrum by using dynamic spectrum sensing and allocation. One of the main components of CR is to find a spectrum hole for data transmission. Spectrum hole can be found by using spectrum sensing, a geolocation database, or by using a beacon signal. In this chapter, the authors describe algorithms for spectrum sensing in the presence of both additive white Gaussian and colored Gaussian noise. The algorithms include blind, non-blind, and cooperative sensing-based methods. The authors have compared the performance of various methods for IEEE 802.22 standard (which is the first standard incorporating CR).

Section 2
Overview of Various Sensing Schemes for Cognitive Radio Networks

the unlicensed users can use the underutilized spectrum without causing any harmful interference to the licensed users. Its key domains are sensing, cognition, and adaptation. The spectrum sensing problem is one of the most challenging issues in cognitive radio systems to detect the available frequency bands. This chapter introduces the concepts of various transmitter detection techniques, namely energy detection, matched filter detection, and cyclostationary feature detection. The chapter also discusses other sensing techniques that are introduced to enhance the detection performance of the conventional energy detector. Additionally, the introduced sensing techniques are implemented using extensive MATLAB simulations and their performances are evaluated and compared in terms of sensing time, detection sensitivity, and ease of implementation. The implementation is based on BPSK and QPSK modulation schemes under various SNR values for AWGN noisy channel with Rayleigh fading.

Cognitive radio is the technology used to solve the problem of spectrum underutilization by performing spectrum sensing, spectrum management, spectrum sharing, and spectrum mobility. The primary goal of cognitive radio is open spectrum sharing. Spectrum is a scarce and valuable natural resource that has to be used very effectively. The static allocation of spectrum to the licensed users will lead to wastage of resources when the spectrum is unused by the licensed user. Spectrum sensing methodology helps in detecting the spectrum holes and enables the unlicensed users to access the unused bands in the licensed spectrum effectively without interfering the licensed users. Cognitive thinking takes wireless communication to the next level by sensing the electromagnetic environment and dynamically adjusts its operating parameters in order to achieve maximum throughput, mitigate interference, facilitate interoperability, etc. The chapter presents the basics of cognitive radio networks, its architecture, its application, and advantages of cognitive radio networks.

Cognitive radio has come a long way in the recent years with the advent of improved algorithms and instrumentation. However, for ease and efficient working of cognitive radio, there is a need to have effective detection of spectrum sensing. The objective of spectrum sensing techniques is to find spectrum holes which can be accessible by the users of cognitive radio. The deployment of suitable sensing techniques reduces undesirable congestion in traffic and enhancement of spectrum usage. All these require sensing techniques whose main goal is oriented towards efficient identification and subsequent deployment of spectrum. This chapter is aimed to give a brief overview of some spectrum sensing techniques. An attempt is made to give the characteristics of the highly deployable sensing schemes. Accordingly, the merits and demerits are comprehensively highlighted. Further, emphasis has been given to relevant future challenges.

Spectrum sensing plays an important role in cognitive radio networks. Based on the bandwidth considered for spectrum sensing, it can be classified into narrowband and wideband spectrum sensing. Traditional spectrum sensing methods are devised for narrow band as it focuses on narrow frequency a range that is the channel bandwidth is lesser than the coherence bandwidth of the channel. To provide more spectral opportunity to cognitive user and to increase the throughput, cognitive radio network needs techniques that exploits spectral opportunities over a wide frequency range. Wideband spectrum sensing techniques aim to sense a channel bandwidth that exceeds the coherence bandwidth of the channel. Narrowband sensing techniques cannot be directly employed to perform wideband spectrum sensing as they make a single binary decision. In this chapter, the advanced spectrum sensing techniques and their taxonomy are discussed in detail.

5G is one of the newest and most acclaimed engineering technology of wireless era. This technology is also known as CRNs, in which spectrum detection plays a key role. Through this chapter, the authors explore multiple antennas-based improved sensing detector (MA_ISD). In the said scheme by using adaptive threshold and multiple antennas patterns, its easy to mitigate sensing failure issue enriches reliability. The presented scheme uses two detectors (TD) concept. The concept of two detectors (i.e., TD scheme has been applied in which multiple antennas are used for electing the quality signals). The considered model upgrades the detection operation and acquires limited or inferior detection time. The first energy detector uses a single adaptive threshold (ED-SAT) while another energy detector employs two adaptive thresholds (ED-TAT). The threshold value is accommodative as it relays on noise variance (σ_ω^2), the behavior of noise variance transforms in accordance with noise signal. Both the detectors run collaboratively and their gain is then supply to a decision device which operates OR functions. In this research work, results shows that with cooperation of two antennas (Nr=2) in multiple antennas-based improved sensing detector (MA_ISD) technique delivers enhancement in detection performance by 24.6%, 53.4%, 37.9%, and 49.6%, in contrast to existing schemes (i.e., EDT-ASS-2015 scheme, ED and cyclo-2010, adaptive SS-2012, and conventional-ED) scheme at -12dB SNR, respectively. During the time, proposed technique also reduces the sensing time in the order of 47.0 ms, 49.0 ms, and 53.2 ms as compared to existing schemes (EDT-ASS-2015, Adaptive SS-2012, and ED and Cyclo-2010) scheme at - 20 dB SNR, respectively.

lower than the Nyquist rate and still be reconstructed with high accuracy. Different sparsifying bases for compressive sensing are presented in this chapter and their performance is compared. The design of these matrices is based on different types of wavelet transforms, including the discrete wavelet transform and the stationary wavelet transform; the latter having shown a clear improvement in performance over the former. The authors present different ways of implementing these transforms in a compressive sensing framework. Additionally, different types of reconstruction methods including the genetic algorithm and the auxiliary function method are also presented and their impact on the overall performance is discussed.

One challenge of a sensing technique is reducing sensing time while ensuring good effective data rate. In fact, once compressive sensing based on sub-Nyquist sampling is adopted, sensing time can be reduced by saving number of samples. This increases the probability of missed detection which causes collisions with primary service and worsens channel imperfections. In such case erasures occur in addition to errors. In this chapter, the authors propose a new technique to correct erasures while keeping sensing time at a desired level. Based on polar code and low complexity decoding algorithm, the proposed technique exhibits for high code rates better performance in terms of bit error rate (BER) compared to two existing techniques based on other codes, namely low-density parity check (LDPC) and BCH.

The role of artificial intelligence techniques and its impact in context of cognitive radio networks has become immeasurable. Artificial intelligence redefines and empowers the decision making and logical capability of computing machines through the evolutionary process of leaning, adapting, and upgrading its knowledge bank accordingly. Significant functionalities of artificial intelligence include sensing, collaborating, learning, evolving, training, dataset, and performing tasks. Cognitive radio enables learning and evolving through contextual data perceived from its immediate surrounding. Cognitive science aims at acquiring knowledge by observing and recording externalities of environment. It allows self-programming and self-learning with added intelligence and enhanced communicational capabilities over wireless medium. Equipped with cognitive technology, the vision of artificial intelligence gets broadened towards optimizing usage of radio spectrum by accessing spectrum availability, thereby reducing channel interferences while communication among licensed and non-licensed users.

Section 5
Practical Use of Cognitive Radio Networks

The world is witnessing widespread roots of ubiquitous computing across disciplines and industries. It is equipped with the ability to monitor anything from anywhere with efficient, easy, and equitable goods that offer services for everyone. This has become possible through usage of wireless technologies, which possess an extensive scope in healthcare domain. However, despite various advantages, wireless technologies are faced with distinct challenges in hospital environment. For instance, wireless devices often tend to cause electromagnetic interference to critical medical devices resulting in malfunctioning. Further, with ubiquitous computing, sensitive data about health state of patients is constantly being shared remotely from one place to another. Therefore, systems in place must address requirements of data security, and thus privacy. For this purpose, the chapter presents a collaborative study on cognitive-radio-based healthcare system, including advantages, architecture, and challenges related to implementation of cognitive radios in hospital environment.

Energy conservation is one of the essential requirements for the betterment of the Earth's future. The main objective of this chapter is to provide an effective framework of smart grid (SG) using the intelligence of cognitive radio network (CRN). The SG connectivity will be enhanced by using the functionality of internet of things (IoT). Using IoT, the devices between the consumers, manufactures, and energy source provider (government) can conserve resources and economy. In this chapter, the requirement of smart grid, the need for IoT technology in SG, and the usage of CRN in SG are all elaborated. The infrastructural design pattern of cognitive radio-IoT-based smart grid is introduced in this chapter.

Internet of things (IoT) is a new area of advanced networking where things or devices share information among each other though collected sensor data and perform tasks together. This chapter is a detailed description of IoT and its architecture, challenges, security, applications, and future scope. Since the devices are heterogeneous in nature, middleware are used for giving them a platform to communicate in homogeneous manner. Service requirement of IoT middleware is discussed further in the chapter. Furthermore, recent technologies integrated with IoT like software-defined networks and cognitive radios are a part of this chapter.

Foreword

Cognitive Radio network is an area of technology that is rapidly growing in importance with the need to utilize the available spectrum more efficiently and the need to provide more effective communications. In Cognitive radio, the wireless node is able to sense its environment, and especially spectrum holes, and change its transmission and reception chains to communicate efficiently without interfering with licensed users. Cognitive radio thus aims to improve the way the radio spectrum is utilized. Spectrum sensing is used to sense the unused spectrum in an opportunistic manner.

In this context, this book can be considered as a platform to introduce readers to the sensing techniques for next-generation radio networks and their applications. The contents of the book will provide the readers a firm understanding of the fundamental concepts of sensing techniques, cognitive radios, wireless networks, cognitive radio sensor networks, simulations and performance evaluation, applications and efficient use of the spectrum using various models including neural networks, and artificial intelligence algorithms.

The mission of this book is to integrate both research as well as education and present details about one of the promising technologies of the sensing techniques for next-generation cognitive radio networks in such a manner that it can be used both as a textbook for undergraduate and graduate courses, as well as a good reference book that guides researchers to solve complex problems and discusses future challenges for open research problems in this area. This is the first such comprehensive book on sensing techniques, challenges and its applications that could be useful for both students and researchers, as well as for industry practitioners and business professionals.

The chapters in the different parts have been organized systematically such that the readers do not get lost with the complexity of the materials presented in the individual chapters. There will be continuity among the chapters and even though each chapter has been contributed by different authors, the editors have ensured that the chapters are written in such a way that the flow of contents across the chapters is more logical and easy to understand. The book can be used as a prescribed book for an advanced undergraduate elective course or a Masters level graduate course on sensing techniques for cognitive radio networks: We suggest parts 2 and 3 for this purpose. With the fundamental knowledge gained from the above-mentioned parts, the remaining parts or chapters of the book can be used as good case studies for a student to pursue further research on sensing techniques, their challenges, and applications to other networks. To summarize, this textbook stands at the forefront of current literature in the field of sensing techniques for cognitive radio networks.

I highly recommend this timely book. I believe it will benefit many readers and be a fine reference.

Dil Muhammad Akbar Hussain
Aalborg University, Denmark

Preface

Cognitive Radio Network (CRN) is future technology, makes use of various techniques (such as spectrum sensing techniques, data sharing techniques, decision making techniques, network management, network optimization, problem detection and resolution techniques, and etc). Spectrum sensing function detects all the available spectrum holes in order to avoid interference. One of the main challenges with cognitive radio networks is how to detect primary user signal with efficiently.

In this book, we present the state-of-the-art on spectrum sensing techniques for the cognitive radio networks operations. This includes various sensing algorithms for configuring radio networks to meet perfect requirements and applications in the real world.

FOCUS

The focus of this book is on radio network operations, current and advanced sensing techniques, how much work has been done till now and gathers recent advances in cognitive radio networks. The sensing techniques presented are general in terms of energy detector, cooperative detection, wavelet-based compressive detection, and etc.

INTENDED AUDIENCE

This book is intended for engineers (that are practicing cognitive radio related developments), researchers, and higher degree students that are working on radio scene analysis for cognitive radios and networks. For researchers, this book serves as a starting point for research in areas such as algorithms for detectors, sensing techniques, or applications of radio networks. It may serve as a comprehensive tutorial or be used as the bases for new research projects. For practitioners, this book contains many algorithms, protocols, and experiments that will prove useful in the designing and deployment of sensing techniques for radio networks.

ORGANIZATION

This book is organized into five parts. The first part provides the introduction of the book and gives a background on cognitive radio networks. The second part of the book presents the motivation and the overview of the various sensing schemes. The third part is arranged into five chapters and presents many improved sensing algorithms/ techniques for the radio networks. This includes multiple antennas based sensing methods, cooperative sensing methods, compressive sensing methods, and polar codes. Further, the fourth part discusses the role of artificial intelligence in radio networks. The fifth part of the book presents cognitive radio network applications and illustrates how radio networks may be used.

Ashish Bagwari
Uttrakhand Technical University, India

Jyotshana Bagwari
Uttrakhand Technical University, India

Geetam Singh Tomar
THDC Institute of Hydropower Engineering and Technology, India

Acknowledgment

We express our deep gratitude to our parents, Mr. Pradeep Kumar Bagwari and Mrs. Saraswati Bagwari, who had been encouraging and supporting us throughout the tenure of our research work, and Grace of Baba Kedarnath to achieve this goal. We also wish to acknowledge our family members i.e. personal and professional (WIT) for their continuous support, help and encouragement without which this research work would not have been completed in time. Last but not the least, special thanks to our lovely nephew Anshul Semwal, niece Anushka Gaur, and our son Agrim Bagwari for their true love.

Ashish Bagwari
Uttrakhand Technical University, India

Jyotshana Bagwari
Uttrakhand Technical University, India

Section 1
Introduction to Cognitive Radio Networks

Chapter 1
An Overview of Cognitive Radio Networks:
A Future Wireless Technology

Pooja Joshi
Uttarakhand Technical University, India

Ashish Bagwari
Uttarakhand Technical University, India

ABSTRACT

The major hurdle that restricts the efficient utilization of radio spectrum is its inflexible and rigid licensing scheme. This constraint limits the emerging wireless systems to operate in static spectrum band. This deficiency compelled the innovators to introduce new paradigm (i.e., cognitive radio [CR] technique to boost productive spectrum utilization). In this chapter, the authors briefly introduce the CR to exploit wireless spectrum opportunistically. Cognitive radio network architecture is also discussed in detail. Authors also mentioned the software-defined radio (SDR) platform as a support to regulate spectrum reuse. IEEE 802.22, the first worldwide standard for wireless regional area network (WRAN), based of CR technology is presented in descriptive way along with its capacity, coverage, MAC layer frame structure. It has been observed that CR strikes its presence in today's modern wireless systems, applications, various projects, and trends, and become their desire.

INTRODUCTION

Nowadays global market has witnessed a rapid growth in various wireless communication technologies. In our day-to-day life, devices like a TV remote, personal digital assistant (PDAs), cellular phones, satellite TV setup box, all are based on wireless communication technology. This observation shows that the number of users subscribing and demanding for wireless services and applications have surpassed the number of users using traditional wired telephone services. Basically, wireless communication systems are based on the transmission of radio (electromagnetic) wave having the frequency of range 3Hz to 300 GHz. Hence this dramatic growth of mobile users and wireless data causing spectrum crunch problem

DOI: 10.4018/978-1-5225-5354-0.ch001

for various upcoming wireless devices and services. Visual networking index (VNI) on February 2017 has reported that mobile data traffic has been raised 18 times over the last five years and the growth rate of global mobile data traffic extended up to estimate of 63% in 2016 (Networking, 2017).

Figure 1 shows the variation in mobile data traffic growth of different regions all over the world, where the Middle East and Africa possess the highest growth in the year 2016 in comparison to the previous year.

VNI also reported that in coming years the worldwide data traffic is expected to raise 49 Exabyte per month by 2021, illustrated in Figure 2.

Therefore, there is a need to make appropriate efforts against the spectrum crunch problem. Traditional spectrum assignment policy, which is defined by the Federal Communication Commission (FCC), is based on command and control model. Under this model, only the authorized user who wins the radio frequency bid can use the spectrum without any interference. However, this static and inflexible spectrum assignment policy which is conducted by the government agencies causes the utilization of radio resource is either inefficient or underutilized. Here, Figure 3 shows the diversity in licensed RF spectrum utilization, the spectrum consumption is cumulated on certain parts of the spectrum while significant spectrum capacity remains unused (Bagwari & Singh, 2012) because the authorized user may not utilize the spectrum at all times in all locations.

Therefore, in 2002, the spectrum policy task force has generated the (Force, 2002) to FCC, which recommended the estimate changes in spectrum policy for public benefits and addressed the several worldwide spectrum issues like, including interference protection, spectral efficiency, effective public safety communications, and international spectrum policies. So, it can be stated that, Data explosion (high dimension data processing) and spectrum underutilization these are the two salient factors to initiates a new communication hypothesis i.e. dynamic spectrum access scheme(DAS), to utilize the existing

Figure 1. Mobile Data Traffic Growth in 2016
Networking, 2017.

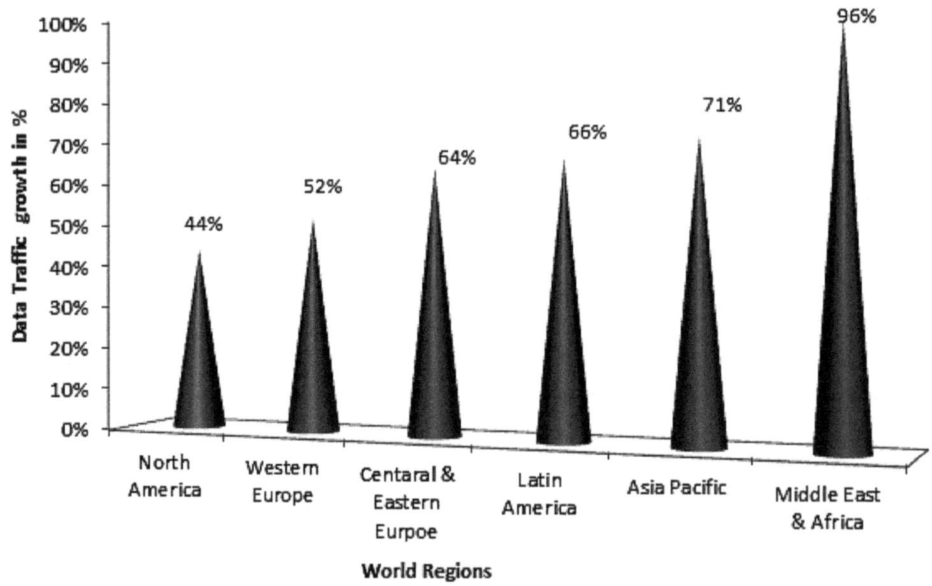

Figure 2.Cisco Forecasts growth of Mobile Data Traffic from 2016-2021
Networking, 2017.

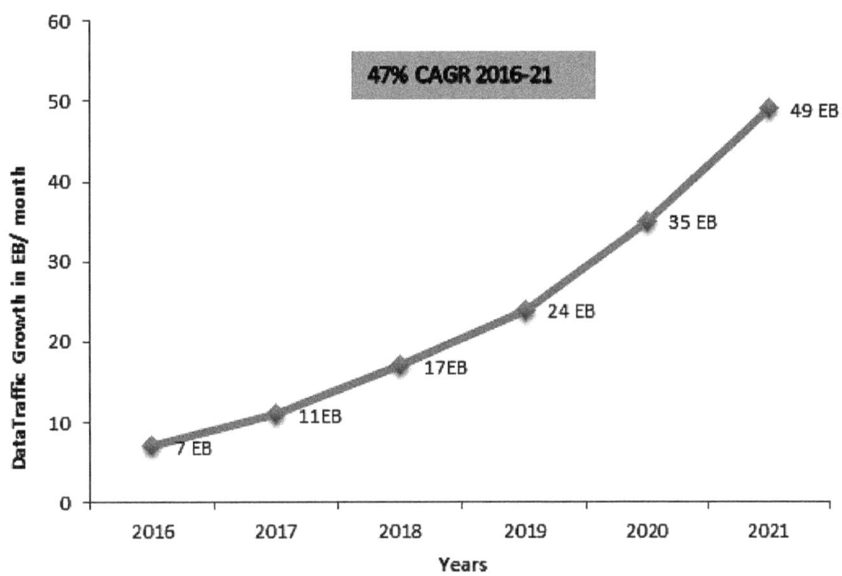

Figure 3. The diversity in licensed RF spectrum utilization

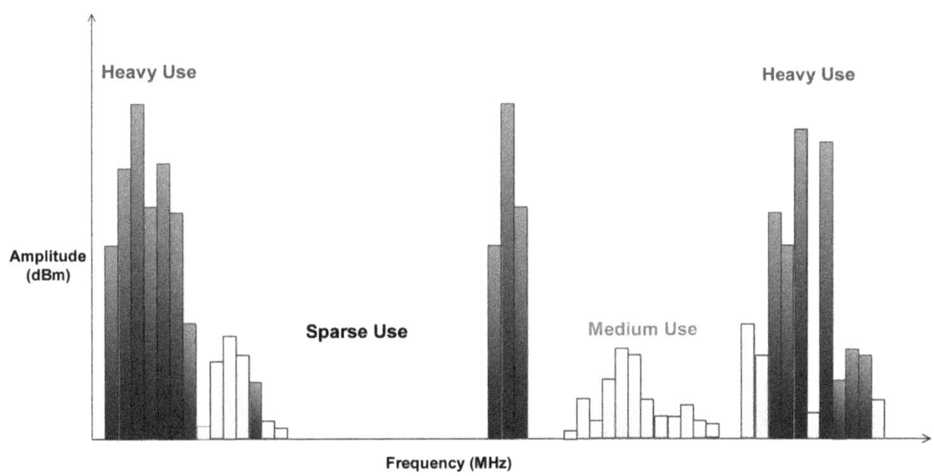

spectrum services more effectively, hence the dynamic utilization of spectrum saw the seed for new key technology i.e. Cognitive radio.

BACKGROUND/ LITERATURE REVIEW

This section briefly mentions the evolution of different wireless technologies based on various key aspects like data rate, standard, modulation, switching schemes etc.

First-Generation Wireless Technology (1G)

The era of 1G wireless communications were introduced in the early of 20th century, when radio phones were introduced for the purpose of military communication only. initially in 1950's telecommunication systems are based on half duplex mode (Push To Talk System), where a single channel is used to enable transmission and to disable reception. Later in this series two channel (full duplex), one for sending and one for receiving is added in 1960's known as IMTS (improved mobile telephone system). In 1970's various private companies have further evolved the existing systems in order to develop their own systems (Badoi, Prasad, Croitoru, & Prasad, 2011), (These private systems are Analogue Mobile Phone System used in America, Total Access Communication system and Nordic Mobile Telephone system used in parts of Europe, total Access Communication system used in Hong Kong and Japan. The independently developed system is basically 1st generation communication system introduced in1982 by Bell Labs known as Advanced Mobile Phone System (AMPS). The key feature and facility of 1G is to divide the geographic area into cell, was served by Base station (BS) to implement the reuse of frequency (Bhalla & Bhalla, 2010; Pereira & Sousa, 2004).

Characteristics:

1. Complete analogy system
2. Amplitude/ Frequency modulation were used
3. Direct dialing device, operator help not required
4. Gives data rate up to 2kbps
5. Cordless handsets

Limitations:

1. Supports Only voice communication not data
2. No roaming
3. Low capacity and not up to the mark voice link
4. Weak security on air interface

Second- Generation Wireless Technology (2G)

The commercial launch of second generation cellular telecom network was introduced in 1991 in Finland; it is based on global system for mobile (GSM) standard. It initiates voice transmission using digital signal up to 64kbps speed and had bandwidth of 30 to 200 KHz.2G standards, GSM and Interim Standard 95(IS-95) are classified on the bases of Time Division Multiple Access (TDMA) and Code Division Multiple Access (CDMA) channel access methods (Bhalla & Bhalla, 2010; Pereira & Sousa, 2004). Further to obtain better data rate GSM carries started developing the service named General packet Radio Service (GPRS), a packet switched network able to transmit data at up to 160 kbps. Next phase after GPRS is called Enhanced Data Rate for GSM Evolution (EDGE), It delivers the data at up to 500kbps and worked with 8PSK modulation. During this era, the internet was becoming popular, so all post 2.5 G, multimedia services started growing and supporting web browsing (Khan & Barman, 2015).

Characteristics:

1. Conversations were digitally encrypted
2. Efficient Spectrum proficiency
3. Enables Roaming
4. Support Both the Data and voice service
5. Security Improved

Limitations:

1. Require strong digital signals to make the mobile system works
2. Difficult to handle video data

Third- Generation Wireless Technology (3G)

The ultimate goal of 3 G, built on universal Mobile Telecommunication System (UMTS) is to provide at least 2Mbps data rate for stationary or mobile user and 384 Kbps in moving vehicle. This 3GPP was also considered as Release 99. The first 3G network was introduced in October 1, 2001 by NTT Do Co Mo in Japan to facilitate the transmission of high volume data while using Internet Protocol (IP) based services. This technology supports wireless multimedia services, web base access, video conferencing, email. 3G W-CDMA air interface connect all the devices like computer, handsets and allow sharing between them via packet –based wireless service. The main 3G standards are W-CDMA (Wideband Code Division Multiple Access), CDMA2000 (this standard recall Evolution-Data Optimized (EV-DO), typically for broadband Internet access (Bhalla & Bhalla, 2010; Pereira & Sousa, 2004; Prinima, & Jyoti, 2016).

Characteristics:

1. High Performance
2. Fast Internet Access
3. Supports services like GPS (Global positioning system), Video Conferencing Video call.
4. Provides Digital Navigation
5. Support TV through internet

Limitations:

1. Requires Compatible 3G Handsets
2. Require High Power
3. Distance from Base Station also an Issue
4. Upgrading cost for 3G devices were also high

Fourth-Generation Wireless Technology(4G) and Fifth Generation System (5G) Using CR

4G considered as the successor of 3G standard, where services were not limited only up to voice and SMS only. 4G was dedicated to provide the services like video chat, mobile TV, Digital Video Broadcasting, HD TV, Multimedia Messaging Service (MMS), gaming services (Bhalla & Bhalla, 2010). Japan did the first successful field trial for 4G in June 2005. 4G is also known as full IP integrated; it’s all about the integration of various technologies and standards anytime and anywhere. LTE (Long Term Evolution) is designated as key technology of 4G which knocked the market door in 2009 (Javed & Siddiqui, 2017; Kanojia, Motwani, Khan, & Yadav, 2012).

Characteristics:

1. Simple Unified Protocol Architecture
2. High Mobility
3. Support 100 Mbps data rate at high mobility and 1Gbps (at low mobility)
4. Supports video conferencing, GPS services
5. Enhanced Scalability
6. Peak Download and Uplink Data rate: 1Gbps and 500Mbps respectively

Limitations:

1. Hardware Complexity
2. Consume more battery power
3. Network Implementation is also an issue.

The limitations faced by previous technologies (1G, 2G, 3G, 4G) has been resolved by the next upcoming generation i.e. Fifth generation of cellular wireless standard (5G) along with cognitive radio(CR) technology. They are believed to be the future technology, is not deployed yet but expected to be launched in market by 2020-25. CR technology offered the possibilities to increase spectrum efficiency by utilizing the free licensed users spectrum holes. It would be the real wireless world without any limitation to internet access and geographical issue using Artificial Intelligence (AI). It delivers the global service portability and expected to provide 65,000 connections at a time because of its extraordinary capabilities. 5G implies the expected frequency band from 3-3000GHz and bandwidth up to 1Gbps and more. It is more powerful to convey worldwide wireless interconnections (WISDOM- Wireless Innovative System for Dynamic Operating Mega communications concept), jointly with fast data rate of the quality of service (QoS) applications. To acquire WISDOM CR technology is the evolved as good solution to implement 5G (Badoi, Prasad, Croitoru, & Prasad, 2011).

COGNITIVE RADIO

Cognitive radio (CR)is an intelligent software radio who is capable to sense its RF surrounding environment and site, able to alter their radio operating parameters like power, frequency, modulation in order to maximize throughput, reduces the severity of interference, accelerate interoperability among primary and secondary users. The primary user having priority to the access network, their access rights are considered as property rights to the channels. Secondary users having low priority, who always trying to make opportunistic use of the same spectrum without harming PU signals (Ibnkahla, 2014). Here some pre-eminent definitions of CR have been mentioned below: In 2001, Mitola defined CR (Mitola, 2001) as follows:

Cognitive radio is a goal-driven framework in which the radio autonomously observes the radio environment, infers context, assesses alternatives, generates plans, supervises multimedia services, and learns from its mistakes. This observe-think-act cycle is radically different from today's handsets that either blast out on the frequency set by the user, or blindly take instructions from the network. Cognitive radio technology thus empowers radios to observe more flexible radio etiquettes than was possible in the past.

Another famous definition cited was by Simon Haykins (Haykin, 2005) while doing a meticulous review on CR, as below:

An intelligent wireless communication system that is aware of its surrounding environment (i.e., outside world), and uses the methodology of understanding-by-building to learn from the environment and adapt its internal states to statistical variations in the incoming RF stimuli by making corresponding changes in certain operating parameters (e.g., transmit-power, carrier frequency, and modulation strategy) in real-time, with two primary objectives in mind:

1. *Highly reliable communications whenever and wherever needed;*
2. *Efficient utilization of the radio spectrum.*

For Opportunistic spectrum access cognitive radios endowed two main attributes, one is cognitive capability and another is re-configurability, these can expound as (Kanti, Bagwari, & Tomar, 2015).

Cognitive Capability

Cognitive capability helps CR to examine the appropriate parameter for communication and real-time interaction with its environment. In 1999, J.Mitola has coined cognitive cycle, an overview of four main steps of this cycle has been illustrated in Figure 4 (Haykin, 2005; Kanti, & Tomar, 2016 ; Akyildiz, Lee, Vuran, & Mohanty, 2006).

1. **Spectrum Sensing:** Monitor and capture the information of spectrum band, detect licensed user's presence and absence in that particular band.
2. **Spectrum Management:** Select best radio frequency channel to promote efficient use.
3. **Spectrum Sharing:** It delivers standard access of spectrum to other users.
4. **Spectrum Mobility:** Switch to another band when licensed user arrives or become active.

Re-Configurability

Re-configurability is that ability of CR which helps to reconfigure the transmitting and receiving parameters without changing their hardware components. In earlier systems the transceiver parameters can't be modified once the system is configured or delivered to the customer, there are several reconfigurable parameters, which are as follows:

- **Operating Frequency:** CR determines the best frequency after sensing the radio environment and changes its own operating frequency for communicating in a dynamic manner.

Figure 4. Cognitive Cycle

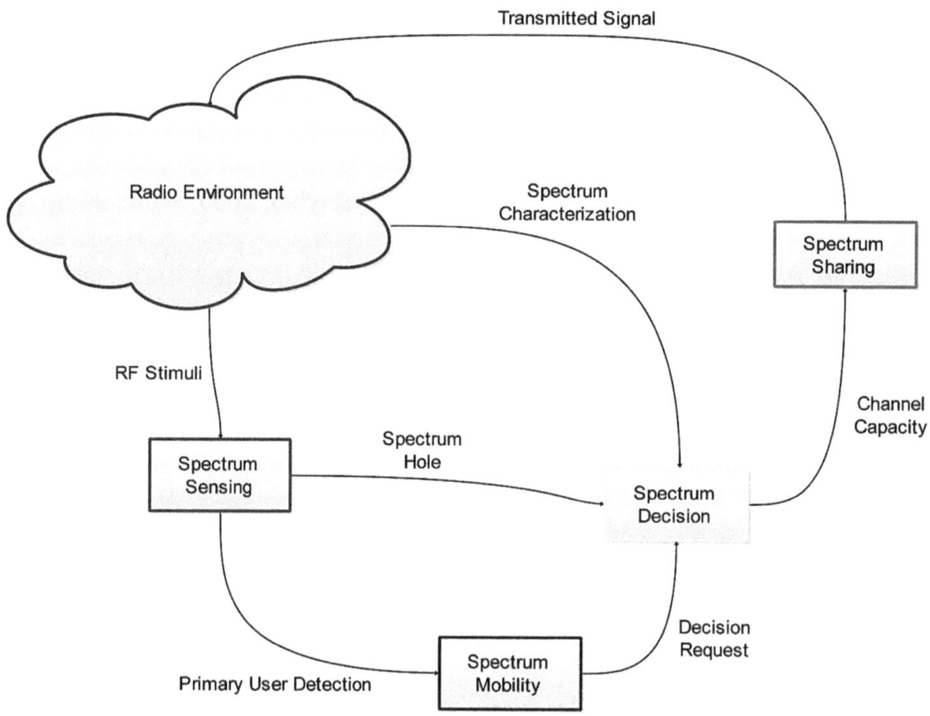

- **Modulation:** It opts the modulation scheme according to the user requirements and channel condition. Selection of modulation schemes varies and depends on delay sensitive and loss sensitive applications.
- **Transmission Power:** This capability of CR is used to operate on low power and energy harvesting mode when high power operations are not required or there is power limit.
- **Communication Technology:** CR provides interoperability function among the heterogeneous communication systems (Akyildiz, Lee, Vuran, & Mohanty, 2006).

Cognitive radio is interpreted as a part of the physical world to provide and use information from surroundings in order to enhance flexibility and to find the best spectrum by using cognitive capability and re-configurability.

In the following subsection, there is a description of the cognitive radio network, their elements, various network access types and Cognitive radio network heterogeneity.

COGNITIVE RADIO NETWORK ARCHITECTURE

Cognitive radio networks provide the favourable environment among the various cognitive radios and users for healthy communication. Current wireless network architecture consists some licensed spectrum bands while other portion of spectrum remains unlicensed so, it is essential to develop communication protocol with clear description of cognitive radio architecture. Cognitive radio networks are more feasible as compared to the single cognitive radio. Figure 5 shows the components of CRN architecture, which can be broadly classified in two parts: Primary network and Cognitive radio network (CRN) (Akyildiz, Lee, Vuran, & Mohanty, 2006).

1. **Primary Network:** Primary network is also referred as traditional network, having exclusive access and right only to a certain band of spectrum.
2. **Cognitive Radio Network (CRN):** This network is also termed as unlicensed network or secondary network because it has no license to operate in desired band. These types of networks can access the spectrum only in an opportunistic way.

The Basic Elements of Primary Network

1. **Primary User (PU):** PU operates on certain spectrum band, for which it is licensed. This access is only controlled by its Base-station. They should not be affected by any other illegal and unauthorized user. PU doesn't require any changes and additional functions to establish harmony with CR users and CR base-station.
2. **Primary Base Station:** Primary Base station (BS) also called licensed BS. Primary base-station may be required to have both legacy and cognitive radio protocols for the primary network access of cognitive radio users.

Figure 5. Cognitive Radio Network Architecture

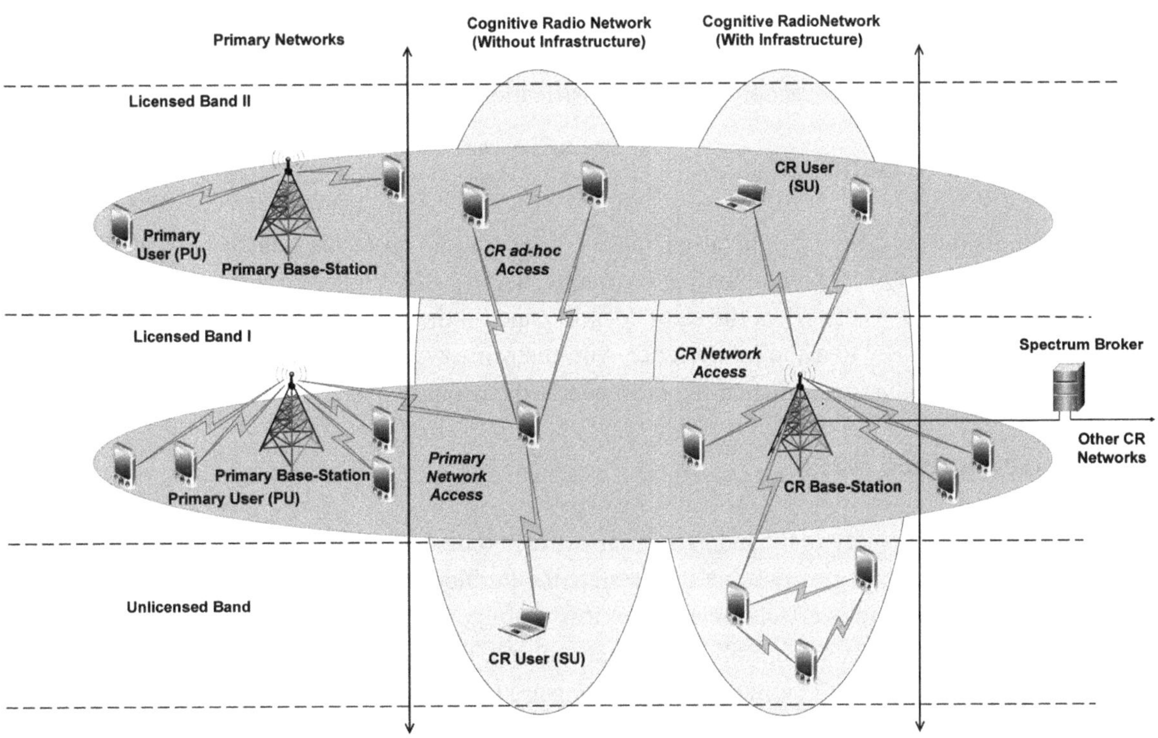

The Elements of CRN

1. **Cognitive Radio User:** These radio users don't have spectrum license, so they behave only in an opportunistic way. Users having some special capabilities like spectrum sensing, spectrum decision and spectrum handoff etc. CR user is capable of communicating with both base-stations and other CR users.
2. **Cognitive Radio Base-Station:** It is a secondary base station having fixed infrastructure components along with cognitive capabilities.
3. **Spectrum Broker:** Spectrum broker acts as a scheduling server that helps to share spectrum resources among the various types of network. It is connected to each network and facilitates them.

Cognitive Radio Network Architecture Having Three Different Types of Network Access

1. **Cognitive Radio Network Access:** In this network, CR user can acquire or access their CR base-station both in licensed and unlicensed spectrum band. Their medium access technique is independent to that of the primary network.
2. **Cognitive Radio Ad-Hoc Access:** This network access provides the Ad-Hoc connection which is required to make communication between cognitive radio users in both licensed and unlicensed band.
3. **Primary Network Access:** In this access type, the cognitive radio user seeks the permission of primary network to access the primary through the licensed band.

CRN HETEROGENEITY

The cognitive radio network acquires the property to operate both in licensed and unlicensed spectrum band, so this section is explained according to its classification (Akyildiz, Lee, Vuran, & Mohanty, 2006).

CRN on Licensed Band

According to the previous Figure 3, licensed bands consist unused spectrum holes and we exploit these spectrum holes through new communication technique, using CRN. Figure 6a represents the existence of CRN with the primary network at the same location and on the same band. This condition itself the brings a challenge to CRN to avoid interference with the primary user. The main aim of CRN is to determine the best available spectrum and the presence of the primary user.

CRN on Unlicensed Band

Present open spectrum policy in the industrial, scientific and medical (ISM) band has introduced magnificent and major technologies but due to interference problem in various heterogeneous networks, the efficiency of ISM band is affected badly. So, it is required to design CRN on unlicensed bands to improve spectrum efficiency. Figure 6b shows that multiple CRN is existing at the same location and using same spectrum band for communication, there is no licensed holder.

Figure 6. Left: CR Network on licensed band; right: CR Network on unlicensed band

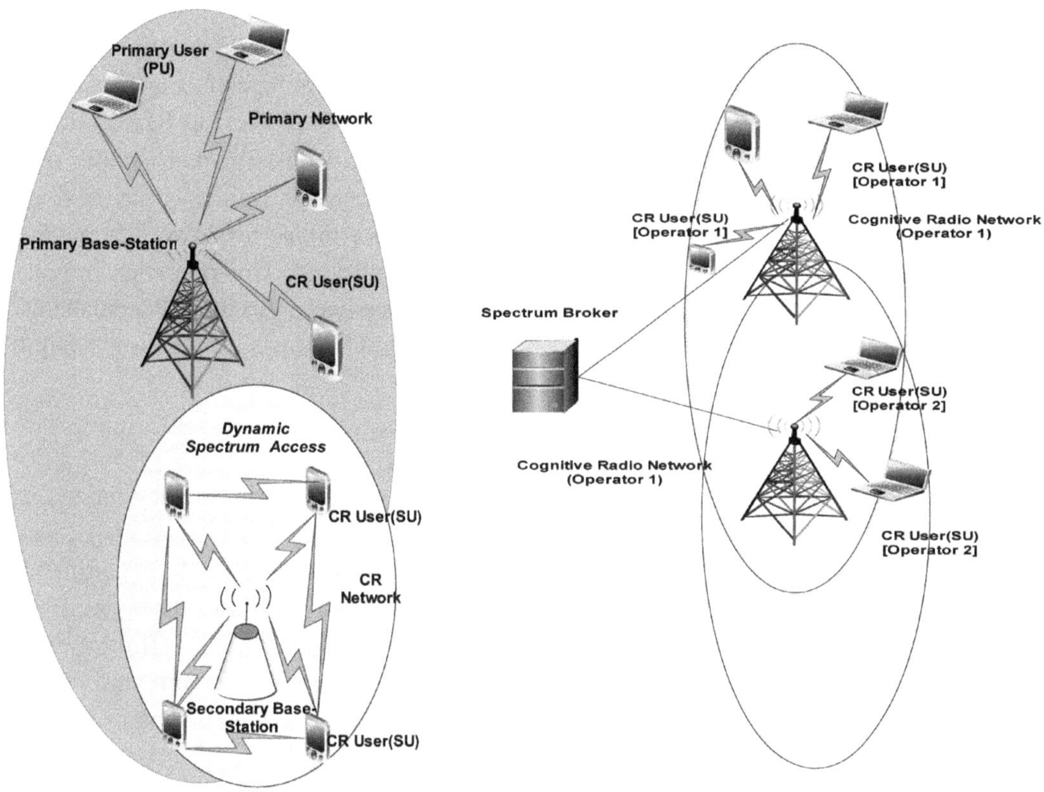

We need a sophisticated and intelligent spectrum sharing algorithm or method for fair spectrum sharing. So that these algorithms will support this type of architecture to improve the efficiency of spectrum usage since all CRN users have same right to access unlicensed band.

So, it is required to design CRN on unlicensed bands to improve spectrum efficiency. Figure 6b shows that multiple CRN is existing at the same location and using same spectrum band for communication, there is no licensed holder. We need a sophisticated and intelligent spectrum sharing algorithm or method for fair spectrum sharing. So that these algorithms will support this type of architecture to improve the efficiency of spectrum usage since all CRN users have same right to access unlicensed band.

EXAMPLES OF EXISTING ARCHITECTURE

The most famous examples of Cognitive radio network architecture are described in this section, here some of them have been considered.

Spectrum Pooling

Today's spectrum for mobile communication is owned by (GPRS, Emergency services, T.V Broadcast, UMTS etc), they all are licensed owners and share their unused spectrum with secondary users. This resource sharing strategy, where licensed owner having highest priority is known as spectrum pooling. Spectrum pooling architecture has been developed in University of Kaelsruhe, based on orthogonal frequency-division multiplexing (OFDM). OFDM is a technique of transmitting multiple digital signals simultaneously over a large number of orthogonal subcarriers. The advantage of using OFDM is to feed sub- carriers (occupied by the licensed user) with zero consuming radio power. The base station periodically broadcast detection frames (frame having no data) in order to determine licensed user or PU. All mobile users perform spectrum sensing during broadcasting of detection frame. These sensing data is collected at the base station by using boosting protocol. All mobile terminals modulate a complex symbol at maximum power in those sub-carriers where licensed user appeared. This operation provides base station an amplified signal on all sub-carriers along with new licensed users. In this architecture, there are some issues like the detection of spectral access in MAC and PHY layer, scheduling, handoff etc.

CORVUS

CORVUS architecture, a group of cognitive radios operates over the large frequency range from tens of MHz to several GHz for the usage of virtual unlicensed spectrum. In this system model, secondary users create their groups known as secondary user group (SUG) for coordinating their communication. This system works in a coordinated manner and this collaborative effort increases the system's ability in identifying and avoiding the primary user (Akyildiz, Lee, Vuran, & Mohanty, 2006; Brodersen, Wolisz, Cabric, Mishra, & Willkomm, 2004). Figure 7 presenting the principle of spectrum pool in CORVUS system. The red shaded Frequency Bands (FB) indicates that the PU is currently using its spectrum and consequently this Frequency-Band cannot be used by any SU. Each member of secondary user group continuously senses the Spectrum pool. The large spectrum pool is divided into a number of sub-channels, these sub-channels are nothing but a part of the frequency band.

Sets of sub-channels are used by the secondary users to form a secondary user link. This architecture consists of two logical control channels as follows:

1. **A Universal Control Channel**: This channel is used to declare the information of existing groups and provides authentic transmission parameters to the newly arriving user.
2. **Logical Group Control Channel**: This channel is used by the members of the group to exchange sensing information and to the establish secondary user link.

There are many more projects and networks which are deployed over the heterogeneous cognitive radio network and its technology, example: The European Dynamic radio for IP services in Vehicular environments (DRiVE), Nautilus project, OCRA network etc. (Akyildiz, Lee, Vuran, & Mohanty, 2006).

ADVANTAGES of CRN

In this present century, cognitive radio is remarkable technology satisfying the current increasing numbers of users. This network consists of a number of radio stations and cognitive radio nodes which gives significant raise to the speed and availability of channels for the new secondary wireless networks (FreeWimaxInfo.com, 2017).

Figure 7. Spectrum Pooling Idea

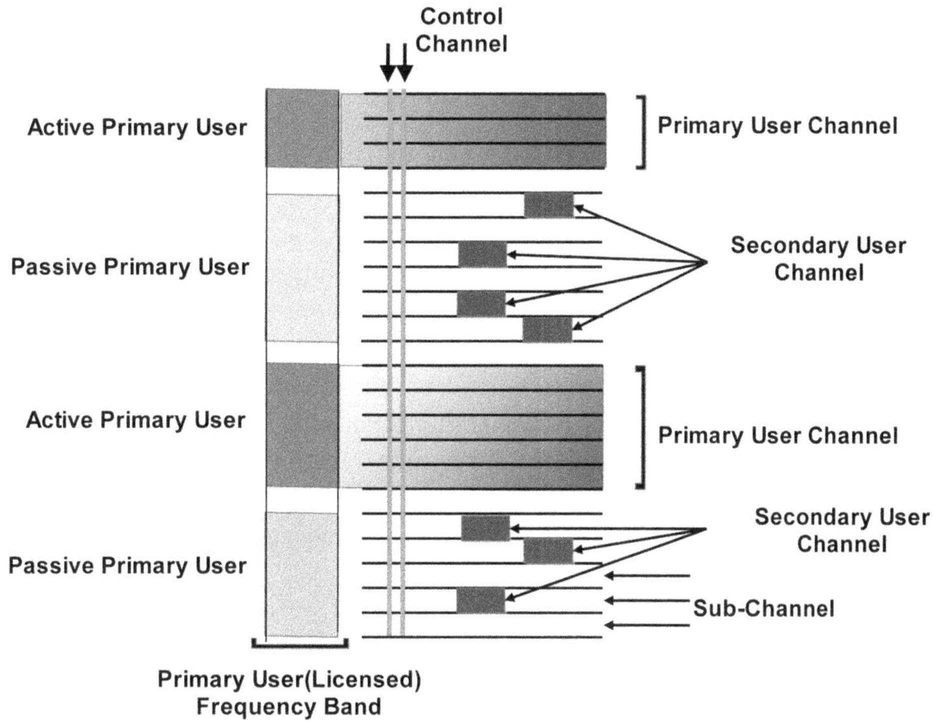

1. **Better Radio Link Performance:** The basic function of the cognitive radio network is spectrum sensing. Cognitive radio spectrum sensing method is capable of providing updated information regarding current transmission done by the primary user and sends information back to the central processing unit (CPU).
2. **Improved Coverage:** CRN helps to relay data from one point to next node. In this way, less power consumption can be reduced without compromising the performance.

SOFTWARE DEFINED RADIO BASICS

Cognitive radio is also known as SDR (software defined radio) deals with precious resource i.e. spectrum space, it is new breed of radio in wireless communication system which act as a self-organized coordinator and controls all the transmission variables and specifications like operating frequency band, protocols, modulation mode dynamically and appropriately (Jondral, 2005; Tomar, Bagwari, & Kanti, 2016). This radio is expert to make changes in the function, which is required/demand by the system at that particular instance of time. The Kansas University Agile Radio (KUAR) (Hossain, Niyato, & Han, 2009) is one of the most famous, flexible, portable and powerful SDR, which enables advance research in wireless cognitive radios. The hardware part of this platform having subsystems namely as Power supply board, control processor board, digital board of programmable signal processor, analog-to-digital and digital-to-analog converters, RF transceiver, antennas. Its software section consists control and management programs namely as Boot (load radio modules used by other components), Policy (controls transmission parameters according to the regulatory rules), Ops, and QoS (measure network parameter).

Classification of Software Radio

Software radio has basically 5 stages, starts from 0 to 4. Figure 8 shows the various stages of software radio which are defined as (Tomar, Bagwari, & Kanti, 2016).

Stage 0: HR (HARDWARE RADIO): In this type of radio, Software can't make any changes to the system.

Stage 1: SCR (SOFTWARE CONTROLLED RADIO): Modulation and frequency band attributes can't be changed without changing the hardware, even though software implements control function.

Stage 2: SDR (SOFTWARE DEFINED RADIO): It has Potential to store large number of air interfaces or waveform, able to cover substantial frequency range. SDR provides various modulation techniques, supports wide-band and narrow-band operation while maintaining security during communication. This System software also supports new modules for new functions; fix the bugs without reloading the entire software set again.

Stage 3: **(ISR) Ideal Software Radio:** This radio is same as SDR, but doesn't perform analog amplification and heterodyne mixing before A/D conversion and after D/A conversion.

Stage 4: **(USR) Ultimate Software Radio:** It is fabricated on chip, mends for comparison purpose, and operates on wide range of operating frequency, without external antenna.

Figure 8. Presenting various categories of software radio from stage 0-4
Tomar, Bagwari, & Kanti, 2016.

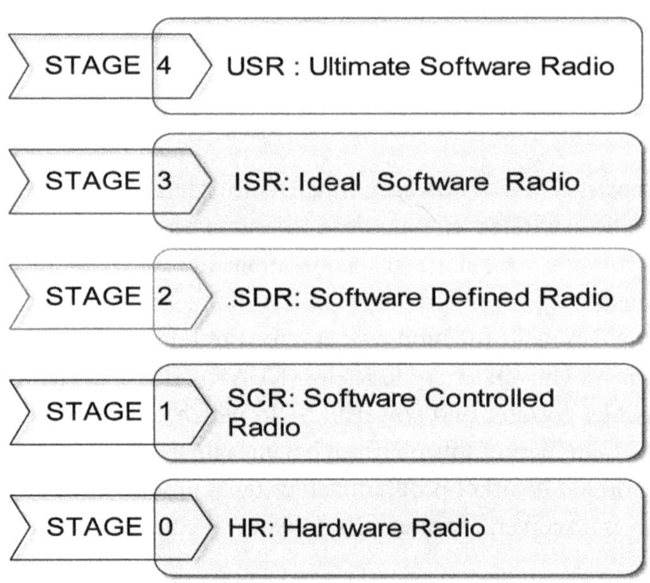

SDR Transceiver

As shown in Figure 9 SDR transceiver has two major components namely as RF front end and baseband processing unit, connected with each other via control bus in RF environment. RF- front end performs functions like- amplification, mixing, and A/D conversion, while baseband unit is as similar as to our existing transceiver, which is responsible for encoding/decoding and for modulation/demodulation of signals. RF-front has aptness to recieve wideband signals because it consists of hardwares like wideband anteena, adaptive filter, power amplifier, these hardware technologies suppots CR to measure real-time spectrum information SDR integrates both the analog and digital techniques together (Hossain, Niyato, & Han, 2009; Wikimedia Foundation Inc. Software Defined Radio Wikipedia, 2008).

RF front end also have some issues/ complications like, signals of different power and bandwidth coming from various locations are transmitted to RF antenna, therefore, it is quite difficult to detect accurate weak signal among various dynamic frequency. RF front end requires advanced high speed A/D converters with over-the-top resolution, achieving this scenario is quite infeasible. Another challenging task is to reduce dynamic range of signal before analog to digital conversion. This type of reduction can be obtained by filtering all strong signal which are present within the spectrum at different locations. This challenge can be satisfied by tunable notch filter, which is again a critical issue in CR network (Cabric, Mishra, & Brodersen, 2004; Cabric & Brodersen, 2005) RFfront end influences the performance of SDR, they are aimed to facilitate high level of performance and flexibility with the help of high quality of filters and oscillators.

This Figure 10 shows the various components, responsible for the activities happening between the antenna and the baseband system. Each unit performs specific task (Akyildiz, Lee, Vuran, & Mohanty, 2006; Dodson, Bradford, & Laneman, 2010). RF filter is used to select the required band by filtering the RF signal from bandpass filter. This signal is forwarded to Low noise amplifier (LNA) which amplify the received signal and minimize its noise components. Then Mixer Generates IF i.e. intermediate frequency, by mixing the recieved signal with local RF frequency. For a given voltage, a specific frequency signal is generated by voltage controlled oscillator(VCO), and further mixed with an incoming signal. At the same time, Phase locked LOOP(PLL) locks the signal on a specific frequency to generate accurate frequency with fine resolution. After this, Channel selection filter Selects the needed channel and rejects the unwanted one, different channel selection filters are used for different recievers.Last unit is Automtic gain control (AGC),which is an amplifier circuit is used for providing a constant output power level over a wide variety of input levels.

Main Functions of SDR (Hossain, Niyato, & Han, 2009):

Figure 9. Components of SDR transreciver
Hossain, Niyato, & Han, 2009.

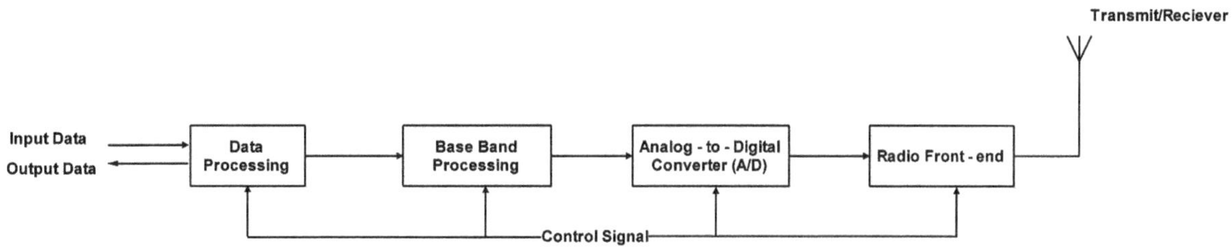

Figure 10. RF-Front end components
Akyildiz, Lee, Vuran, & Mohanty, 2006.

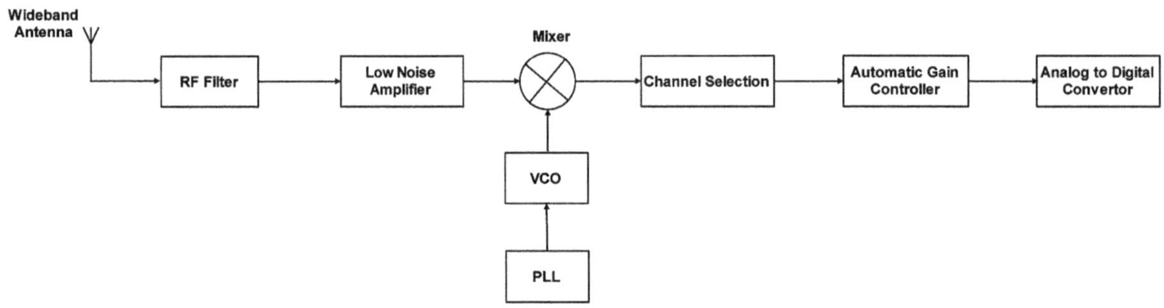

1. **Multiband Operation Support:** SDR embrace different wireless access systems (TV band, cellular band, ISM band) for data transmission over different frequency spectrum.
2. **Multi-Standard Support:** It supports different standards (GSM, WCDMA, CDMA2000, Wi-Fi, Wi-Max) and different air interfaces with in same standard (IEEE 802.11a, 802.11b, 802.11g, Wi-Fi standard).
3. **Multiservice Support:** Cellular telephony services and broad band internet access supported by SDR.
4. **Multichannel Support:** Its transceiver works on various frequency bands simultaneously.

So, we can say that SDR is a completely software configured radio, capable of any type of standards up gradation in various applications with different protocols used in the wireless network. Its transceiver is very much skilled to welcome the job of mobile phones, web browser (wireless), telephone, fax (wireless), pager, GPS, etc from any location. These radios are committed to keeping up high bandwidth applications along with good service quality SDR components and its programming, supports versatility and a high degree of freedom in radio discipline (Bagwari & Tomar, 2013; Bagwari, Kanti, & Tomar, 2016).

Next section i.e 1.8 belongs to one of the most representative standards of cognitive radio architecture: IEEE 802.22 along with their capacity, coverage details, and MAC layer frames.

IEEE 802.22 WRAN SYSTEM

The IEEE 802.22 working group was formed to develop a WRAN system based on cognitive radio technology to provide broadband connectivity, especially for lower population density areas to utilize VHF/UHF TV band between 54 to 864 MHz.TV spectrums usages differ over very larger timescale.

The TV band Notice of Proposed Rule Making (NPRM), released in May 2004, which stated that how we can operate unlicensed radios in the television broadcast band without causing any interference to incumbent services. Table 1 illustrates the different wireless data communication standards along with their Range, Frequency, and Data rate introduced by IEEE 802 LAN/MAN standard committee (Stevenson, Chouinard, Lei, Hu, Shellhammer, & Caldwell, 2009).

Table 1. Different wireless data communication standards

S. No.	Standard	Network Description	Channel Capacity	Range	Frequency/ Spectrum	Data Rate
1.	IEEE 802.15	Wireless Personal area network (WPAN) (Bluetooth), (UWB),(Zig-Bee)	720Kbps for Bluetooth, 20-250 Kbps for Zig-Bee	< 10 m	2.4 GHz for 802.15.1, 915 MHz for 802.15.4	< 10 Mbps
2.	IEEE 802.11	Wireless Local area network (WLAN) (Wi-Fi)	11-54 Mbps shared	< 150 m	2.4 GHZ for 802.11b/g 5.2 GHz for 802.11a	11-54 Mbps
3.	IEEE 802.16	Wireless Metropolitan area network (WMAN) (Wi-Max)	Share Up to 70 Mbps	< 5 km	2-11 GHz for 802.16a 11-60 GHz for 802.16	11 Mbps
4.	IEEE 802.20	Wireless Wide area network (WWAN) (MobileFi)	Up to 2 Mbps each	< 15 km	3.5 GHz	
5.	IEEE 802.22	Wireless Regional area network (WRAN)	18 to 24 Mbps	< 100 km	54-862 MHz	37 Mbps

Cognitive radio is smart enough to determine the unused section of the spectrum which is also called white space (unused fragment of time). CR adopts two methods for spectral awareness with the help of IEEE 802.22, first one is, spectrum sensing: in this method we sense or observe the spectrum to identify the occupancy of channel and second method isgeo-location: we combine the location information of CR with the database of licensed transmitter, for targeting those channels which can be reused.

System Coverage

IEEE 802.22 cell is formed with the help of base station (BS): it is wireless access point act as a transceiver in wireless network and customer premises equipment (CPEs): type communication equipment located on customer's home and business site. BS controls all the CPE's. It provides fixed point to multipoint and point to point (mesh) connectivity. In point to multipoint wireless air interference, transmission is coordinated by BS, they work on the principle of the master (BS) and slave (CPE). Where BS acts as a professionally installed entity, which is armed with GPS receiver. BS transmits data to various CPEs in the downward direction and receives data from CPEs through the uplink. It also operates like medium access controller within the cell while managing the distributed sensing.

BS instruct various CPEs to measure different TV channel signal to noise ratio along with their locations and then take the decision of adjusting modulation, bandwidth, coding and so on, as per the feedback is given by CPEs . Fixed point to the multipoint connection face the problem of hidden incumbent. This problem can be handled by introducing alternative architecture i.e. point to point (mesh) network in which all CPEs communicate directly with each other, but for establishing mesh network channel should be predefined for BS/CPEs which is not feasible in case of IEEE802.22 system practically, as it shares the spectrum band dynamically with licensed user. In mesh network all BS/CPE's try to hog the available bandwidth because they are operating in close proximity, this may increase interference among 802.22 system. This interference may render the system useless. To satisfy bandwidth hungriness system proposed channel bonding which maximizes the utilization of spectrum and Maximum utility graph coloring method for fair division of spectrum band, interference degradation (Liang, Hoang, & Chen, 2008; Sengupta, Chatterjee, & Kwiat, 2009).

System Capacity

The IEEE 802.22 system specifies spectral efficiency of range 0.5b/s/Hz up to 5 b/s/Hz. If we consider an average of 3 b/s/Hz, this would correspond to a total PHY data rate of 18 mbps in a 6MHz TV channel. It is an international standard which operates in any regulatory regime,while there is an on-going debate to accommodate the various TV channel bandwidth of 7 and 8 MHz also. In order to achieve minimum data rate per CPE, a total of 12 simultaneously users have been considered which leads to a required minimum peak throughput rate edge of coverage of 1.5 Mbps and 384 kbps per CPE in the downward and upward direction respectively. This level of performance is similar to DSL services.

IEE802.22 WRAN Base station coverage range is 100 km, which is quite remarkable in comparison of earlier IEEE 802 standards. Around mid of 2006, the first draft of IEEE 802.22 was announced. IEEE802.22 System's coverage and capacity information are mentioned below in Table 2. With the help of their various parameters and specifications. Wide variety of modulation scheme like quadratic phase shift keying (QPASK), Quadrature amplitude modulation (16-QAM) and 64-QAM has been used within the orthogonal frequency division multiplexing Access (OFDMA). OFDM is an exceptionally popular advanced technology, which is used in 4G (LTE). OFDM is used to provide efficient allocation of sub-carriers to match the requirements of various CPEs (Akyildiz, Lee, Vuran, & Mohanty, 2006; Corderio, Challapali, Birru, & Shankar, 2006; Poole, 2017).

Table 2. Parameter Values for CRs as per IEEE 802.22 Standards

Parameters	Specification
Frequency Rang	54 to 862 Mega Hertz
Channel Bandwidth (RF)	6, (7, 8) Mega Hertz
Average Spectrum Efficiency	3 bits/ second/hertz
Channel Capacity	18 mega bit /second
System capacity /subscriber	Downlink: 1.5Mega byte per second Uplink: 384 kilo byte per second
Channel capacity	18 Mbps
Payload Modulation	QPASK(30 km), 16-QAM (22 km), 64-QAM(15 km),
Transmit Effective Isotropic Radiated Power	Default 4 W for CPEs
Cyclic Prefix Modes	1/4, 1/8, 1/16, 1/32
Radius of Coverage For WRAN System	30.7 km
Multiple access	OFDMA

MAC Layer

MAC layer assembled with synchronous time structure having super frame (SF) structure, SF structure is made up of 16 number of frames with fixed duration of 10 ms each First frame consists SF Permeable, frame permeable, and SCH, while rest of the 15 frames only have permeable as shows in Figure 11. MAC frame is composed of Downstream (DS) sub-frame and upstream (US) Sub-frame.

1. The Downstream (DS) sub-frame Consists:
    a. Frame Control Header (FCH): Contains the size of the DS/US MAP fields and US/DS;
    b. Channel descriptors;
    c. DS/US MAP: Gives the scheduling information for CPE bursts.
2. The Upstream (US) sub-frame contains:
    a. Information of BS's range;
    b. Total bandwidth requested;
    c. The subscriber can alert the BS about the sudden appearance of the incumbent by notifying, UCS (the urgent coexistence situation);
    d. Self-coexistence window (SCW);
    e. Time Buffers: To absorb propagation delay.

Every super-frame contains 26–42 OFDM symbols, bandwidth from 6 to 8 MHz and cyclic prefixes from 1/4 to 1/32. Each OFDM symbol consists of maximum 60 sub-channels, where each sub-channel has 28 subcarriers.

An extended version of SCH i.e. CBP (Coexistence Beacon Protocol), enables WRAN BS to manage and operate in an intelligent manner with neighbor WRAN cell under coexistence situation. This protocol capable of finding another neighbor 802.22 system and coordinate with them.BS sends special permeable and super control header to those TV channel bands (up to 3 continuous channels) which are participating in communication without affecting incumbent's requirements (Zhang, Zheng, & Chen, 2016; Stevenson, Chouinard, Lei, Hu, Shellhammer, & Caldwell, 2009). CPE makes its association with BS and communicates with any of these channels. The current 802.22 draft supports multiple MAC frames and hence can provide good range, multipath diversity, and better system performance (Cavalcant & Ghosh, 2008).

Channel Sensing

MAC layer is also responsible for in-band and out-of-band measurement. The in-band measurement conducts the sensing of actual channels used by the BS and CPE's. Out-of-band performs the sensing for remaining channels (Corderio, Challapali, Birru, & Shankar, 2006). To protect the incumbent, temporarily transmission break is provided by the BS and CPE. To stop the data transmission 802.22 draft MAC employs the quite period mechanism it is comprised of two different time scales: fast and fine sensing as shown in Figure 12.

1. **Fast Sensing:** In fast sensing, the time taken by the CPEs to measure the energy of affected channel, is too short, under 1ms/channel. If the energy measured by CPE is below the threshold, then BS decides whether to perform fine sensing or not.

Figure 11. IEEE 802.22 Super-frames and Hierarchical Frame Structure
Stevenson, Chouinard, Lei, Hu, Shellhammer, & Caldwell, 2009.

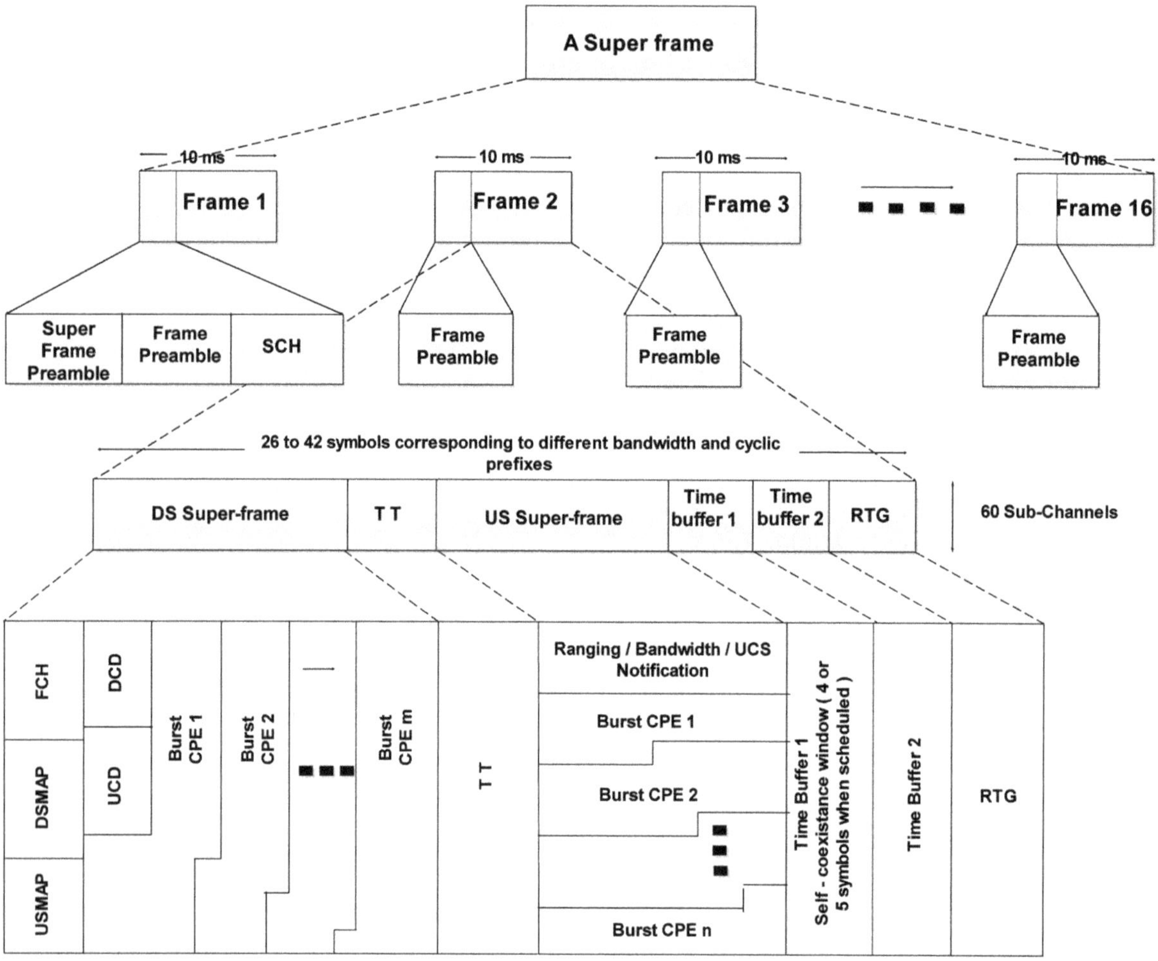

2. **Fine Sensing:** It is another mechanism which takes more time approximately (25 ms per channel and more), also be considered as detailed sensing of signals used/transmitted by PU. Progress report of this mechanism depends on the outcomes of the previous (fast sensing) mechanism. Sensing mechanism plays a major role in identifying and to protect incumbent transmitter (analog TV, digital TV, and low power licensed users such as wireless microphones) from interference.

PROJECTS AND TRENDS IN COGNITIVE RADIO

The main purpose of introducing cognitive radio technology is to get better spectrum utilization. Numerous agencies were already studied the conventional usage of radio spectrum and proposed many projects like Adaptive Reconfigurable Access and Generic Interfaces for Optimization in Radio Networks (ARAGORN).

Figure 12. Two Stage Quite Period Mechanism
Corderio, Challapali, Birru, & Shankar, 2006.

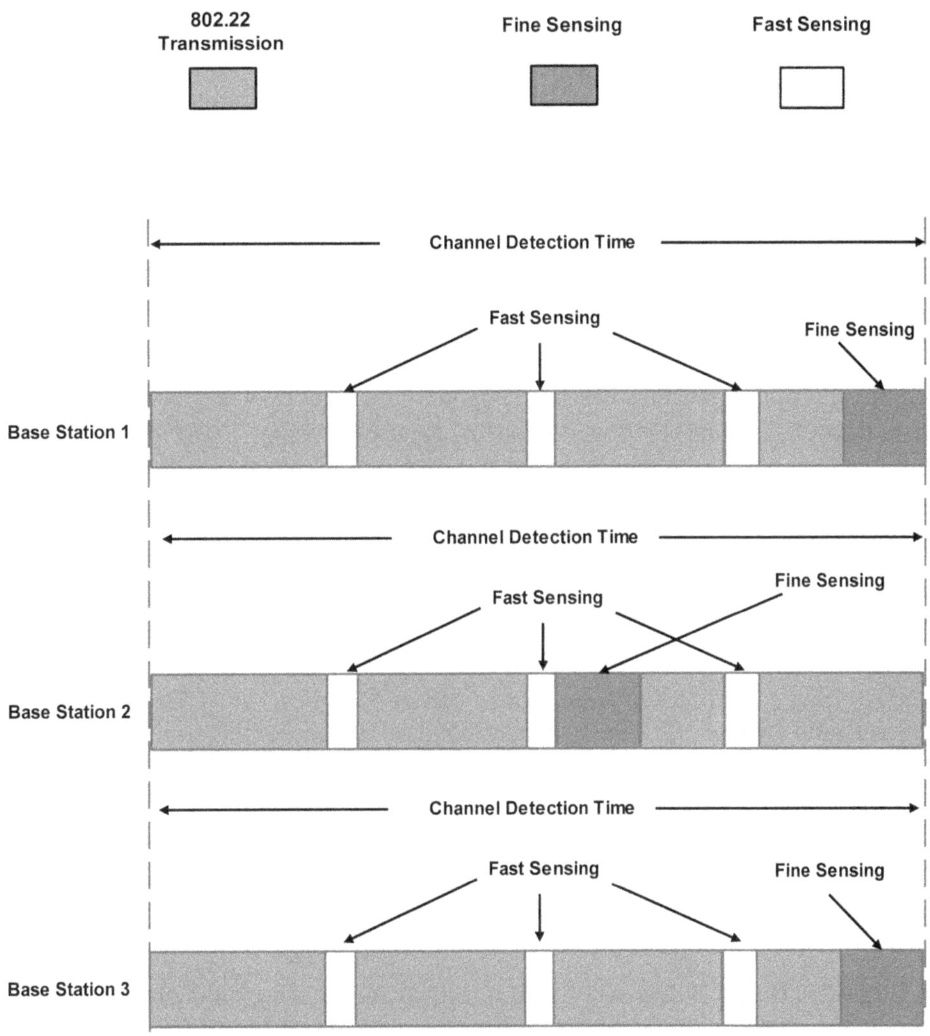

- **Adaptive Reconfigurable Access and Generic Interfaces for Optimization in Radio Networks (ARAGORN):** This project aimed to proliferate efficiency and performance of the present day as well as future systems using cognitive radio network paradigms. The experiment of this project is executed in 2.4 GHz band, aimed to combine cognitive radio module and related software module (Pandit & Kumar, 2017).
- **European Research on Ultra-Wideband (EUWB):** This project has made technical advancement in the scientific area using cognitive signalling, multiband antenna, and short–ranged ultra-band communication (Giorgetti, Chiani, Dardari, Piesiewicz, & Bruck, 2008).
- **Physical Layer for Dynamic Spectrum Access and Cognitive Radio (PHYDYAS):** This project based on filter bank-based multicarrier (FBMC) technique which provides high spectrum resolution, independent sub-channels, while maintaining the high data rate ability. The short term to

develop and demonstrate algorithms for single and multi-antenna terminals, scalability and adaptively, and multiple access (Pandit & Kumar, 2017; Di Benedetto, Hua, Kaiser, & Wang, 2008).

- **End-to-End Efficiency (E3):** This project aimed to integrate cognitive wireless network in beyond 3G (B3G) world.E3 ensures the interoperability, flexibility, and scalability between existing legacy and future wireless systems, manage the overall system complexity, convergence across access technologies, business domains, regulatory domains and geographical regions (Koenig, Nolte, Gebert, Demestichas, Stavroulaki, & Saatsakis, 2009).
- **Spectrum Pricing for Cognitive Radio:** This project actually initiates the dynamic spectrum management using cognitive radio system, European Commission funded to utilize the UHF TV spectrum for mobile communication (Marques, Rodriguez, Delaere, Delahaye, Lecroart, Gundlach, ... & Noguet, 2013).

These are Five European schemes based on cognitive technology were recommended with a fully budget of more than €50 million (Di Benedetto, Hua, Kaiser, & Wang, 2008).

APPLICATIONS OF COGNITIVE RADIO

Today's century is definitely the "century of speed" especially in the wireless communication where we have achieved best usage of the spectrum with the help of cognitive radio in pre-existing wireless networks.

- **Cognitive Radio in 5-G:** Cellular wireless standard 5G in combination with CR is presented as a future technology. At one side where CR increasing the spectrum efficiency by using free licensed spectrum. On other side 5G offering the whole wireless world interconnection jointly with very high data transmission rate and quality of service applications. The WISDOM (Wireless Innovative System for Dynamic Operating Mega communications) concept is actually utilized in 5G with the help of CR to handle the network load efficiently (Badoi, Prasad, Croitoru, & Prasad, 2011).
- **Green Cognitive Radio Networks:** Green communications is a growing research area in wireless communication. The basic objective of green radio concept is to introduce a radio architecture that reduces the total energy used in operating a radio access network. Various case strategies have been discussed that how to harvest energy, how to minimize power use in wireless base stations, how to maximize the utility and performance of CRN. Green cognitive radio requires optimization of dynamic spectrum access as well as optimal utilization of green energy (Han, Harrold, Armour, Krikidis, Videv, Grant, ... & Le, 2011; Huang, Han, & Ansari, 2015).
- **Wireless Sensor Network-Aided Cognitive Radio:** The main objective of using the cognitive technique in wireless sensor network is to circumvent the limitations of conventional wireless sensor networks. CR-WSN is a new paradigm that has the capability of handling bursty data, packet loss reduction, power waste deduction, high degree of buffer management, and has better communication quality by utilizing the spectrum efficiently.CR-WSN is good options for various applications like Military and Public Security Applications, healthcare, real-time surveillance, vehicular networks etc. (Joshi, Nam, & Kim, 2013).

CONCLUSION

CISCO reported that rapid growth in mobile traffic in last five year and worldwide data traffic is expected to raise 49 Exabyte per month by 2021 so there is need to harness underutilized spectrum. Cognitive radio will be a remarkable future technology which will integrate itself with next generation wireless system i.e 5g, and will also emerged as the best model to tackle underutilization and spectrum scarcity issue. SDR is a supportive platform which brings the versatility, a high degree of freedom in radio discipline. IEEE 802.22 standard is required for the development and effective implementation of cognitive radio and its network. This effectual approach is aimed to improve spectrum utilization in rural areas economically.

ACKNOWLEDGMENT

We thank our parents for supporting and motivating for this work because without their blessings and God's grace this work would not be possible.

REFERENCES

Akyildiz, I. F., Lee, W. Y., Vuran, M. C., & Mohanty, S. (2006). NeXt generation/dynamic spectrum access/cognitive radio wireless networks: A survey. *Computer Networks*, *50*(13), 2127–2159. doi:10.1016/j.comnet.2006.05.001

Badoi, C. I., Prasad, N., Croitoru, V., & Prasad, R. (2011). 5G based on cognitive radio. *Wireless Personal Communications*, *57*(3), 441–464. doi:10.100711277-010-0082-9

Bagwari, A., Kanti, J., & Tomar, G. S. (2016). *New Cooperative Spectrum Detection Technique in Cognitive Radio Networks*. Academic Press.

Bagwari, A., & Singh, B. (2012, November). Comparative performance evaluation of spectrum sensing techniques for cognitive radio networks. In *Computational Intelligence and Communication Networks (CICN), 2012 Fourth International Conference on* (pp. 98-105). IEEE. 10.1109/CICN.2012.66

Bagwari, A., & Tomar, G. S. (2013). Cooperative spectrum sensing with adaptive double-threshold based energy detector in cognitive radio networks. *Wireless Personal Communications*, *73*(3), 1005–1019. doi:10.100711277-013-1244-3

Bhalla, M. R., & Bhalla, A. V. (2010). Generations of mobile wireless technology: A survey. *International Journal of Computers and Applications*, *5*(4).

Brodersen, R. W., Wolisz, A., Cabric, D., Mishra, S. M., & Willkomm, D. (2004). *Corvus: a cognitive radio approach for usage of virtual unlicensed spectrum*. Berkeley Wireless Research Center (BWRC) White paper, 1-21.

Cabric, D., & Brodersen, R. W. (2005, September). Physical layer design issues unique to cognitive radio systems. In *Personal, Indoor and Mobile Radio Communications, 2005. PIMRC 2005. IEEE 16th International Symposium on* (Vol. 2, pp. 759-763). IEEE. 10.1109/PIMRC.2005.1651545

Cabric, D., Mishra, S. M., & Brodersen, R. W. (2004, November). Implementation issues in spectrum sensing for cognitive radios. In *Signals, systems and computers, 2004. Conference record of the thirty-eighth Asilomar conference on* (Vol. 1, pp. 772-776). Ieee. 10.1109/ACSSC.2004.1399240

Cavalcanti, D., & Ghosh, M. (2008, May). Cognitive radio networks: Enabling new wireless broadband opportunities. In *Cognitive Radio Oriented Wireless Networks and Communications, 2008. CrownCom 2008. 3rd International Conference on* (pp. 1-6). IEEE.

Corderio, C., Challapali, K., Birru, D., & Shankar, S. (2006). IEEE 802.22: An introduction to the first wireless standard based on cognitive radios. *Journal of Communication*, *1*(1), 38–47.

Di Benedetto, M., Hua, Y., Kaiser, T., & Wang, X. (2008). Cognitive Radio Technology. *IEEE Signal Processing Magazine*, *25*(6), 10–198. doi:10.1109/MSP.2008.930499

Dodson, N., Bradford, G. J., & Laneman, J. N. (2010, November). A high performance rf transceiver implementation. In *Proceedings of the SDR 10 Technical Conference and Product Exposition* (pp. 652-658). Academic Press.

Force, S. (2002). Spectrum policy task force report. Federal Communications Commission ET Docket 02, vol. 135.

Freewimaxinfo.com. (2017). *Advantages of cognitive radio networks*. Available at: http://freewimaxinfo.com/advantages-of-cognitive-radio-networks.html

Giorgetti, A., Chiani, M., Dardari, D., Piesiewicz, R., & Bruck, G. H. (2008, September). The cognitive radio paradigm for ultra-wideband systems: The *European project EUWB. In Ultra-Wideband, 2008. ICUWB 2008. IEEE International Conference on* (Vol. 2, pp. 169-172). IEEE.

Han, C., Harrold, T., Armour, S., Krikidis, I., Videv, S., Grant, P. M., ... Le, T. A. (2011). Green radio: Radio techniques to enable energy-efficient wireless networks. *IEEE Communications Magazine*, *49*(6), 46–54. doi:10.1109/MCOM.2011.5783984

Haykin, S. (2005). Cognitive radio: Brain-empowered wireless communications. *IEEE Journal on Selected Areas in Communications*, *23*(2), 201–220. doi:10.1109/JSAC.2004.839380

Hossain, E., Niyato, D., & Han, Z. (2009). *Dynamic spectrum access and management in cognitive radio networks*. Cambridge University Press. Retrieved from http://freewimaxinfo.com/advantages-of-cognitive-radio-networks.html

Huang, X., Han, T., & Ansari, N. (2015). On green-energy-powered cognitive radio networks. *IEEE Communications Surveys and Tutorials*, *17*(2), 827–842. doi:10.1109/COMST.2014.2387697

Ibnkahla, M. (2014). *Cooperative Cognitive Radio Networks: The Complete Spectrum Cycle*. CRC Press.

Javed, M., & Siddiqui, A. T. (2017). Transformation of Mobile Communication Network from 1G to 4G and 5G. *International Journal of Advanced Research in Computer Science*, *8*(3).

Jondral, F. K. (2005). Software-defined radio: Basics and evolution to cognitive radio. *EURASIP Journal on Wireless Communications and Networking*, (3): 275–283.

Joshi, G. P., Nam, S. Y., & Kim, S. W. (2013). Cognitive radio wireless sensor networks: Applications, challenges and research trends. *Sensors (Basel), 13*(9), 11196–11228. doi:10.3390130911196 PMID:23974152

Kanojia, D., Motwani, M., Khan, M. J., & Yadav, M. (2012). Beyond 3G: 4G Mobile Communication. *International Journal of Innovations in Engineering and Technology, 1*(2).

Kanti, J., Bagwari, A., & Tomar, G.S. (2015). Novel Spectrum detector for IEEE 802.22 Wireless Regional Area Network. *International Journal of Smart Device and Appliance, 3*(2), 9-25.

Kanti, J., & Tomar, G. S. (2016). Various Sensing Techniques in Cognitive Radio Networks: A Review. *Spectrum, 9*(1).

Khan, M. H., & Barman, P. C. (2015). 5G-Future Generation Technologies Of Wireless Communication Revolution 2020. *American Journal of Engineering Research, 4*, 206–215.

Koenig, W., Nolte, K., Gebert, J., Demestichas, P., Stavroulaki, V., & Saatsakis, A. (2009, September). Introducing cognitive systems in the B3G world: The E 3 approach to the" invisible" network of the future. In *Personal, Indoor and Mobile Radio Communications, 2009 IEEE 20th International Symposium on* (pp. 1-5). IEEE.

Liang, Y. C., Hoang, A. T., & Chen, H. H. (2008). Cognitive radio on TV bands: A new approach to provide wireless connectivity for rural areas. *IEEE Wireless Communications, 15*(3).

Marques, P., Rodriguez, J., Delaere, S., Delahaye, P., Lecroart, B., Gundlach, M., . . . Noguet, D. (2013, June). Shared use of radio spectrum in the eu: From research projects to standards. In *Communications Workshops (ICC), 2013 IEEE International Conference on* (pp. 1249-1254). IEEE.

Mitola Iii, J. (2001). Cognitive radio for flexible mobile multimedia communications. *Mobile Networks and Applications, 6*(5), 435–441. doi:10.1023/A:1011426600077

Networking, C. V. (2017). *Global mobile data traffic forecast update, 2016–2021.* Cisco white paper.

Pandi, N., & Kumar, A. (2017). *A Review on Cognitive Radio for Next Generation Cellular Network and its Challenges*. Academic Press.

Pereira, V., & Sousa, T. (2004). *Evolution of Mobile Communications: from 1G to 4G.* Department of Informatics Engineering of the University of Coimbra.

Poole, I. (2017). *IEEE 802.22 Physical Layer / Air Interface.* Available at:http://www.radio-electronics.com/info/wireless/ieee-802-22/physical-layer-air-interface.php

Prinima & Jyoti. (2016). Evolution of Mobile Communication Network: from 1G to 5G. *International Journal of Innovative Research in Computer and Communication Engineering, 4*(4).

Sengupta, S., Chatterjee, M., & Kwiat, K. (2009). *Interference aware spectrum allocation in IEEE 802.22 wireless mesh networks.* Air Force Research Lab.

Stevenson, C. R., Chouinard, G., Lei, Z., Hu, W., Shellhammer, S. J., & Caldwell, W. (2009). IEEE 802.22: The first cognitive radio wireless regional area network standard. *IEEE Communications Magazine*, *47*(1), 130–138. doi:10.1109/MCOM.2009.4752688

Tomar, G., Bagwari, A., & Kanti, J. (Eds.). (2016). *Introduction to Cognitive Radio Networks and Applications*. CRC Press. doi:10.1201/9781315367545

Wikimedia Foundation Inc. (2008). *Software Defined Radio*. Wikipedia.

Zhang, Y., Zheng, J., & Chen, H. H. (Eds.). (2016). *Cognitive radio networks: architectures, protocols, and standards*. CRC Press.

Chapter 2
Cognitive Radio Networks:
IEEE 802.22 Standards

Abhijeet Bishnu
Indian Institute of Technology Indore, India

Vimal Bhatia
Indian Institute of Technology Indore, India

ABSTRACT

Cognitive radio (CR) offers a novel way for effective usage of wireless spectrum by using dynamic spectrum sensing and allocation. One of the main components of CR is to find a spectrum hole for data transmission. Spectrum hole can be found by using spectrum sensing, a geolocation database, or by using a beacon signal. In this chapter, the authors describe algorithms for spectrum sensing in the presence of both additive white Gaussian and colored Gaussian noise. The algorithms include blind, non-blind, and cooperative sensing-based methods. The authors have compared the performance of various methods for IEEE 802.22 standard (which is the first standard incorporating CR).

INTRODUCTION

Telecommunication regulation authority of each country allocates spectrum to licensed users, also known as primary users (PUs), on a long-term basis and for large topographical regions. However, it has been found that a large amount of the allocated spectrum remains unutilized (Mishra, 2014, Sharswat, Godara, Bishnu, & Bhatia, n.d). The inefficient utilization of the limited spectrum leads to the development of dynamic spectrum access techniques. The users who do not have a license for the spectrum, also known as secondary users (SUs), are allowed to use the unutilized licensed spectrum on a temporary basis. In recent years, the researchers have been considering more comprehensive and flexible uses of the available spectrum through the use of cognitive radio (CR) technology (Wang, 2011). Figure 1 shows the contiguous bandwidth available in three different locations of Indore, India on television (TV) bandwidth. It is observed from Figure 1 that about 85-90% of the band is available which is not used for TV transmission.

DOI: 10.4018/978-1-5225-5354-0.ch002

Figure 1. Contiguous bandwidth available for (a) Mundla Nayta, (b) Harnya Khedi, and (c) Simrol

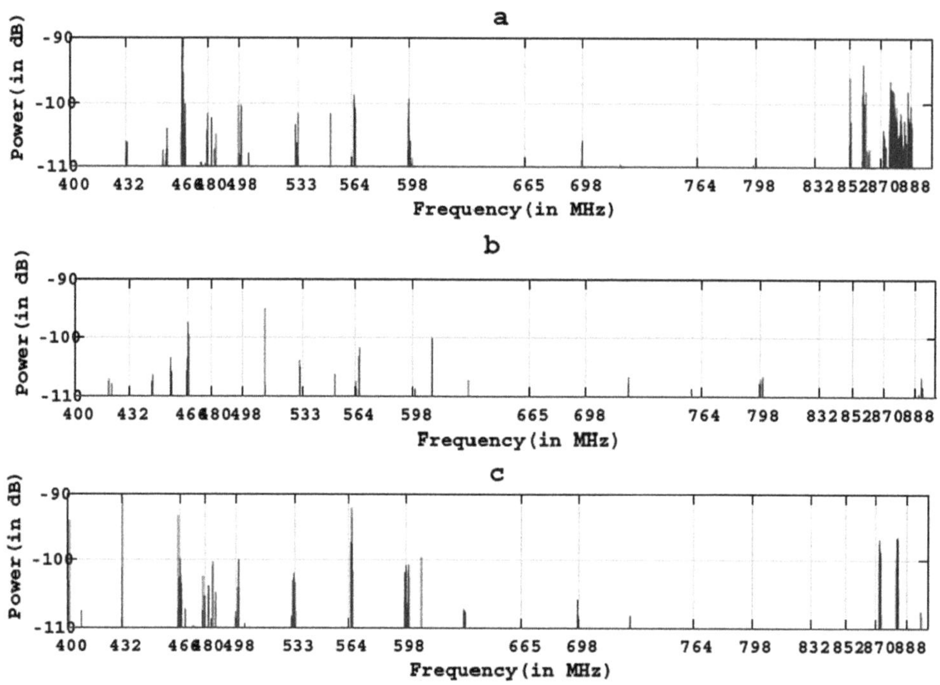

Cognitive radio offers a novel way for effective usage of spectrum by using dynamic spectrum allocation. It is defined as a radio that can alter its transmitter's parameters according to the interactions with the environment in which it operates. It differs from conventional radio devices as a CR provides *cognitive capability* and *reconfigurability* to users. Cognitive capability refers to the capability to sense and collect information from the nearby environment, such as information about transmission bandwidth, frequency, modulation power, etc. With this capability, secondary users can recognize the best available spectrum (Wang, 2011). Reconfigurability refers to the capability to rapidly adjust the operational parameters according to the sensed information in order to attain the optimal performance (Wang, 2011). By exploiting the spectrum opportunistically, the CR enables secondary users to sense the available spectrum, choose the best available channel, coordinate spectrum access with other users, and vacate the channel when a primary user returns back to access the same channel. In CR based dynamic spectrum access, the secondary users (SUs) can utilize the spectrum opportunistically without interfering with the PUs using three approaches: underlay, overlay, and interweave techniques (Haykin, 2015).

The IEEE 802.22 wireless regional area networks (WRAN) is the first wireless standard to include interweave CR in its specification where primary users are digital television (DTV) and wireless microphone, and secondary users are customer premises equipment (CPE) (Bishnu, 2015). According to federal communication commission (FCC) for in-band spectrum sensing (SS), primary users after their appearance should be detected within 2 s with the probabilities of false detection and misdetection no greater than 0.1 for maximum protection of primary users (Kim, 2010). To meet above mentioned requirements, in-band sensing must be performed once every 2 s and a detection method (e.g., energy or feature detection (Cordeiro, 2006)) that yields the best performance should be selected. Since SS is conducted during quiet periods within which communications between secondary users are suspended. Hence, the impact of SS on secondary user's quality-of-service impairment should also be considered.

SS techniques can be classified into blind, non-blind and cooperative techniques. In blind techniques, such as energy detection-based SS, statistical covariance based SS, eigenvalue based SS, principal component analysis based SS; a priori knowledge of PU signal characteristics is not needed. While, in non-blind techniques such as pilot tones, cyclic prefix, power spectral density, and cyclostationary based SS, SUs have to know some features of the PUs. The performance of most of the algorithms degrades in the presence of colored Gaussian noise. There are few algorithms proposed in the literature for SS in the colored noise which basically depends on the correlation coefficient of the noise samples. There is another technique for SS known as cooperative sensing which circumvents various issues like fading, noise uncertainty, and shadowing. Cooperative sensing also reduces the probabilities of misdetection and false alarm.

SYSTEM MODEL

Let *y(t)* be the continuous-time signal received by a SU sampled with period T_s (higher than Nyquist rate), and denoted is as $y[n] = y(nTs)$. In order to detect the PU's signal, the binary hypothesis detection model (Wang, 2011) can be written as:

$$H_0 : y\left[n\right] = w\left[n\right] \tag{1}$$

$$H_1 : y\left[n\right] = x\left[n\right] + w\left[n\right] \tag{2}$$

where $x[n]$ and $w[n]$ are the transmitted PU signal samples and additive white Gaussian noise (AWGN) respectively. Without loss of generality, it is assumed that $x[n]$ and $w[n]$ is uncorrelated. The PU signal is not present under null hypothesis, H_0 and present under alternative hypothesis, H_1.

SS is coducted based on the statistics of sensing segments having M sensing vectors with L (smoothing factor) consecutive output received samples vectors.

$$\mathrm{y}_i = \left[y\left[i\right], y\left[i+1\right], \ldots, y\left[i-L+1\right]\right]^T \tag{3}$$

where $(.)^T$ is the transpose of (.). The received sample covariance matrix for large M can be approximated as:

$$\boldsymbol{R}_{yy} = \frac{1}{M}\sum_{i=1}^{M} y_i y_i^T \tag{4}$$

Therefore, the received sample covariance matrix of kth sub-segment can be calculated as:

$$\boldsymbol{R}_{yy,k} = \frac{1}{M}\sum_{i=(k-1)M+1}^{(k-1)M+M} y_i y_i^T \tag{5}$$

Similarly, the transmitted sample covariance matrix is given as:

$$\boldsymbol{R}_{xx} = \frac{1}{M}\sum_{i=1}^{M} x_i x_i^T \tag{6}$$

and, the noise sample covariance matrix is given as:

$$\boldsymbol{R}_{ww} = \frac{1}{M}\sum_{i=1}^{M} w_i w_i^T \tag{7}$$

Two probabilities are important to evaluate the performance of SS; detection probability $P_d = P(H_1 \mid y[n] = x[n] + w[n])$, and false alarm probability $P_f = P(H_1 \mid y[n] = w[n])$. In cognitive radio, P_f relates to the spectrum utilization, and P_d relates to the interference to PU (Hou, 2014).

SPECTRUM SENSING UNDER WHITE GAUSSIAN NOISE

SS under white Gaussian noise can be mainly classified into blind, non-blind and cooperative techniques.

Blind Spectrum Sensing

In blind SS technique, *a priori* knowledge of characteristics of PUs signal is not needed to be known. It can be further classified into energy detection technique, statistical covariance matrix based technique and eigenvalue based technique. However, the eigenvalue based techniques depend on the statistical covariance matrix of the received samples. The blind SS algorithms are given as:

1. Energy Detection

Energy detection (ED) is the simplest technique to detect the signal. However, the performance of ED degrades at low signal-to-noise ratio (SNR). This method is based on the energy of the N_s received samples as:

$$T_{ED} = \frac{1}{N_s}\sum_{i=1}^{N_s} \left|y[i]\right|^2 \tag{8}$$

The probability density function (PDF) of T_{ED} follows a Chi-squared distribution of degree N_s and can be approximated as Gaussian distribution with mean, $\mu_{ED,}$ and variance, σ_{ED}^2. On the basis of PDF of T_{ED} and P_f, threshold is calculated under H_0 as (Sobron, 2015):

$$\gamma_{ED} = Q^{-1}\left(P_f\right)\sigma_{ED} + \mu_{ED} \tag{9}$$

where Q^{-1} is the inverse Q function ($Q\left(t\right) = 1 / \sqrt{2\pi}\int_t^{\infty} e^{-\left(\frac{x^2}{2}\right)} dx$) and γ_{ED} is the detection threshold. If $T_{ED} > \gamma_{ED}$, then PUs signal exist otherwise not.

2. Three Event Energy Detection

In three event energy detection (3EED), energy in three consecutive slots, E_{i-1}, E_i, E_{i+1} are estimated. First check the value of E_i, if E_i is greater than threshold, then PU exists. If E_i is lower than threshold, then additional check is performed in the preceding sensing slot, *i*-1. Again if E_{i-1} is greater than the threshold, then PU exists. If E_{i-1} is lower than threshold, then additional check is performed in the next sensing slot, *i*+1. Hence, if the energy exceeds the threshold in any of the three sensing intervals, then PU is present otherwise not. This method increases the computational complexity as compared to ED.

On the basis of PDF of energy and P_f, threshold is calculated under H_0 as (Vladeanu, 2016):

$$\gamma_{3EED} = \left[Q^{-1}\left(1 + \sqrt[3]{P_f - 1}\right)\sqrt[2]{2N_s} + N_s\right]\sigma_w^2 \tag{10}$$

where γ_{3EED} is the detection threshold and σ_w^2 is the variance of noise.

3. Covariance Absolute Value

In covariance absolute value (CAV) based detection; first, the covariance matrix of received samples is calculated. Then, two test statistics are obtained from sample covariance matrix and finally, the PU signal exists or not depends on the two test statistics ratio and threshold, γ_{CAV}. The two test statistics are given as (Zheng, 2009):

$$T_1 = \sum_{i=1}^{M}\sum_{j=1}^{M}\left|r_{ij}\right| \tag{11}$$

$$T_2 = \sum_{i=1}^{M}\left|r_{ii}\right| \tag{12}$$

where r_{ij} is the element of $\boldsymbol{R}_{yy}$. The threshold γ_{CAV} under H_0 can be calculated as:

$$\gamma_{CAV} = \frac{1+\left(L-1\right)\sqrt{\frac{2}{N_s \pi}}}{1-Q^{-1}\left(P_f\right)\sqrt{\frac{2}{N_s}}} \tag{13}$$

where N_s is the number of samples used to calculate covariance matrix of the received signal. It is noted that γ_{CAV} does not depend on noise variance. If $T_1/T_2 > \gamma_{CAV}$, then PUs signal exists, otherwise not.

4. Maximum Eigenvalue Detection

In maximum eigenvalue (ME) based detection, maximum eigenvalue of received sample covariance matrix is used as test statistics. Random matrix theory is used to analyze the method and set the threshold. The test statistics is given as (Zeng, 2008):

$$T_{ME} = \lambda_{\max} \tag{14}$$

where $\lambda_{\max}$ is the maximum eigenvalue of received sample covariance matrix. The threshold γ_{ME}, under H_0 is calculated as:

$$\gamma_{ME} = \frac{\left(\sqrt{N_s}+\sqrt{L}\right)^2}{N_s}\left(1+\frac{\left(\sqrt{N_s}+\sqrt{L}\right)^{-\frac{2}{3}}}{\left(N_s L\right)^{\frac{1}{6}}}F_1^{-1}\left(1-P_f\right)\right) \tag{15}$$

where F_1 is the cumulative distribution function (CDF) of the Tracy-Widom distribution of order 1. The distribution of maximum eigenvalue follows the Tracy-Widom distribution of order 1 for real data and Tracy-Widom distribution of order 2 for complex data. If $T_{ME} > \gamma_{ME}\sigma_w^2$, the PUs signal exists, otherwise not.

5. Maximum-Minimum Eigenvalue Detection

In maximum-minimum eigenvalue (MME) detection, the ratio of maximum eigenvalue to minimum eigenvalue of received sample covariance matrix is used as test statistics. The test statistics is given as (Zeng, 2009):

$$T_{MME} = \frac{\lambda_{\max}}{\lambda_{\min}} \tag{16}$$

where $\lambda_{\min}$ is the minimum eigenvalue of received sample covariance matrix. The threshold γ_{MME}, under H_0 is calculated as:

$$\gamma_{MME} = \frac{\left(\sqrt{N_s} + \sqrt{L}\right)^2}{\left(\sqrt{N_s} - \sqrt{L}\right)^2} \left(1 + \frac{\left(\sqrt{N_s} + \sqrt{L}\right)^{-\frac{2}{3}}}{\left(N_s L\right)^{\frac{1}{6}}} F_1^{-1}\left(1 - P_f\right) \right) \tag{17}$$

If $T_{MME} > \gamma_{MME}$, the PUs signal exists, otherwise not. It is noted that the threshold is independent from noise variance and hence it is robust than ED under noise uncertainty problem.

6. Energy With Minimum Eigenvalue Detection

In energy with minimum eigenvalue (EME) detection, the ratio of received samples energy to minimum eigenvalue of received sample covariance matrix is used as test statistics. The test statistics is given as (Zeng, 2009):

$$T_{EME} = \frac{T_{ED}}{\lambda_{\min}} \tag{18}$$

The threshold γ_{EME}, under H_0 is calculated as:

$$\gamma_{EME} = \left(\sqrt{\frac{2}{N_s}} Q^{-1}\left(P_f\right) + 1 \right) \frac{N_s}{\left(\sqrt{N_s} - \sqrt{L}\right)^2} \tag{19}$$

If $T_{EME} > \gamma_{EME}$, the PUs signal exists, otherwise not. It is noted that again the threshold is independent from noise variance and hence it is robust than ED under noise uncertainty problem.

7. Arithmetic-to-Geometric Mean Detector

In arithmetic-to-geometric mean (AGM) detector, a generalized likelihood test ratio is used in problem which finds an unstructured estimate of $\boldsymbol{R}_{yy}$. This detector calculates the AGM of eigenvalues of received sample covariance matrix. The test statistics is given as (Font-Segura, 2010):

$$T_{AGM} = \frac{\frac{1}{L}\sum_{l=1}^{L} \lambda_l}{\left(\prod_{l=1}^{L} \lambda_l\right)^{1/L}} \tag{20}$$

where λ_l is the eigenvalues of $\boldsymbol{R}_{yy}$. The threshold γ_{AGM}, under H_0 determined from the given probability of false alarm and the PDF of T_{AGM}. If $T_{AGM} > \gamma_{AGM}$, the PUs signal exists, otherwise not. Again, the threshold is independent from noise variance.

8. Scaled Largest Eigenvalue Detection

In scaled largest eigenvalue (SLE) based detection, scaled version of the largest eigenvalue of received sample covariance matrix is used as test statistics. It is shown that in (Wei, 2011) that the detection performance of SLE is better than MME and EME. The test statistics is given as (Wei, 2011):

$$T_{SLE} = \frac{\lambda_{\max}}{Tr\left(\boldsymbol{R}_{yy}\right)} \tag{21}$$

where Tr(.) is the trace of (.). The threshold γ_{SLE}, under H_0 is calculated as:

$$1 - H\left(\xi\right) = P_f \tag{22}$$

where $H\left(\xi\right)$ is the CDF of T_{SLE} with a polynomial equation in ξ [36]. If $T_{SLE} > \gamma_{SLE}$, the PUs signal exists, otherwise not.

9. Function of Matrix Based Detection

In function of matrix based detection (FMD), the trace operation which is monotonically increasing property of function of matrix is used as test statistics. The test statistics is given as (Lin, 2012):

$$T_{FMD} = Tr\left(\frac{1}{K}\sum_{k=1}^{K}\boldsymbol{R}_{yy,k}\right) \tag{23}$$

The threshold γ_{FMD}, under H_0 is calculated as:

$$\gamma_{FMD} = \left(1 + Q^{-1}\left(P_f\right)\sqrt{\frac{2}{KN_s}}\right)L\sigma_w^2 \tag{24}$$

If $T_{FMD} > \gamma_{FMD}$, the PUs signal exists, otherwise not.

10. Feature Template Matching

In feature template matching (FTM) detection, the principal eigenvector of the sample covariance matrix of both transmitted and received sample is used. The test statistics is given as (Zhang, 2013):

$$T_{FTM} = \left| \varphi_{xx,1}^{T} \varphi_{yy,1} \right| \tag{25}$$

where $\varphi_{xx,1}$ and $\varphi_{yy,1}$ are the principle eigenvector of $\boldsymbol{R}_{xx}$ and $\boldsymbol{R}_{yy}$ respectively, and |x| represents the absolute value of x. With the help of initial samples of the received signal and by some threshold $\varphi_{xx,1}$ is calculated. The threshold γ_{FTM}, under H_0 is calculated with the help of given P_f and PDF of T_{FTM}. If $T_{FTM} > \gamma_{FTM}$, the PUs signal exist otherwise not.

11. Principal Component Analysis

Principal component analysis (PCA) is an analysis technique which is used to lower the dimensionality of a data set having a large number of variables, while maintaining as much as possible of the variation present in the data set (Jolliffe, 1986). This reduction is acheived by transforming to a new set of variables, the principal components (PCs), which are uncorrelated and ordered so that the first few PCs keep most of the variation present in the original data set (Bhatti, 2012). In PCA based detection, the test statistics consists of the sum of the *P* largest eigenvalues of $\boldsymbol{R}_{yy}$ as (Bhatti, 2012):

$$T_{PCA} = \frac{1}{P} \sum_{p=1}^{P} \lambda_p \tag{26}$$

where $\lambda_1 \geq \lambda_2 \ldots \geq \lambda_p$. The threshold γ_{PCA}, under H_0 is calculated with the help of given P_f and PDF of T_{FTM}. If $T_{PCA} > \gamma_{PCA}$, the PUs signal exist otherwise not.

Non-Blind Spectrum Sensing

In non-blind SS, transmitted signal or its feature is known *a priori* to the receiver. Non-blind SS mainly consists of pilot, cyclic prefix, power spectral density, and cyclostationary based SS for orthogonal frequency division multiplexing (OFDM) based system. The non-blind SS are given as:

1. Pilot Tones Based SS

In this method, SS algorithm is based on the symbol cross-correlation of two time-domain OFDM symbols (TDSC). The TDSC algorithm uses the property that there is a non-zero constant term inserted in the TDSC if the two OFDM symbols have the same pilot symbols in frequency-domain (Chen, 2009). Thus, the proposed SS method can be applied to any OFDM system which allocates pilot symbols in the frequency domain. The TDSC algorithm has two approaches as:

a. Neyman-Pearson Approach

The test statistics of this approach is given as (Chen, 2009):

$$T_{NPPT} = \left| C(\nu) \right| \tag{27}$$

where $C(\nu)$ is the accumulated TDSC function, and ν is the symbol index difference of two OFDM symbols. $C(\nu)$ is represented as (Kim, 2007):

$$C(\nu) = \xi(\nu) \text{ under } H_0 \tag{28}$$

$$C(\nu) = e(v)\Lambda + \xi(\nu) \text{ under } H_1 \tag{29}$$

where $e(v)$ shows a phase rotation caused by the carrier frequency offset, Λ is the average received signal power in the positions of the pilot tone divided by N^2 (N is the number of subcarrier in OFDM system), and $\xi(\nu)$ follow the normal distribution.

b. Maximal Ratio Combining Approach

The test statistics of this approach is given as (Chen, 2009):

$$T_{MRCPT} = \left| \sum_{\nu} S_{\nu} S_{\nu+d} R(\nu, \nu + d) \right| \tag{30}$$

where S_{ν} is the number of TDSC function of two OFDM symbols that are accumulated and added, and $R(\nu, \nu + d)$ is the conjugate product of two accumulated TDSC function. The threshold γ_{PT}, under H_0 is calculated as:

$$\gamma_{PT} = \sqrt{-\sigma_{H_0}^2 \ln(P_f)} \tag{31}$$

If $T_{NPPT}(T_{MRCPT}) > \gamma_{PT}$, the PUs signal exists, otherwise not.

2. Cyclic Prefix Based SS

In cyclic prefix (CP) based SS, the correlation of two parts of the received samples which are N samples apart is used as test statistics which is given as (Lei, 2010):

$$T_{CP} = \sum_{d=1}^{W} \frac{w[d] w^*[N + d]}{|y[d]|^2} \text{ under } H_0 \tag{32}$$

$$T_{CP} = \sum_{d=1}^{W} \frac{y[d]\, y^*[N+d]}{\left|y[d]\right|^2} \text{ under } H_1 \tag{33}$$

where W is the observation window and (.)* represents the conjugate of (.). The threshold γ_{CP}, under H_0 is calculated as:

$$\gamma_{CP} = \frac{W}{2} P^{-1}_{k=2,\lambda_0=\frac{2c^2}{W}} \left(1 - P_f\right) \tag{34}$$

where $P^{-1}_{k,\lambda}(y)$ indicates the inverse of the non-central chi-square CDF with k-degrees of freedom and non-centrality parameter λ_0 at a particular P_f. If $T_{CP} > \gamma_{CP}$, the PUs signal exists, otherwise not.

3. Joint Sensing Scheme

In this scheme, the detection of OFDM signals for CR, based on CP and pilot tones (PT) embedded in OFDM signals (Chen, 2012). The test statistic is given as (Chen, 2012):

$$T_{CPPT} = \frac{m_{1,R}}{\sigma^2_{0,R}} \Re(R) + \frac{m_{1,C_v}}{\sigma^2_{0,C_v}} \Re(C_v) \tag{35}$$

where $m_{1,R}$ is the mean of time domain cross-correlation (R) of the received samples under H_1, $\sigma^2_{0,R}$ is the variance of R under H_0, m_{1,C_v} is the mean of C_v under H_1, σ^2_{0,C_v} is the variance of C_v under H_0 (C_v is the test statistics based on pilot tones), and $\Re$ (.) represents the real part of (.). The threshold γ_{CPPT}, under H_0 is calculated as:

$$\gamma_{CPPT} = Q^{-1}\left(P_f\right)\sqrt{\sigma^2_{0,T_{CPPT}}} \tag{36}$$

where $\sigma^2_{0,T_{CPPT}}$ represents the variance of T_{CPPT} under H_0. If $T_{CPPT} > \gamma_{CPPT}$, the PUs signal exist otherwise not.

4. Periodogram Based Detection

This method is based on the correlation of periodogram of received signal with the *a priori* known spectral features of the PU (Quan, 2009), and its test statistic is given as:

$$T_{PG} = \frac{1}{n} \sum_{k=0}^{n-1} S_Y^{(n)}(k)\, S_X^{(n)}(k) \tag{37}$$

where n represents n-point discrete Fourier transform, $S_X\left(k\right)$ and $S_X\left(k\right)$ are the power spectral density of the transmitted PU signal and received signal respectively. The threshold γ_{PG} is based on the PDF of T_{PG} at a particular P_f under H_0. If $T_{PG} > \gamma_{PG}$, the PUs signal exists, otherwise not.

5. Cyclostationary

Cylostationary is the statistical property of signal which varies periodically with time.

a. Spectral Correlation Based Detection

In spectral correlation based detection (SCD), first the spectral correlation function of the received signal is calculated by searching the whole detected bandwidth as given by (Han, 2006):

$$S_{yT}^{\alpha}\left(f\right)_{\Delta t} = \frac{1}{\Delta t}\int_{-\Delta t/2}^{\Delta t/2}\frac{1}{\sqrt{T}}Y_T\left(t, f+\frac{\alpha}{2}\right)\frac{1}{\sqrt{T}}Y_T^{*}\left(t, f-\frac{\alpha}{2}\right)dt \tag{38}$$

where Δt represents the interval time, T is the cyclostationary period, α is the cyclic frequency parameter, and $Y_T(.,.)$ is given as (Lunden, 2007):

$$Y_T\left(t,v\right) = \int_{t-T/2}^{t+T/2} y\left(u\right)e^{-j2\pi vu}du \tag{39}$$

Then the obtained spectral correlation function of the received signal is used to look for the distinctive cyclic frequencies. If unique cyclic frequencies are found, then PUs signal exist otherwise not.

b. Multiple Cyclic Frequency Based Detection

Multiple cyclic frequency (MCF) based detection is based on the fact that a signal which is cyclostationary at the symbol frequency is typically cyclostationary at all integer multiples of symbol frequency as well. The test statistic is given as (Lunden, 2007):

$$T_{MCF} = \max_{\alpha \in A} Mr_{yy}\left(\alpha\right)\Sigma_{yy}^{-1}\left(\alpha\right)r_{yy}^{T}\left(\alpha\right) \tag{40}$$

where A denote the set of cyclic frequency, $r_{yy}\left(\alpha\right)$ is given as:

$$r_{yy}\left(\alpha\right) = \left[\Re\{R_{yy}(\alpha,\tau_1\},\ldots,\Re\{R_{yy}(\alpha,\tau_N\},\Im\{R_{yy}(\alpha,\tau_1\},\ldots,\Im\{R_{yy}(\alpha,\tau_N\}\right]. \tag{41}$$

$R_{yy}\left(\alpha,\tau\right)$ is given as:

$$R_{yy}\left(\alpha,\tau\right)=\frac{1}{M}\sum_{t=1}^{M}y\left(t\right)y^{*}\left(t+\tau\right)e^{-j2\pi\alpha t} \tag{42}$$

$\Sigma_{yy}\left(\alpha\right)$ is given as:

$$\Sigma_{yy}\left(\alpha\right)=\begin{bmatrix}\Re\left\{\frac{Q+Q^{*}}{2}\right\} & \Im\left\{\frac{Q-Q^{*}}{2}\right\}\\ \Im\left\{\frac{Q+Q^{*}}{2}\right\} & \Re\left\{\frac{Q-Q^{*}}{2}\right\}\end{bmatrix} \tag{43}$$

where the (m,n)th entries of covariance matrix is given as:

$$Q\left(m,n\right)=S_{f\tau_m f\tau_n}\left(2\alpha,\alpha\right) \tag{44}$$

$$Q^{*}\left(m,n\right)=S^{*}_{f\tau_m f\tau_n}\left(0,-\alpha\right) \tag{45}$$

where $S_{f\tau_m f\tau_n}$ and $S^{*}_{f\tau_m f\tau_n}$ denote the unconjugated and conjugated cyclic spectra of $f\left(t,\tau\right)=y\left(t\right)y^{*}\left(t+\tau\right)$ respectively. The threshold γ_{MCF}, under H_0 is calculated with the help of given P_f and PDF of T_{MCF}. If $T_{MCF} > \gamma_{MCF}$, the PUs signal exists, otherwise not.

c. Cycle Frequency Domain Profile Based Detection

In this method, the cycle frequency domain profile (CDP) is used for signal detection and preprocessing for signal detection (Kim, 2007). Signal features are obtained from CDP using a threshold-test method. Hidden Markov model has been used to process obtained signal feature for classification. The test statistic is given as (Kim, 2007):

$$T_{CDP}=\frac{\max\left(I\left(\alpha\right)\right)}{\sqrt{\left(\sum_{\alpha=0}^{M-1}I^{2}\left(\alpha\right)\right)/M}} \tag{46}$$

where $I\left(\alpha\right)\triangleq\max_{f}\left|C_{y}^{\alpha}\left(f\right)\right|$ and:

$$C_{y}^{\alpha}\left(f\right)=\frac{S_{y}^{\alpha}\left(f\right)}{\left[S\left(f+\frac{\alpha}{2}\right)S\left(f-\frac{\alpha}{2}\right)\right]^{1/2}} \tag{47}$$

is the spectral coherence and $S_y^{\alpha}\left(f\right)$ is the spectral correlation function. The threshold γ_{CDP} defined by (46) under H_0. If $T_{CDP} > \gamma_{CDP}$, then PUs signal exists, otherwise not.

SPECTRUM SENSING UNDER COLORED GAUSSIAN NOISE

The algorithms that we have discussed so far are performed well in the presence of AWGN. However, performance of most of the algorithms degrades in the presence of colored or correlated noise. The algorithms which are based on CP, PT and cyclostationary are mainly depend on the cross-correlation function and hence perform well at low signal-to-noise ratio (SNR) in the presence of uncorrelated samples. However, in the presence of correlated samples due to the colored noise, their performance degrades at low SNR. Similarly the algorithms which are based on eigenvalue of the received sample covariance matrix perform well at low SNR in the presence of uncorrelated samples or AWGN. However, the performance of eigenvalue based algorithms degrades and sometimes fails to detect PU's signal in the presence of correlated signals due to the swelling effect of covariance matrix (Horev, 2017). The received signal is correlated because of filtering and oversampling (Sharma, 2013). When the received signal is passed through a filter (low pass filter or shaping filter) at the receiver's input, then the white noise is also passed through the same filter and because of the transfer function of the shaping filter, output of the filter results in correlated signal. Similarly, at the receiver the analog-to-digital converter samples the received analog signal at greater than twice the bandwidth of the signal and leads to correlation in the output sampled data (Sharma, 2013). There are few algorithms which are based on eigenvalue of the received sample covariance matrix for the correlated samples which are as follows:

1. Standard Condition Number Based Detection

The standard condition number (SCN) of a matrix is the ratio of maximum eigenvalue to minimum eigenvalue. The test statistic is given by (Sharma, 2013):

$$T_{SCN} = \frac{\tilde{b}}{\tilde{a}} \tag{48}$$

is also known as SCN of the received sample covariance matrix. Here,

$$\tilde{b} = 1 + \beta + 2\mu\beta + 2\sqrt{\beta}\sqrt{\left(1+\mu\right)\left(1+\mu\beta\right)}. \tag{49}$$

And:

$$\tilde{a} = 1 + \beta + 2\mu\beta - 2\sqrt{\beta}\sqrt{\left(1+\mu\right)\left(1+\mu\beta\right)} \tag{50}$$

β<1, $\mu = \left(\sqrt{\sigma_2} - \sqrt{\sigma_1}\right)^2 / 4\sigma_1\sigma_2$, $\sigma_2 = 1 + \rho / 1 - \rho$, $\sigma_1 = 1 - \rho / 1 + \rho$, and ρ (<1) is the correlation coefficient of the colored noise. The threshold γ_{SCN} defined by (48) under H_0. If $T_{SCN} > \gamma_{SCN}$, then PUs signal exists, otherwise not.

2. Maximum Eigenvalue Based Detection

In this detection, the test statistic depends on the maximum eigenvalue of the received sample covariance matrix in the presence of colored noise as given by (Sharma, 2014):

$$T_{MECN} = \lambda_{\max}^{c} \tag{51}$$

where $\lambda_{\max}^{c}$ is the maximum eigenvalue of $\mathbf{R}_{yy}$ in the presence of colored noise. The threshold γ_{MECN} under H_0 can be calculated as (Sharma, 2014):

$$\gamma_{MECN} = \frac{\sigma_c F_1^{-1}\left(1 - P_f\right) + N_s^{2/3}\nu_c}{N_s^{2/3}} \tag{52}$$

ν_c is defined as (Sharma, 2014);

$$\nu_c = \frac{1}{c}\left(1 + \frac{c}{2\pi\mu\delta}\int_{\sigma_1}^{\sigma_2} \frac{1}{\left(1 - \lambda c\right)\lambda}\sqrt{-\lambda^2 - 1 + 2\lambda\mu_1 d\lambda}\right) \tag{53}$$

$$\sigma_c^3 = \frac{1}{c^3}\left(1 + \frac{c^3}{2\pi\mu\delta}\int_{\sigma_1}^{\sigma_2} \frac{\lambda}{\left(1 - \lambda c\right)^3}\sqrt{-\lambda^2 - 1 + 2\lambda\mu_1 d\lambda}\right) \tag{54}$$

Here, $c \in [0, 1/\lambda_{\max}^{c}]$, $\delta > 1$, $\sigma_1 \le \lambda \le \sigma_2$, and $\mu_1 = \left(1 + \rho^2\right)/\left(1 - \rho^2\right)$. If $T_{MECN} > \gamma_{MECN}$, then PUs signal exists, otherwise not.

COOPERATIVE SENSING

As discussed in previous section, SS is heavily sensitive to noise uncertainty, shadowing, and fading. To circumvent this issue, cooperative sensing has been proposed in literature. The cooperative sensing decreases the probabilities of false alarm and misdetection considerably (Yucek, 2009). Additionally, cooperation can solve hidden problem of PU and also decreases the sensing time (Ganesan, 2005). Developing efficient information sharing algorithm and increased complexity are the main challenges of

SS (Guo, 2006). Cooperative SS is most effective when cooperative CR's observes independent fading or shadowing (Liu, 2006). In (Peh, 2007), it is proved that cooperating with all users in the network does not necessarily attain the optimum performance and thus, cognitive users with highest PU's SNR are chosen for cooperation. Cooperative SS can be implemented in three ways as centralized sensing, distributed sensing and external sensing.

1. Centralized Sensing

In centralized sensing, a central unit also called as access point gathers sensing information from cognitive devices, recognize the available spectrum, and broadcast this information to others cognitive devices.

a. Signal-to-Noise Ratio Based Detection

In this detection method (Visotsky, 2005), the existence of PU is checked by the observed SNR at each sensing node U_i, (i=1, 2, …, n where n is the number of sensing device or SU). The SS problem can be cast as:

$$H_0 = \boldsymbol{SNR}_U \sim N\left(\mu_0 x1, \sigma_w^2 \Sigma\right) \tag{55}$$

$$H_1 = \boldsymbol{SNR}_U \sim N\left(\mu_1 x1, \sigma_w^2 \Sigma\right) \tag{56}$$

where $\boldsymbol{SNR}_U$ is the vector of SNR_{U_i} of dimension nx1, 1 is a vector of ones with dimension nx1, $\Sigma = \bar{\Sigma} / \sigma_w^2$ (where **£** is the covariance matrix of the received samples), $\mu_0 = \mu\left(\bar{R}\right)$, and $\mu_1 = \mu\left(\bar{R} + \delta\right)$ with δ >0 ($\bar{R}$ is the distance from the PU transmitter at which SU transmissions are statistically unlikely to cause interference to the PU receivers). The test statistic is given as:

$$T_{SNR} = \sum_{i=1}^{n} B_i \tag{57}$$

Here, B_i=1, if i^{th} SU detects the PU otherwise 0. The threshold γ_{SNR} under H_0 is based on a particular P_f. The information of each B_i is sent to central unit where central unit decides whether a PUs signal is present or not.

b. Cluster Based Detection

In cooperative SS, all the sensors send their sensing observation to the central unit through fading channels, which leads to degradation in the sensing performance. Cluster based SS improves the sensing performance by selecting all the sensors into a few clusters and selecting the most appropriate sensor in each cluster to report to the central unit (Sun, 2007). The steps for the cluster based SS are as follows (Sun, 2007):

1. Every sensor j in cluster i gathers energy $E_{i,j}$ and dispatches observation $G_{i,j}$ to the cluster head, where $G_{i,j}$ is related to $E_{i,j}$ by a function Ω.
2. Then each cluster head receives observation and makes a cluster decision C_i according to some fusion function Φ.
3. The cluster decision C_i is reported to the central unit and then a final decision is made according to some fusion function φ.

2. Distributed Sensing

In distributed sensing, sensor nodes share their sensing observations with each other nodes. Distributed sensing is more superior to centralized sensing as distributed sensing does not require backbone structure and also reduces the cost (Yucek, 2009). A SU receives decision from rest of the sensors or SU's and decides H_1 if any of the sensor decision is H_1. This fusion rule is known as OR-rule (OR is logical OR operator) (Ghasemi, 2005).

3. External Sensing

There is another method which is used for acquiring spectrum information is external sensing. In external sensing, an external agent collects the information of spectrum occupancy and broadcast this information to other CR. External sensing helps to solve overcoming hidden PUs and the uncertainty due to fading and shadowing. Additionally spectrum efficiency is also increased as the CRs do not spend time in SS. External sensing is also one of the method which is proposed for identifying PUs in IEEE 802.22 standard.

PERFORMANCE EVALUATION

From all the above various algorithms for SS, the most popular and studied SS are covariance and eigenvalue based SS. Hence, in this section we evaluate the performance of covariance and eigenvalue of received sample covariance matrix based algorithms, we used real world measurement of DTV signal (Tawil, 2006) captured with duration of about 25 seconds in Washington, DC, USA as a PU's signal. Received SNR and communication channel between transmitter and receiver are unknown. For all results, 1000 Monte Carlo simulations are performed with $L = 10$, $K = 10$, $M = 1000$, and $N_s = 100000$. To obtain various SNR level, both white Gaussian noise and colored Gaussian noise is added to the primary signal as given in (Shellhammer, 2006).

Figure 2 shows the P_d vs SNR of blind algorithms in the presence of white Gaussian noise at $P_f = 0.05$. It is observed from Figure 2 that the FMD based detection has best performance while Energy detection has worst performance in terms of PU's signal detection at low SNR. PCA and ME based algorithms also perform better with small performance degradation (2-3 dB) than FMD. The performance of rest of the algorithms except FTM and EME are same with high performance degradation (7 dB) than FMD.

The performance of FTM and EME are similar with very high performance loss (10 dB) than FMD. Figure 3 shows the P_d vs P_f (also known as receiver operating characteristic (ROC)) of blind algorithms in the presence of white Gaussian noise at -25 dB SNR. It is observed from Figure 3 that FMD, PCA, and ME based algorithms are nearly same performance and much better performance than rest of other algorithms as they detect the signal at very low P_f. The ROC performances of rest of the algorithms are similar and poor than FMD, PCA and ME based algorithms.

Figure 4 shows the P_d vs SNR of blind algorithms in the presence of colored Gaussian noise at P_f = 0.05. It is observed from Figure 4 that the FMD based detection has best performance while many of algorithms including CAV, ME, MME, and SLE fail to detect the signal. MECN and SCN based algorithm which are proposed for colored noise have much better performance than their corresponding ME and MME based algorithm for white noise respectively. There is small performance degradation (2 dB) of MECN and SCN than FMD, while there is high performance degradation (8 dB) of PCA and FTM based algorithm than FMD. Energy and EME based detection require more SNR than SNR required for FMD to detect the signal. Figure 5 shows the ROC of blind algorithms in the presence of white Gaussian noise at -18 dB SNR. It is observed from Figure 5 that FMD, MECN, and SCN based algorithms are nearly same performance and much better performance than rest of other algorithms as they detect the signal at very low P_f. The ROC performance of PCA and FTM is better than that of EME and energy based detection.

Figure 2. P_d vs SNR for blind algorithms in the presence of white Gaussian noise at P_f = 0.05.

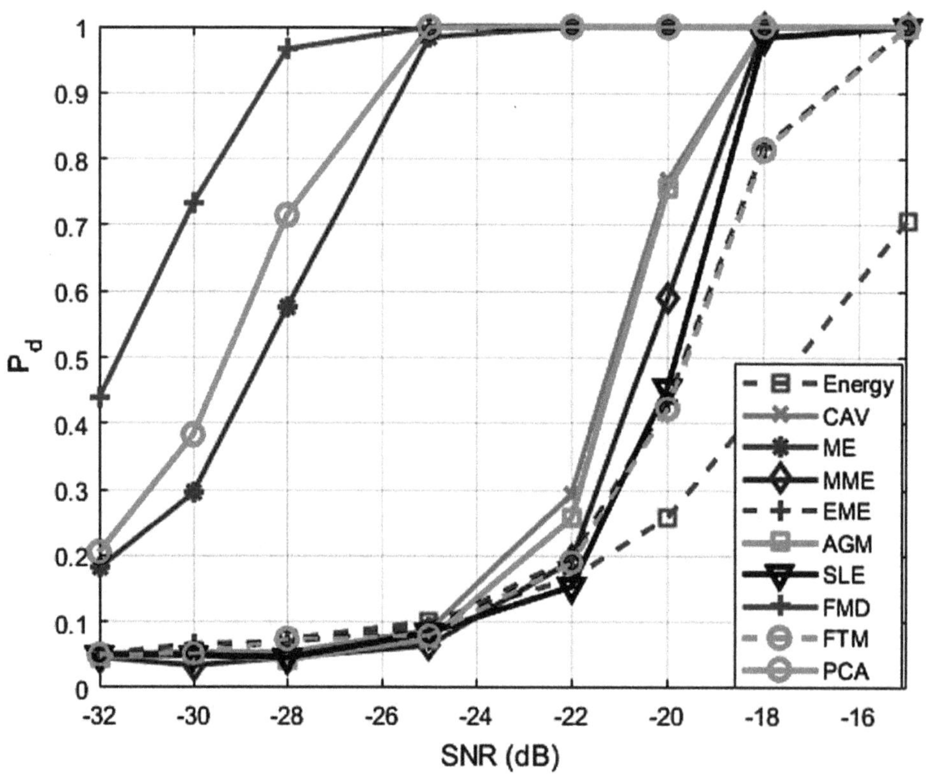

Figure 3. P_d vs P_f of blind algorithms in the presence of white Gaussian noise at -25 dB SNR.

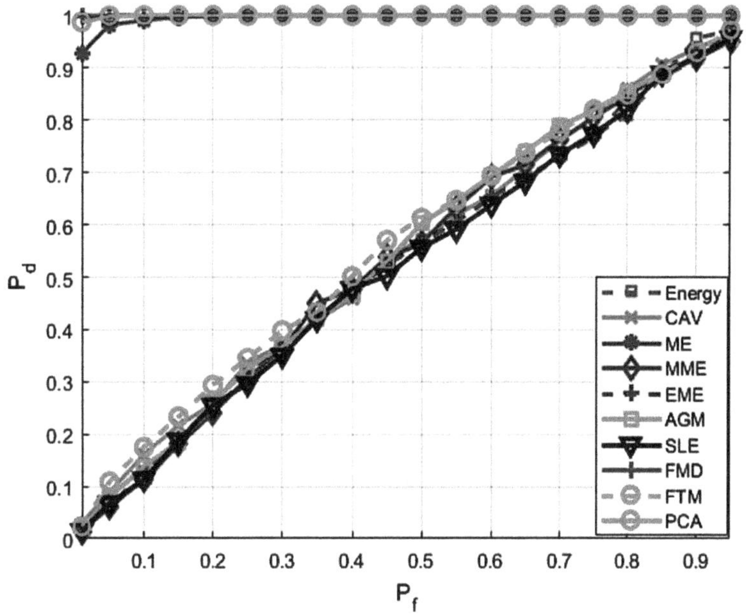

In terms of computational complexity, energy detection based algorithms are least complex, CAV requires O(ML) computation, FMD requires O((M+1)L), and rest of other algorithms of blind method nearly require O(ML+L^3) computation. The non-blind methods generally depend on the cross-correlation function and hence all non-blind methods have nearly same computational complexity and it is much less than the calculation of the eigenvalues of a matrix.

SPECTRUM SENSING FOR IEEE 802.22

IEEE 802.22 is the first wireless standard which is based on CR to provide wireless access to rural areas where PU's are DTV and wireless microphone, and SU's are CPE. Thus for maximum protection of PU's, PU should be observed within 2 s of their appearance by SU's with the probability of misdetection (P_{md}) and P_f no greater than 0.1 (Chen, 2012). The constrained on P_{md} focuses on the protection of PUs by accurately detecting their re-appearance otherwise SUs can interfere the operation of PUs. On the other hand, the constrained on P_f aim to boost the QoS of SUs by avoiding frequent channel switching. For large P_f, the SUs will vacate the in-band channel and move to another channel even in the absence of PUs. Once the channel vacate then it will be sensed or reused after 10 minutes (Stevenson, 2006) and hence the collection of idle channels will be soon depleted.

To meet the requirement constrained on P_{md} and P_f, IEEE 802.22 provides two step sensing mechanism which includes fast and fine sensing during quiet period. Fast sensing employed energy detection, while fine sensing employed either feature detection or eigenvalue based algorithms. For one time sensing, fast sensing takes small period of time but its performance is sensitive to noise variance, while fine sensing consume much amount of time but its performance is much better than fast sensing. Hence, an efficient sensing algorithm requires for fine sensing which capture the trade-off between low P_{md}, P_{fa},

Figure 4. P_d vs SNR for blind algorithms in the presence of colored Gaussian noise at $P_f = 0.05$.

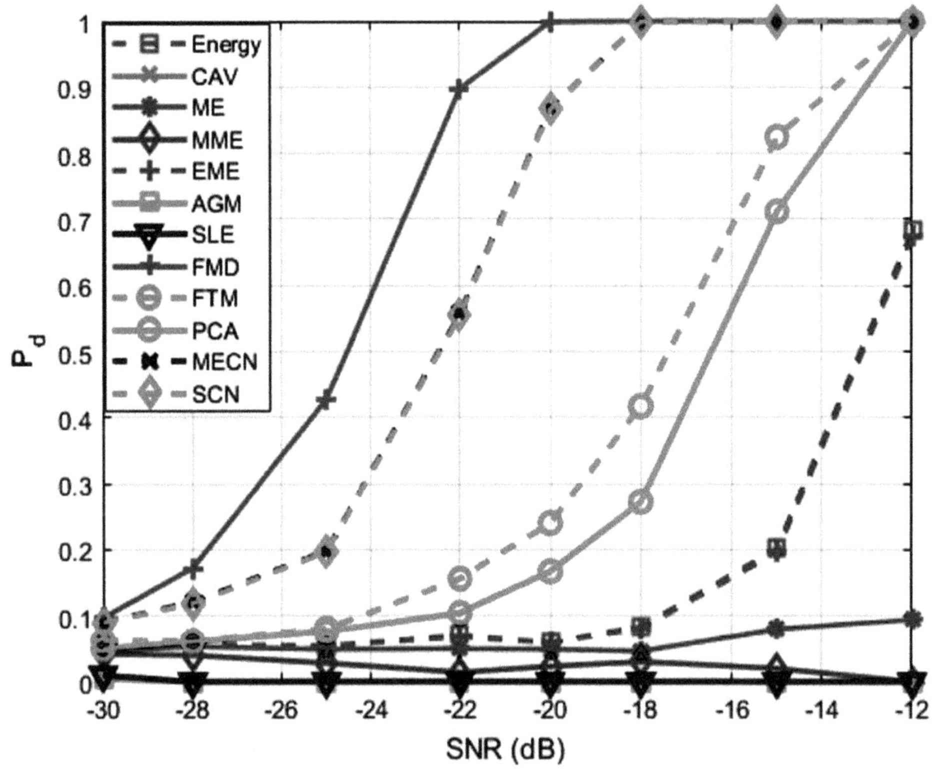

and computational complexity. For fine sensing, there are non-blind methods which are generally based on cross-correlation and eigenvalue based methods. The performance of non-blind methods may degrade due to the correlated samples because of filtering or/and oversampling. In eigenvalue based methods, it is observed from all the figures that FMD, PCA, and ME based algorithms perform much better than other algorithms in the presence of white Gaussian noise and FMD, SCN, and MECN based algorithms perform much better in the presence of colored Gaussian noise. However, the threshold of these algorithms except SCN depends on the noise variance and hence the above mentioned algorithms except SCN are not robust. The threshold of SCN depends on the correlation coefficient of noise samples, and hence SCN is more robust than above algorithms because the correlation coefficient varies much slowly than noise variance. MME and EME based algorithms are those algorithms whose threshold is independent from noise variance and hence MME and EME are robust algorithms. It is observed from Figure 2 that MME performs better than EME in the presence of white Gaussian noise, and it is also observed from Figure 4 that SCN (which is nothing but MME in the presence of colored noise) performs much better than EME. Thus, MME in the presence of white Gaussian noise and SCN in the presence of colored Gaussian noise are best for fine sensing for IEEE 802.22.

Figure 5. P_d vs P_f of blind algorithms in the presence of colored Gaussian noise at -18 dB SNR.

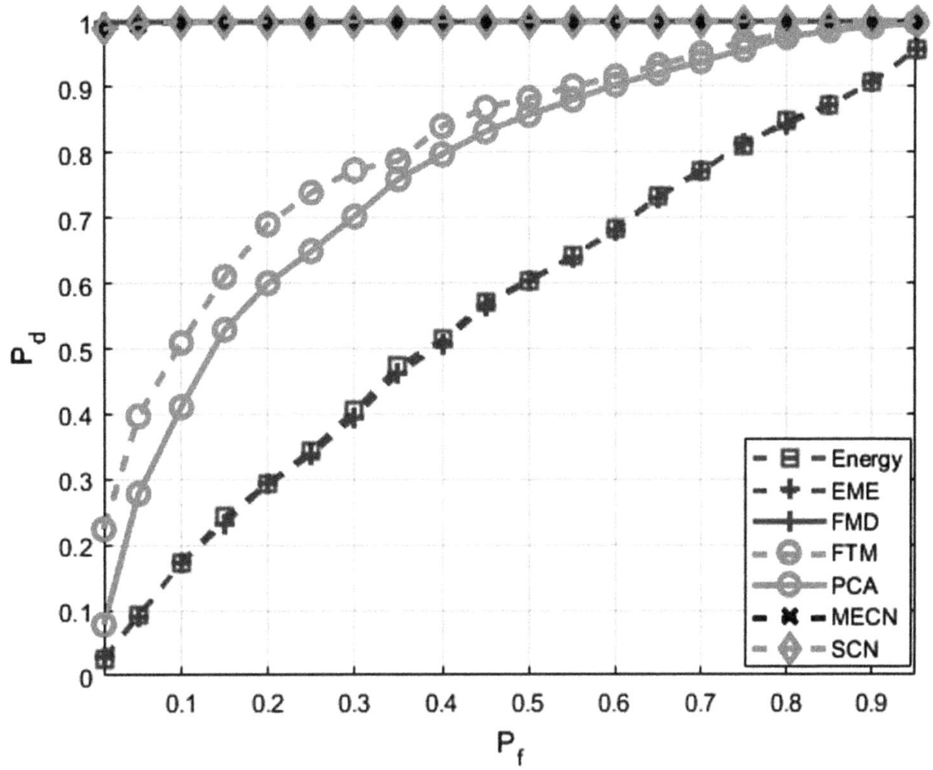

CONCLUSION

In this chapter, we presented a review study of various spectrum sensing algorithms under both white Gaussian noise and colored Gaussian noise. We have also presented the cooperative spectrum sensing which is more beneficial than local spectrum sensing. By analyzing the performance of different algorithms, we suggest that the maximum to minimum eigenvalue based sensing is appropriate in the presence of white Gaussian noise, while standard condition number based sensing is appropriate in the presence of colored Gaussian noise for fine sensing in the IEEE 802.22 standard.

REFERENCES

Bhatti, F. A., Rowe, G. B., & Sowerby, K. W. (2012). Spectrum Sensing using Principal Component Analysis. In *Proceedings of IEEE Wireless Communications and Networking Conferences* (pp. 725-730). Paris, France: IEEE.

Bishnu, A., & Bhatia, V. (2015). On performance analysis of IEEE 802.22 (PHY) for COST-207 channel models. In *Proceedings of IEEE Conference on Standards for Communications and Networking* (pp. 229-234). Tokyo, Japan: IEEE. 10.1109/CSCN.2015.7390449

Chen, H.-S., Gao, W., & Daut, D. G. (2009). Spectrum Sensing for OFDM Systems Employing Pilot Tones. *IEEE Transactions on Wireless Communications*, *8*(12), 5862–5870. doi:10.1109/TWC.2009.12.080777

Chen, Z., Luan, T., & Zhang, X.-D. (2012). Sensing orthogonal frequency division multiplexing systems for cognitive radio with cyclic prefix and pilot tones. *IET Communications*, *6*(1), 97–106. doi:10.1049/iet-com.2010.1038

Cordeiro, C., Challapali, K., & Ghosh, M. (2006). Cognitive PHY and MAC Layers for Dynamic Spectrum Access and Sharing of TV Bands. *Proceedings of First International Workshop on Technology and Policy for Accessing Spectrum*. 10.1145/1234388.1234391

Font-Segura, J., & Wang, X. (2010). GLRT-Based Spectrum Sensing for Cognitive Radio with Prior Information. *IEEE Transactions on Communications*, *58*(7), 2137–2146. doi:10.1109/TCOMM.2010.07.090556

Ganesan, G., & Li, Y. (2005). Cooperative spectrum sensing in cognitive radio networks. In *Proceedings of IEEE Symposium on New Frontiers Dynamic Spectrum Access Networks* (pp. 137-143). Baltimore, MD: IEEE. 10.1109/DYSPAN.2005.1542628

Ghasemi, A., & Sousa, E. S. (2005). Collaborative Spectrum Sensing for Opportunistic Access in Fading Environments. In *Proceedings of IEEE Symposium on New Frontiers Dynamic Spectrum Access Networks* (pp. 131-136). Baltimore, MD: IEEE. 10.1109/DYSPAN.2005.1542627

Guo, C., Zhang, T., Zeng, Z., & Feng, C. (2006). Investigation on Spectrum Sharing Technology Based on Cognitive Radio. In *Proceedings of IEEE International Conference on Communications and Networking in China* (pp. 1-5). Beijing, China: IEEE. 10.1109/CHINACOM.2006.344901

Han, N., Shon, S., Chung, J. H., & Kim, J. M. (2006). Spectral Correlation Based Signal Detection Method for Spectrum Sensing in IEEE 802.22 WRAN Systems. In *Proceedings of IEEE International Conference on Advanced Communication Technology* (vol. 3). Gangwon-Do, South Korea: IEEE.

Haykin, S., & Setoodeh, P. (2015). Cognitive Radio Networks: The Spectrum Supply Chain Paradigm. *IEEE Transactions on Cognitive Communications and Networking*, *1*(1), 3–28. doi:10.1109/TCCN.2015.2488627

Horev, I., Yger, F., & Sugiyama, M. (2017). Geometry-aware principal component analysis for symmetric positive definite matrices. *Springer Machine Learning*, *106*(4), 93–522.

Hou, S., & Qiu, R. C. (2014). Kernel Feature Template Matching for Spectrum Sensing. *IEEE Transactions on Vehicular Technology*, *63*(5), 2258–2271. doi:10.1109/TVT.2013.2290866

Jolliffe, I. T. (1986). *Principal Component Analysis*. Springer-Verlag. doi:10.1007/978-1-4757-1904-8

Kim, H., & Shin, K. G. (2010). In-Band Spectrum Sensing in IEEE 802.22 WRANs for Incumbent Protection. *IEEE Transactions on Mobile Computing*, *9*(12), 1767–1779. doi:10.1109/TMC.2010.169

Kim, K., Akbar, I. A., Bae, K. K., Um, J.-S., Spooner, C.-M., & Reed, J. H. (2007). Cyclostationary Approaches to Signal Detection and Classification in Cognitive Radio. In *Proceedings of IEEE Symposium on New Frontiers Dynamic Spectrum Access Networks* (pp. 212-215). Dublin, Ireland: IEEE. 10.1109/DYSPAN.2007.35

Lei, Z., & Chin, F. P. S. (2010). Sensing OFDM Systems Under Frequency-Selective Fading Channels. *IEEE Transactions on Vehicular Technology*, *59*(4), 1960–1968. doi:10.1109/TVT.2010.2042139

Lin, F., Qiu, R. C., Hu, Z., Hou, S., Browning, J. P., & Wicks, M. C. (2012). Generalized FMD Detection for Spectrum Sensing under Low Signal-to-Noise Ratio. *IEEE Communications Letters*, *16*(5), 604–607. doi:10.1109/LCOMM.2012.030512.112164

Liu, X., & Sai Shankar, N. (2006). Sensing-based opportunistic channel access. *Mobile Networks and Applications*, *11*(4), 577–591. doi:10.100711036-006-7323-x

Lunden, J., Koivunen, V., Huttunen, A., & Poor, H. V. (2007). Spectrum Sensing in Cognitive Radios Based on Multiple Cyclic Frequencies. In *Proceedings of IEEE International Conference on Cognitive Radio Oriented Wireless Network and Communications* (pp. 37-43). Orlando, FL: IEEE. 10.1109/CROWNCOM.2007.4549769

Mishra, A. K., & Johnson, D. L. (2014). *White Space Communications: Advances, Developments and Engineering Challenges*. Springer.

Peh, E., & Liang, Y.-C. (2007). Optimization for Cooperative Sensing in Cognitive Radio Networks. In *Proceedings of IEEE Wireless Communications and Networking Conference* (pp. 27-32). Paris, France: IEEE. 10.1109/WCNC.2007.11

Quan, Z., Shellhammer, S. T., Zhang, W., & Sayed, A. H. (2009). Spectrum Sensing by Cognitive Radios at Very Low SNR. In *Proceedings of IEEE Global Communications* (pp. 1-6). Honolulu, HI: IEEE. 10.1109/GLOCOM.2009.5426262

Sahai, A., & Cabric, D. (2005). Spectrum sensing: fundamental limits and practical challenges. In *Proceedings of IEEE Symposium on New Frontiers Dynamic Spectrum Access Networks*. Baltimore, MD: IEEE.

Sharma, S. K., Chatzinotas, S., & Ottersten, B. (2013). Eigenvalue-Based Sensing and SNR Estimation for Cognitive Radio in Presence of Noise Correlation. *IEEE Transactions on Vehicular Technology*, *62*(8), 3671–3683. doi:10.1109/TVT.2013.2260834

Sharma, S. K., Chatzinotas, S., & Ottersten, B. (2014). Maximum Eigenvalue Detection for Spectrum Sensing Under Correlated Noise. In *Proceedings of IEEE International Conference on Acoustic, Speech, and Signal Processing* (pp. 7268-7272). Florence, Italy: IEEE. 10.1109/ICASSP.2014.6855011

Sharswat, V., Godara, P., Bishnu, A., & Bhatia, V. (in press). Measurement and Identification of White Space using Low Cost SDR for Rural Wireless Broadband. In *IEEE International Conference on Wireless Communications, Vehicular Technology, Information Theory, Aerospace and Electronic System*. Hyderabad, India: IEEE.

Shellhammer, S., Tawil, V., Chouinard, G., Muterspaugh, M., & Ghosh, M., (2006). *Spectrum Sensing Simulation Model*. IEEE 802.22-06/0028r10, 2006.

Sobron, I., Diniz, P. S. R., Martins, W. A., & Velez, M. (2015). Energy Detection Technique for Adaptive Spectrum Sensing. *IEEE Transactions on Communications*, *63*(3), 617–627. doi:10.1109/TCOMM.2015.2394436

Stevenson C. R., Cordeiro, C., Sofer, E., & Chouinard, G., (2006). *Functional Requirements for the 802.22 WRAN Standard*. IEEE 802.22-05/0007r47, Jan. 2006.

Sun, C., Zhang, W., & Letaief, K. B. (2007). Cluster-Based Cooperative Spectrum Sensing in Cognitive Radio Systems. In *Proceedings of IEEE International Conference on Communications* (pp. 2511-2515). Glasgow, UK: IEEE. 10.1109/ICC.2007.415

Tawil, V. (2006). *51 Captured DTV Signal*. Available: http://grouper.ieee.org/groups/802/22/Meetingsdocuments/2006May

Visotsky, E., Kuffner, S., & Peterson, R. (2005). On Collaborative Detection of TV Transmissions in Support of Dynamic Spectrum Sharing. In *Proceedings of IEEE Symposium on New Frontiers Dynamic Spectrum Access Networks* (pp. 338-345). Baltimore, MD: IEEE. 10.1109/DYSPAN.2005.1542650

Vladeanu, C., Nastase, C.-V., & Martian, A. (2016). Energy Detection Algorithm for Spectrum Sensing Using Three Consecutive Sensing Events. *IEEE Wireless Communications Letters*, *5*(3), 284–287. doi:10.1109/LWC.2016.2543723

Wang, B., & Liu, K. J. R. (2011). Advances in Cognitive Radio Networks: A Survey. *IEEE Journal of Selected Topics in Signal Processing*, *5*(1), 5–23. doi:10.1109/JSTSP.2010.2093210

Wei, L., & Tirkkonen, O. (2011). Analysis of Scaled Largest Eigenvalue Based Detection for Spectrum Sensing. In *Proceedings of IEEE International Conference on Communications* (pp. 1-5). Kyoto, Japan: IEEE. 10.1109/icc.2011.5962520

Yucek, T., & Arslan, H. (2009). A Survey of Spectrum Sensing Algorithms for Cognitive Radio Applications. *IEEE Communications Surveys and Tutorials*, *11*(1), 116–130. doi:10.1109/SURV.2009.090109

Zeng, Y., Koh, C. L., & Liang, Y.-C. (2008). Maximum Eigenvalue Detection: Theory and Application. In *Proceedings of IEEE International Conference on Communications* (pp. 4160-4164). Beijing, China: IEEE.

Zeng, Y., & Liang, Y.-C. (2009). Spectrum-Sensing Algorithms for Cognitive Radio Based on Statistical Covariances. *IEEE Transactions on Vehicular Technology*, *58*(4), 1804–1815. doi:10.1109/TVT.2008.2005267

Zeng, Y., & Liang, Y.-C. (2009). Eigenvalue–Based Spectrum Sensing Algorithms for Cognitive Radio. *IEEE Transactions on Communications*, *57*(6), 1784–1793. doi:10.1109/TCOMM.2009.06.070402

Zhang, P., & Qiu, R. (2013). GLRT-Based Spectrum Sensing with Blindly Learned Feature under Rank-1 Assumption. *IEEE Transactions on Communications*, *61*(1), 87–96. doi:10.1109/TCOMM.2012.100912.120162

Section 2
Overview of Various Sensing Schemes for Cognitive Radio Networks

Chapter 3
A Quick Overview of Different Spectrum Sensing Techniques

Pooja Joshi
Uttarakhand Technical University, India

Ashish Bagwari
Uttarakhand Technical University, India

Ashish Negi
Govind Ballabh Pant Engineering College, India

ABSTRACT

The next generation of emerging wireless technology is dealing with spectrum shortage. For appropriate and practical implementation of latest wireless technologies, the sufficient amount of frequency is needed. Cognitive radio (CR) is introduced as a proficient nominee to manage spectral undersupply problem, as it rapidly increases the use of underutilize spectrum via spectrum sensing. This chapter introduces brief start about spectrum holes in addition to spectrum sensing framework. Further, the chapter explains the issues in spectrum sensing and how the cooperative sensing technique is fit to overcome these issues like shadow fading and receiver uncertainty. Consequently, the various non-cooperative sensing techniques are also discussed including their test statics. The advantages and disadvantages of different sensing techniques is exhibited at the end.

INTRODUCTION

There is always being an immense scope for all the engineers and scientists to create advance, utilize and the technology in best possible way for the betterment of society. In this series, the conventional spectrum allocation regulatory framework have major constrain i.e. under utilization of spectral band, because most of the frequency bands are exclusively provided only to the licensed user and restrict them from any kind of intersystem interference. On the other hand from all over the world nearly 6 billion people have access to mobile phones via using only selected portion of radio frequency due to its fundamental physical limit. These issues motivated the scientists to derive new wireless technology i.e. cognitive radio

DOI: 10.4018/978-1-5225-5354-0.ch003

which allows interference in the territory of licensed user by the unlicensed users'. Therefore, Cognitive radios must have to sense the spectrum first in order to obtain information about the spectrum usage and geographical condition of the existing licensed user. In cognitive radio network spectrum sensing is one of the most decisive tasks, the un-licensed users required to make constant check on the activities performed by the primary user on spectrum band. Secondary user continuously try to grab spectrum hole when primary becomes silent, but again emergence of primary user derives secondary user to quit the frequency band as soon as possible in order to avoid interference. The process of finding spectrum hole is known as spectrum sensing. Spectrum sensing plays one of the major roles in cognitive radio networks. In this chapter, we have discussed spectrum sensing techniques and various sensing types.

BACKGROUND

To select effective sensing method for QoS in cognitive radio (Giweli, Shahrestani, & Cheung, 2016) had discussed, examine various non cooperative sensing methods precisely, and compare them with each other. Further, focused to identify the factors, which are essential to enhance the performance of CR system. In their study, they had also highlighted the requirement of real-time selection mechanism of the proper sensing method.

- **Limitations:** Although the author factually describe the selection of sensing method for improving the Qos in Cognitive radio network (CRN) but simultaneously it has been observed that, in their study cooperative sensing techniques and their issues were not addressed. Secondly, Test statics of different non-cooperative sensing methods and current existing challenges in sensing techniques have not been discussed.

A comparative study of spectrum sensing Techniques in Cognitive Radio Networks (Khan & Nakagawa, 2013), have discussed different sensing techniques to verify the presence of primary user (PU) and had concluded that, more than 99% accurate decision related to the presence and absence of PU should be there in order to achieve effective spectrum sensing. Their study also revealed the appropriate sensing technique adoption for densely populated and for less populated licensed spectrum band.

- **Limitations:** During the study authors didn't explain the spectrum sensing framework and issues of cooperative spectrum sensing techniques. Eigen value Based Detection which is capable to maintain minimal probability of false alarm is not discussed, neither Covariance Based Detection nor Waveform Based techniques were considered. The present issues related to sensing techniques like Rate -Reliability adjustment, Multipath fading and Hidden Terminal Problem were also missing.

Spectrum Sensing in Cognitive Radio Networks: Requirements, Challenges and Design Trade-offs (Ghasemi & Sousa, 2008) the authors addressed a range of issues related to the design of spectrum sensing functionality for dynamic spectrum access. Most of the time author aimed towards the implementation, sensing challenges like (channel uncertainty, noise uncertainty aggregate-interference uncertainty) and regulatory challenges like (sensing periodicity, detection sensitivity), of spectrum sensing in cognitive radio system. In their study they investigate that how these issues are controlled by user-level cooperation, system-level coordination among the cognitive radios and different cognitive radio networks respectively.

- **Limitations:** During the study author explored only energy detection method having low implementation and computational complexities, because this method is optimal to use when there is only power of the random Gaussian noise is known to the receiver. Since there are more sensing techniques like Eigen value Based Detection, Covariance Based Detection, Waveform Based, Cyclostationary Based, Matched Filter techniques could be discussed with their test statics equations.

SPECTRUM HOLES

Spectrum holes are those frequency bands which are assigned to licensed users, but in some location for certain period of times not used by licensed users. Figure 1 illustrate the spectrum hole concept. CR enables the use of these spectrum holes which is also called "white space" for unlicensed user (SU). if this band is again used by PU in near future, then secondary cognitive radio user either switch to another spectrum hole or become silent via alternating its modulation scheme and transmission power (Mitola, 2000; Bhuvaneswari, 2016)

1. **Temporal Spectrum Holes:** These holes appear when PU halts its transmission for some period of time, and SU has access on same the spectrum band for transmission.
2. **Spatial Spectrum Holes:** These holes appears when PU doing its transmission at specific geographic location. Spectrum outside from this location is used by the SU's for communication.

Next section is all about the spectrum sensing framework along with its different input and output signals and their specific role. Further some spectrum issues has been addressed, where cooperative sensing is derived as a solutions of these issues.

Figure 1. Spectrum Hole Concept

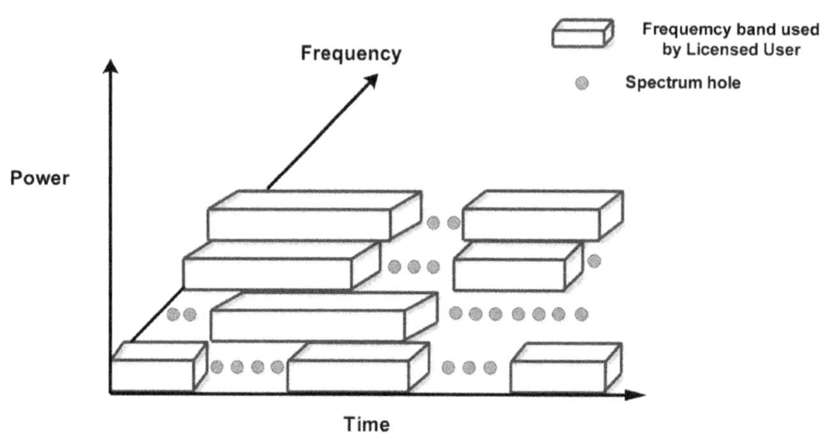

FRAMEWORK: HOW TO SENSE SPECTRUM

The framework of spectrum sensing is discussed by analyzing the various inputs given and outputs driven (Shellhammer, 2008; Holland, Bogucka, & Medeisis, 2015) are illustrated in Figure 2.

1. **Input Signal:**
   a. **Channel Number:** The input channel number signal specifies the channel to be sensed.
   b. **Channel Bandwidth:** Different countries have different bandwidth like in US, channel bandwidth is 6Mz, while other countries prefer 7, 8 MHz, and so bandwidth input signal selects the particular bandwidth.
   c. **Signal Type:** Signal type input selects one type out of these following signals: NTSC/ ATSC/ Microphones (wireless) / DVB-T/ ISDB-T signal types.
   d. **Sensing Mode:** Basically there are three sensing modes: (1) mode 0, activate signal present decision; (2) Mode 1 (optional) activates the signal present decision and confidence matrix, and; (3) Mode 3 (optional) derives Field strength estimate and field strength estimate accuracy.
   e. **Max Pfa (Maximum Probability of False Alarm):** This signal basically identifies the maximum allowed probability of false alarm, only in the presence of noise.
   f. **Sensing Mode and Signal Type:** These signals are used to interpret others output.
2. **Output Signal:**
   a. **Sensing Mode and Signal Type:** Both these signals are used to understand other outputs, so they are just moved through from the input.
   b. **Signal Present Decision:** This output signal gives the simplest and most compact sensing results on the other hand it leads the significant loss of data also. This signal generate binary value to check the presence of specified channel type whose value should be above then the mentioned threshold value (power level).
   c. **Confidence Matrix:** Confidence Matrix reflects, that how much confidence the spectrum sensor have in the signal present decisions, the confidence matrix would be high or low depends on the test statistic, which exceeds significantly or hardly to the threshold value respectively.

Figure 2. spectrum sensing framework
Shellhammer, 2008.

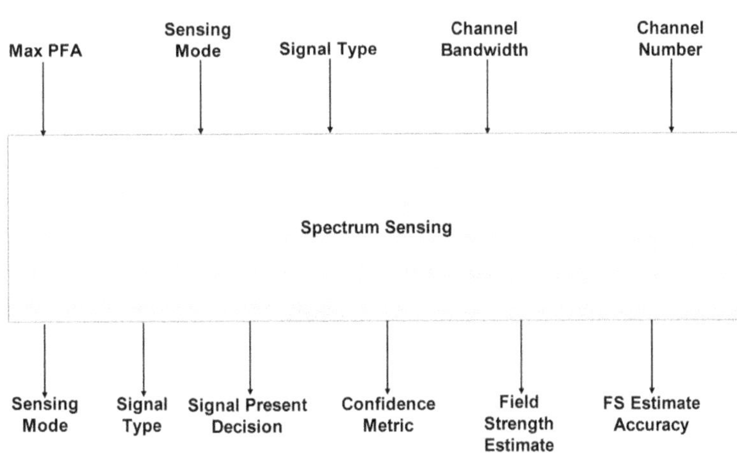

d. **Field Strength Estimate:** This particular signal is dedicated to estimates the signal's electromagnetic field strength, useful to avoid interference with PU
e. **Field Strength Estimate Accuracy:** This signal measures the accuracy of FS estimation.

ISSUES IN SPECTRUM SENSING

Spectrum sensing performance can be affected by the fundamental characteristic of wireless network like multipath fading, noise uncertainty and shadowing as shown in Figure 3. If received PU signal faces deep fading or some obstacle in its path then it become difficult for the secondary user to detect received PU signal even if it is in range of SU. SU 3 is shadowed over the secondary base station (SBS) and SU 1, SU 2 are shadowed over the sensing channel. This situation will cause interference in PU transmission, if SU transmitter fails to detect the presence of the PU transmitter. This problem has been resolved with the help of cooperative spectrum sensing, explained in the next section (Lu, Zhou, Onunkwo, & Li, 2012).

COOPERATIVE SPECTRUM SENSING TECHNIQUES

In cooperative spectrum sensing technique, each cognitive Radio (CR) shares their information periodically to detect a primary user and submit their spectrum observation report to the central controller. This report helps controller to create overall network's spectrum occupancy map by using bitwise-OR operation. All Radio nodes usually combine their local sensing data and help to those individual radios which are not in range and facing problem to determine the exact spectrum occupancy.

Factors like shadowing, fading may also affect the detection of SNR; it's become hard to sense weak signals. For sensing, below SNR wall signal, imposing sensitivity requirements will not gives solution as it may accelerate hardware cost and implementation complexity .so, to achieve performance gain we should adopt corporative mechanism. Figure 4 shows the scenario to improve sensitivity with the help of cooperative sensing.

Figure 3. Receiver uncertainty and shadow fading

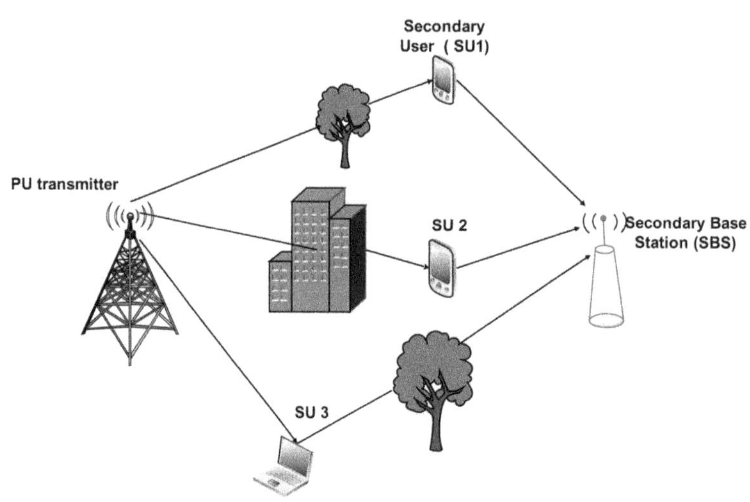

Figure 4. Improvement in Sensitivity with Cooperative sensing

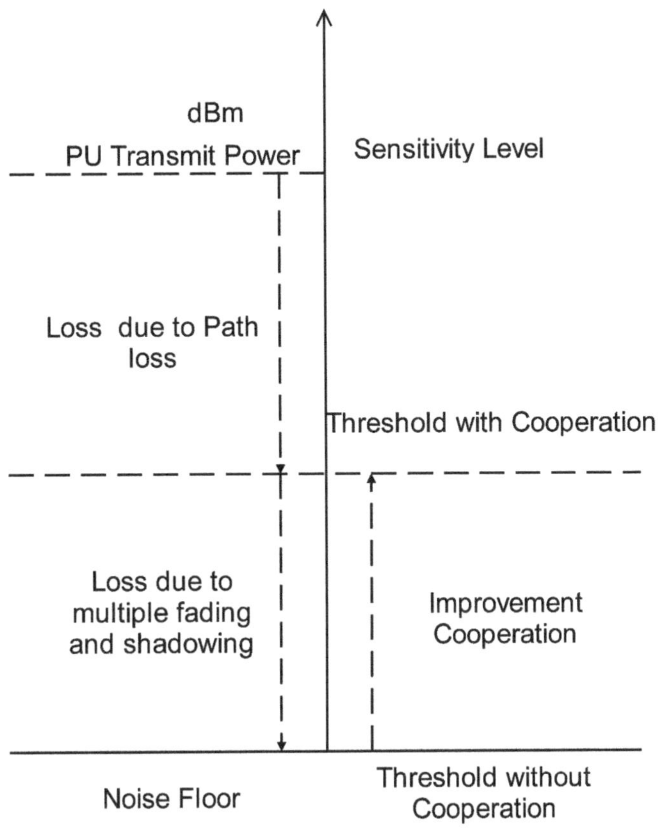

It is helpful to escalate the performance without raising cost and complexity. These problems can be relieved under the corporative sensing mechanism and Being in this mechanism all CR user have opportunity to increase their throughput by transmitting more data as they need not to spend much time to sense, which is already being reduced due to corporation (Mishra, Sahai, & Brodersen, 2006; Tandra & Sahai, 2008).

Requirements of Cooperative SS

1. **Control Channel:** In order to make some modesty among the different elements participating in corporative spectrum sensing, control channel is required. Control channel is responsible for overall system bandwidth quantity.
2. **System Synchronization:** All nodes should work in proper sink, to know when to keep channel free for transmission, how to schedule sensing time, there must be proper synchronization to achieve maximum throughput in system.
3. **Suitable Geo-location:** prime geographical spread, provides performance gain and minimize hidden node problem.

Classification of Cooperative Spectrum Sensing Technique

1. **Centralized:** In centralized sensing approach, all the sensing information is collected by the central processor node which is also known as fusion node, from the other CR's present within the network. Figure 5 depicts how different CR nodes sending the information to Fusion Centre for further processing in order to determine the available frequency and the occupied one.

Fusion centre also consider all the functions performed by different nodes like, some nodes busy to detect channel signal level, while others are looking for suitable alternative in case of channel change request arrive.

2. **Distributed:** In this approach, all radio node have supremacy to achieve ad-hoc approach, where they sense, gather and share information with each other with in their coverage. Figure 6 shows that how various CR nodes distributing the information to each other via reporting channels.
3. **Relay Assisted:** In this approach, relay CR's act as mediator between secondary user and FC to diminish the effect of fading. Hence provide diversity and accuracy to sense spectrum. Figure 7 shows the simple diagram of relay assisted approach in which, the information of all the primary unused channels are detected by the SUs. Some SUs are relay assisted which further report and sends the collected information to the fusion centre for further processing.

Advantages of Cooperative SS

1. **Resolves Hidden Node Issue:** Large number of CR users delivers collective robustness in corporative sensing to ensure the detection of primary user for those CR's facing shadowing or fading problem.

Figure 5. Centralized Approach

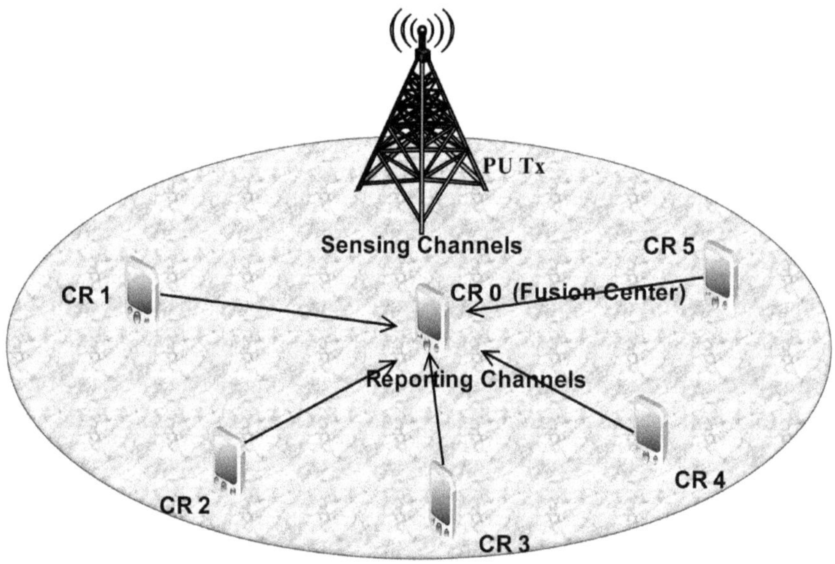

Figure 6. Distributed Approaches

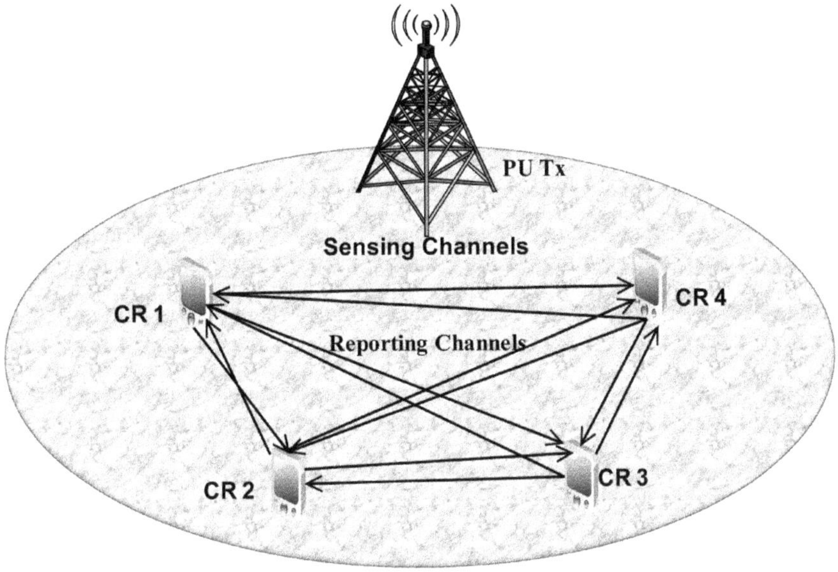

Figure 7. Relay-Assisted Approach

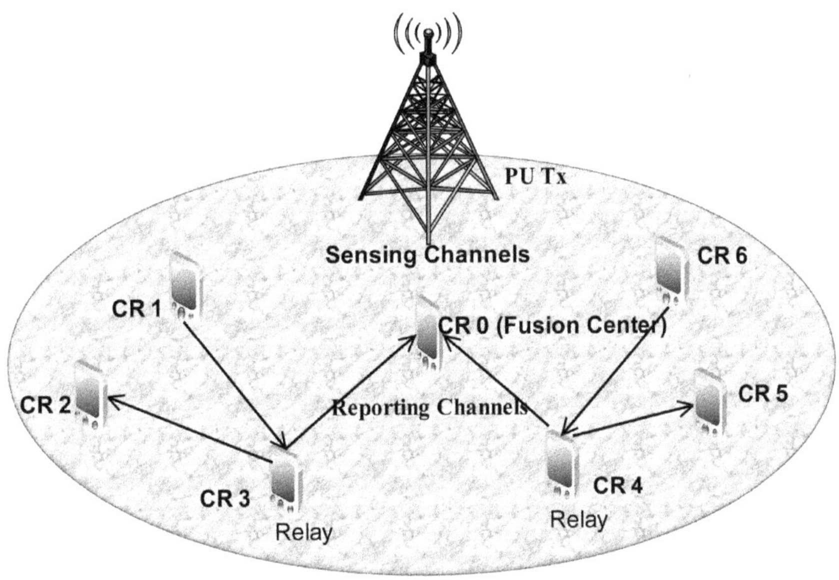

2. **Better Accuracy and Reliability:** Number of spectrum sensing nodes provides accurate and reliable sensing for channel selection under any uncertain condition (Poole, 2017).

Disadvantages of Cooperative SS (Kaabouch, 2014)

1. Required dedicated control channel which forward information to the fusion centre.

2. This technique also demands consequent coarse synchronization in the electromagnetic environment to avoid modification during the spectrum sensing phase.
3. The increase in number of terminals also hikes the cost of the system.

Challenges

1. Combining the results of various users having varying sensing time is also a critical issue in cooperative sensing.
2. In the cooperative sensing technique the development of efficient information sharing algorithms is also a challenge.

NON-COOPERATIVE SPECTRUM SENSING TECHNIQUES

Spectrum sensing technique basically categorized into narrowband and wideband sensing. The cooperative technique has already discussed in earlier section belongs to narrowband spectrum like as Non cooperative technique which is broadly divided into two approaches blind sensing and signal specific sensing as illustrated in Figure 8.

Figure 8. Classification of Non-cooperative Sensing Technique

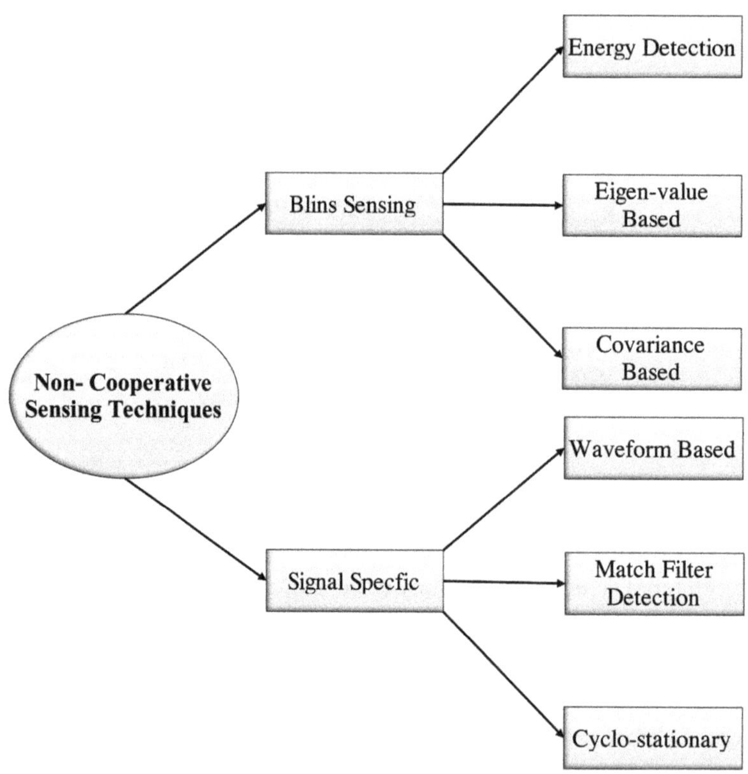

Blind Sensing

Blind sensing approach doesn't require any prior information of PU's system parameters. Signal specific approach demands prior knowledge of PU signal. In this technique each node act as autonomous unit to do the analysis of primary unused channel and performs spectrum occupancy measurement.

The spectrum sensing problem can be explained by Classical binary hypothesis (Holland, Bogucka, & Medeisis, 2015) as:

$$H_0 : y = v \tag{1}$$

$$H_1 : y = x + v \tag{2}$$

In Equation 1:

y is the received signal at SU's receiver.
v is a zero mean additive white Gaussian noise sample.
H_0 hypothesis shows absence of PU.

Where in Equation 2:

x is the transmitted PU signal.
H_1 hypothesis shows presence of PU.

To make choice between H_0, H_1 hypothesis, Test statistic [T(y)] is compared with predefined threshold (λ).

Mathematically this can be expressed as:

$$T\left(y\right) < \lambda, H_0 \tag{3}$$

$$T\left(y\right) \geq \lambda, H_1 \tag{4}$$

Energy Detection

Cognitive radio sensing is much more attracted towards energy detection because it is simple, coherent and an optimal method, because it does not require any information of primary user's signal. The main advantage of energy detection is low implementation and computational complexities (Shellhammer, 2008) therefore the sensing speed and processing cost is relatively very less. This technique is very sensitive to noise uncertainties and depends on the information related to accurate noise variance (σ) or noise power, if the evaluation of noise power is incorrect then there is high probability of false alarm (Tandra & Sahai, 2005; Sahai & Danijela, 2005) An energy detector technique have some drawbacks like low sensing speed, unable to discriminate between modulated signals, noise and interference, not so fit to differentiate between primary and secondary users, threshold for detection is very vulnerable for the in band interference and to noise intensity, and Doesn't work for spread spectrum signals (Zeng & Liang, 2009).

The test statistics is the output of integrator which estimates the energy of the received signal, the test statistics for typical energy detection is expressed as:

$$T\left(y\right) = \frac{1}{\sigma^2} y^2 \tag{5}$$

Test statics is the energy corresponding to received signal which is measured and compared with the threshold to identify the presence of primary user (Atapattu, Tellambura, & Jiang, 2014). Figure 9 represents the block diagram of energy detector.

The performance of the energy detector-based sensing algorithm analyzed by metrics known as probabilities (Holland, Bogucka, & Medeisis, 2015):

1. Probability of detection(P_D)
2. Probability of false alarm (P_f)

Probability of detection (P_D) is used to detect the presence of a PU signal in its presence. Mathematically expressed as:

$$P_D = P_r\left(M > \lambda \mid H_0\right) \tag{6}$$

Probability of false alarm (P_f) is used to detect the presence of a PU signal when it is absent. Mathematically formulated as:

$$P_F = P_r\left(M > \lambda \mid H_1\right) \tag{7}$$

Figure 9. Energy Detector

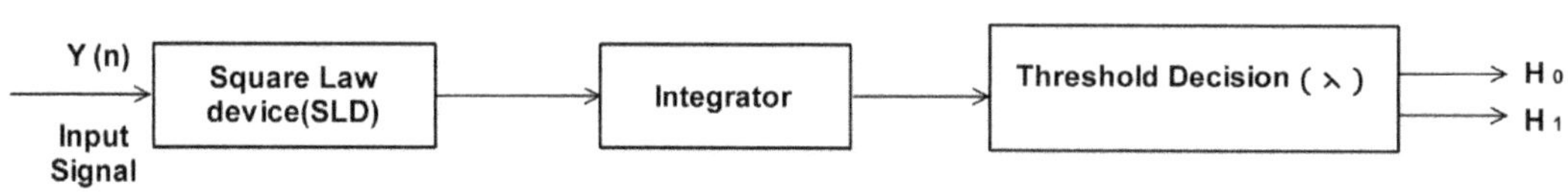

Eigen Value Based Detection

In this sensing method, the detection of primary user's signal is based on the Eigen value of the covariance matrix of the received signal. The ratio of maximum Eigen value to minimum Eigen value of covariance matrix is used to identify the existence of primary user. To determine the threshold, we can perform quantization of this ratio, based on the latest random matrix theories (RMT)(Zeng & Liang, 2007). The major objective of the technique is to achieve optimal threshold that reduces sensing faults. Eigen based technique is very much capable to maintain minimal probability of false alarm, which creates chances to access more channel for the secondary users. This technique consists four methods, which includes MED(maximum Eigen value detector), MME(maximum-minimum Eigen value detector), EME (energy with minimum Eigen value detector), and GLRT detector (He, Ratnarajah, Xue, Yousif, & Sellathurai, 2014) .

The main drawback of this technique is computational complexity of covariance matrix computation and the Eigen value decomposition. The threshold selection is also a critical task (Ma, Li, & Juang, 2009; Zeng & Liang, 2009).

Covariance Based Detection

Covariance sensing technique (Zeng & Liang, 2009)is based on statistical covariance of the received signal. The auto correlation of noise and signal are different, therefore, this discrimination is used in this method to isolate signal component from background noise. This method executes in two stage: 1) Compute the sample of covariance matrix of the received signal, and; 2) Extract two test statistics from the sampled covariance matrix; compare the ratio of these two statistics with the threshold. This method doesn't require any information regarding signal, noise power, channel, in advance. There is no compulsion for synchronization either.

Signal Specific Sensing

In this sensing technique, prior knowledge of primary users signal is required. This technique is further classified as waveform based, matched filter, cyclostationary based etc. Few of them are explained below.

Waveform Based

Waveform sensing technique captures the incoming signal and tests its exactness with PU. This technique check the pattern of these signal through various methods such as preambles, spreading sequence, regularly transmitted pilot pattern, midambles etc (Cabric, Tkachenko, & Brodersen, 2006). These patterns are utilized to assist synchronization in wireless system and for some other purposes. Transmission of preamble pattern occurs before the beginning of each slot where the midamble pattern has been sent in the middle of slot. This method also known as coherent sensing, we implement this technique only to those system, having known signal patterns, so that the received signal can be compared with known signal parameters. Waveform sensing requires short measurement or sensing time, however it is very receptive to error in synchronization. Performance of this method can be improved via increasing the length of the known signal pattern (Tang, 2005).

The received signal y(n) can be expressed as:

$$y(n) = x(n) + v(n) \tag{8}$$

where

$x(n)$ is signal transmitted by PU.
$v(n)$ is a zero mean additive white Gaussian noise sample.
$y(n)$ is the received signal at SU receiver.
n is the sample index.

Wave form-based sensing matrix M can be computed from the mentioned expression:

$$M = \text{Re}\left[\sum_{n=1}^{N} y(n) \times x(n)^{*}\right] \tag{9}$$

where * represents conjugate operation. In the absence of PU, matrix value will be:

$$M = \text{Re}\left[\sum_{n=1}^{N} v(n) \times x(n)^{*}\right] \textit{if } H_0 \tag{10}$$

In the presence of PU, matrix becomes:

$$M = \sum_{n=1}^{N} \left|x(n)\right|^{2} + \text{Re}\left[\sum_{n=1}^{N} v(n) \times x(n)^{*}\right] \textit{if } H_1 \tag{11}$$

By comparing the decision metric M with fixed threshold λ, we can easily make decision on the detection of the presence of PU signal.

Matched Filter

This is an optimum technique, if there is complex knowledge of transmitted signal. It is coherent detection technique in which SU should have correct knowledge of PU signal's properties like (order and type of modulation, pulse shaping, bandwidth, packet format etc). Basically in matched filter is an optimum method because it maximizes SNR. As compared to other detectors it needs small number of samples for detection and due to coherent in nature it performs fast detection. Figure10 shows matched filter detector. The test statistics (Ibnkahla, 2014) of match filter detection is expressed as:

$$T(y) = R\left[x^{H} * y\right] \tag{12}$$

where R is the real part and $()^{H}$ is Hermitian operator.

Figure 10. Matched Filter Detector
Shellhammer, 2008.

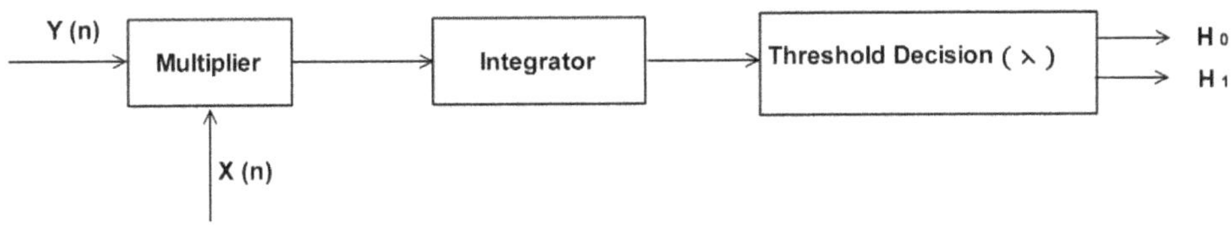

Cyclostationary Based

This detection harness cyclostationary properties of received signals to detect primary user's signals. Spectral correlation Function (SCF) is vital parameter for the PU signals. In the absence of SCF, Stationary noise or Interference signals are rejected by Cyclostationary spectrum sensing method. This technique prefers CSD(cyclic spectral density instead of PSD(power spectral density).

Cyclostationary spectrum sensing is capable to reject noise, and work much better than other techniques at low SNR value. The main disadvantage of this method is that, it requires high computationally complexity and cost. Figure 11 shows cyclostationary detector. The performance of Cyclostationary spectrum sensing declines when lesser number of samples are used (Sahai & Danijela, 2005; Aparna & Jayasheela, 2012).

COMPARISON OF NON COOPERATIVE SPECTRUM SENSING TECHNIQUES

A comparative analysis of each technique has been possessed by mentioning their some advantages and disadvantages, shown in Table 1.

CURRENT EXISTING CHALLENGES IN SPECTRUM SENSING TECHNIQUE

1. **Rate -Reliability Adjustment:** When PU again becomes active in its band then the CR users required to pause its transmission and moved to new idle band and because of this band switching, CR users have to bear the quality of service. On the other hand all the peers (CR) behave in a coordination manner in order to balance frequency transition and increase the communication

Figure 11. Cyclo-stationary detector
Ibnkahla, 2014.

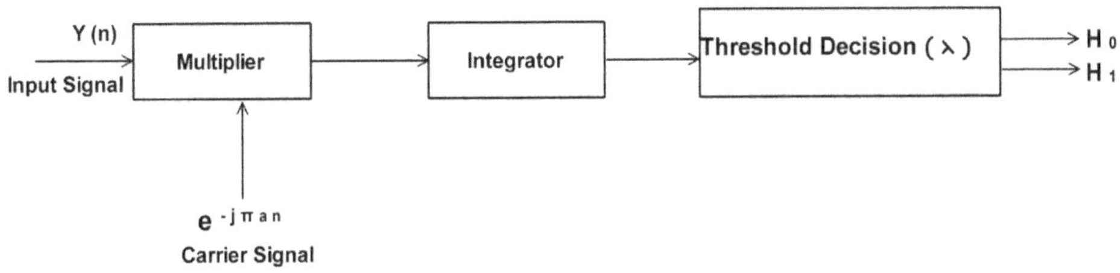

Table 1. Comparison of Non cooperative Spectrum Sensing Techniques

Sensing Techniques	Advantages	Disadvantages
Energy Detection	• Not require prior information of signal, • Low- cost and less power consumption, • Low computational complexity	• Unable to distinguish between PU and SU, • Low sensing accuracy • Can't perform at low SNR
Eigen-value Detection	• Require less sensing time • Low computational complexity • Minimal probability of false alarm • Low power consumption • No synchronization is required	• Very poor in terms of reliability and accuracy • Overcome problematic noise uncertainty • Large sensing time in comparison of energy detection
Covariance Detection	• Do not require information of signal noise power. • Synchronization is not required.	• Overcome problematic noise uncertainty. • Large sensing time in comparison of energy detection.
Waveform Detection	• Reduces noise uncertainty. • Unlike match filter. • No synchronization requirement.	• Complex in terms of computation • Large sensing time.
Matched Filter Detection	• Optimal sensing performance. • Low computational complexity. • Low cost. • Achieve high processing gain in less time. • Maximize the received SNR.	• Acquire Prior knowledge of primary user's signal and channel is difficult. • Require dedicated sensing receiver for synchronizing with each SU.
Cyclostationary Detection	• Robust in low SNR and interference. • High sensing accuracy. • Communication flexibility is good. • Noise rejection capability.	• Incomplete information of PU. • Sensing time duration is very large. • Computational complexity and cost is also high.

reliability via spreading the data transmission over the spectrum. So that, if PU again claims to its band, it hardly affect the CR user's band. At receiver end, Orthogonal-Frequency-Division (OFDM) can be utilized for combining the data from various frequency chunks to construct authentic CR link. OFDM is very much capable for managing non-adjacent frequency bands, but this method also required extra spectral resources for periodic sensing of these additional frequency bands, which affect the data rate of CR users badly (Ghasemi & Sousa, 2008).

2. **Robustness to Noise Models:** Since a CR is proposed to operate in interference and fading surrounding, moreover the noise distribution can only be known upto some measure. At sensing detector side the SNR value can be low as -1dB, so it is necessary to design robust detectors which are agile enough to handle these changeability or uncertainty (Wang, Yang, Zhao, & Zhang, 2009). Presently Goodness-of-Fit Tests (GoFT) is considered as one of the choice for spectrum sensing, requires limited knowledge about noise distribution. Designing of GoFT is an interesting challenge.
3. **Reporting Time - Sensing Time:** Sensing time increases the probability of detection i.e. P_D and decreases the probability of false alarm i.e. P_F, but simultaneously this benefit affect the data transmission time i.e. T_t. Therefore the sensing throughput of CR's will be affected or degraded, which needs to be optimized (Liang, Zeng, Peh, & Hoang, 2008).

4. **Multipath Fading and Hidden Terminal Problem:** The biggest challenge in spectrum sensing faced by CR users, caused due to multipath fading is known as hidden node problem, where Primary user suffers with unwanted interference caused by CR because it is not aware of PU's presence or fails to detect victim incumbent's signal.

CONCLUSION

A quick overview of spectrum hole concept is mentioned in this chapter. A survey is conducted on different sensing techniques, mainly focused on two category named as cooperative and non-cooperative. Cooperative sensing techniques doesn't demand any prior signal information of licensed user. It also overcomes the false alarm, hidden node problem. Moreover the performance of this technique is much better in comparison of other classical spectrum sensing techniques. Further in non-cooperative blind sensing technique, we have observed that, energy detection technique receives tremendous attention despite having low SNR because of its ease of implementation. On the other hand match filter detection technique acquire less sensing time with better SNR, but this technique demands more power and having high complexity in comparison of energy detection. Theoretical analysis on test statistic and threshold has been mentioned. After reviewing all these techniques, it can be stated that different sensing techniques has its own advantages and disadvantages, so they can be implemented on the basis of their application.

REFERENCES

Aparna, P. S., & Jayasheela, M. (2012). Cyclostationary feature detection in cognitive radio using different modulation schemes. *International Journal of Computers and Applications*, *47*(21).

Atapattu, S., Tellambura, C., & Jiang, H. (2014). *Energy detection for spectrum sensing in cognitive radio*. New York, NY: Springer. doi:10.1007/978-1-4939-0494-5_2

Bhuvaneswari, P. T. V. (2016). Spectrum Sensing in Cognitive Radio Networks: A Survey. *Introduction to Cognitive Radio Networks and Applications, 83.*

Cabric, D., Tkachenko, A., & Brodersen, R. W. (2006, October). Spectrum sensing measurements of pilot, energy, and collaborative detection. In Military communications conference, 2006. MILCOM 2006. IEEE (pp. 1-7). IEEE. doi:10.1109/MILCOM.2006.301994

Ghasemi, A., & Sousa, E. S. (2008). Spectrum sensing in cognitive radio networks: Requirements, challenges and design trade-offs. *IEEE Communications Magazine*, *46*(4), 32–39. doi:10.1109/MCOM.2008.4481338

Giweli, N., Shahrestani, S. A., & Cheung, H. N. (2016). Selection of spectrum sensing method to enhance Qos in cognitive radio networks. *International Journal of Wireless and Mobile Networks*, *8*(1), 39–50. doi:10.5121/ijwmn.2016.8104

He, Y., Ratnarajah, T., Xue, J., Yousif, E. H., & Sellathurai, M. (2014, May). Optimal decision threshold for eigenvalue-based spectrum sensing techniques. In *Acoustics, Speech and Signal Processing (ICASSP), 2014 IEEE International Conference on* (pp. 7734-7738). IEEE. 10.1109/ICASSP.2014.6855105

Holland, O., Bogucka, H., & Medeisis, A. (2015). *Opportunistic spectrum sharing and white space access: The practical reality*. John Wiley & Sons. doi:10.1002/9781119057246

Hossain, E., Niyato, D., & Han, Z. (2009). *Dynamic spectrum access and management in cognitive radio networks*. Cambridge University Press. Retrieved from http://www.radio-electronics.com/info/rf-technology-design/cognitive-radio-cr/cooperative-spectrum-sensing.php'

Ibnkahla, M. (2014). *Cooperative Cognitive Radio Networks: The Complete Spectrum Cycle*. CRC Press.

Kaabouch, N. (Ed.). (2014). *Handbook of Research on Software-Defined and Cognitive Radio Technologies for Dynamic Spectrum Management*. IGI Global.

Khan, F., & Nakagawa, K. (2013, June). Comparative study of spectrum sensing techniques in cognitive radio networks. In *Computer and Information Technology (WCCIT), 2013 World Congress on* (pp. 1-8). IEEE. 10.1109/WCCIT.2013.6618728

Liang, Y. C., Zeng, Y., Peh, E. C., & Hoang, A. T. (2008). Sensing-throughput tradeoff for cognitive radio networks. *IEEE Transactions on Wireless Communications*, *7*(4), 1326–1337. doi:10.1109/TWC.2008.060869

Lu, L., Zhou, X., Onunkwo, U., & Li, G. Y. (2012). Ten years of research in spectrum sensing and sharing in cognitive radio. *EURASIP Journal on Wireless Communications and Networking*, *2012*(1), 28. doi:10.1186/1687-1499-2012-28

Ma, J., Li, G. Y., & Juang, B. H. (2009). Signal processing in cognitive radio. *Proceedings of the IEEE*, *97*(5), 805–823. doi:10.1109/JPROC.2009.2015707

Mishra, S. M., Sahai, A., & Brodersen, R. W. (2006, June). Cooperative sensing among cognitive radios. In *Communications, 2006. ICC'06. IEEE International conference on* (Vol. 4, pp. 1658-1663). IEEE. 10.1109/ICC.2006.254957

Mitola, J. (2000). *Cognitive radio---an integrated agent architecture for software defined radio*. Academic Press.

Poole, I. (2017). *IEEE 802.22 rf-technology-design/cognitive-radio-cr/cooperative-spectrum-sensing*. Available at: http://www.radio-electronics.com/info/rf-technology-design/cognitive-radio-cr/cooperative-spectrum-sensing.php

Sahai, A., & Danijela, C. (2005). Spectrum sensing: fundamental limits and practical challenges. *In Proc. International Symposium on New Frontiers in Dynamic Spectrum Access Networks.2005. DySPAN 2005 first IEEE international symposium on* (vol. 41). IEEE.

Shellhammer, S. J. (2008). Spectrum sensing in IEEE 802.22. *IAPR Wksp. Cognitive Info. Processing*, 9-10.

Tandra, R., & Sahai, A. (2005, June). Fundamental limits on detection in low SNR under noise uncertainty. In *Wireless Networks, Communications and Mobile Computing, 2005 International Conference on* (Vol. 1, pp. 464-469). IEEE. 10.1109/WIRLES.2005.1549453

Tandra, R., & Sahai, A. (2008). SNR walls for signal detection. *IEEE Journal of Selected Topics in Signal Processing*, *2*(1), 4–17. doi:10.1109/JSTSP.2007.914879

Tang, H. (2005, November). Some physical layer issues of wide-band cognitive radio systems. In *New frontiers in dynamic spectrum access networks, 2005. DySPAN 2005. 2005 first IEEE international symposium on* (pp. 151-159). IEEE.

Wang, H., Yang, E. H., Zhao, Z., & Zhang, W. (2009). Spectrum sensing in cognitive radio using goodness of fit testing. *IEEE Transactions on Wireless Communications*, *8*(11), 5427–5430. doi:10.1109/TWC.2009.081586

Zeng, Y., & Liang, Y. C. (2007, September). Maximum-minimum eigenvalue detection for cognitive radio. In *Personal, Indoor and Mobile Radio Communications, 2007. PIMRC 2007. IEEE 18th International Symposium on* (pp. 1-5). IEEE.

Zeng, Y., & Liang, Y. C. (2009). Eigenvalue-based spectrum sensing algorithms for cognitive radio. *IEEE Transactions on Communications*, *57*(6).

Zeng, Y., & Liang, Y. C. (2009). Spectrum-sensing algorithms for cognitive radio based on statistical covariances. *IEEE Transactions on Vehicular Technology*, *58*(4), 1804–1815.

Chapter 4
Basic Spectrum Sensing Techniques

Bhuvaneswari P. T. V.
Anna University, India

Bino J.
Anna University, India

ABSTRACT

Cognitive radio network (CRN) is an upcoming networking technology that can utilize both radio spectrum and wireless resources efficiently based on the information gathered from the past experience. There are two types of users in CRN, namely primary and secondary. PUs (PU) have the license to operate in certain spectrum band while the secondary (SU) or cognitive radio (CR) users do not have the license to operate in the desired band. However, they can opportunistically utilize the unused frequency bands. Spectrum sensing, spectrum management, spectrum sharing, and spectrum mobility are the four major functions of cognitive radio systems. The main objective of spectrum sensing is to provide better spectrum access to CR users, without causing any harmful interference to PUs. Sensing accuracy is considered as the most important factor to determine the performance of cognitive radio network. In this chapter, the challenges and requirement involved in spectrum sensing are detailed. Further, various spectrum sensing basic techniques are also discussed in detail.

INTRODUCTION

Spectrum Sensing in Cognitive Radio Network

In CRN, the spectrum sensing function, measures the metrics related to operating environment. They are white spaces, noise, signal power, temperature and interference status. Whenever the frequency band is detected to be unused over a period of time by the PU located in a particular position, then it can be assigned to the CR user. This function can be executed in four different dimensions namely time, frequency, space and code (Ali, & Hamouda, 2017).

Various tasks executed by spectrum sensing are:

DOI: 10.4018/978-1-5225-5354-0.ch004

- White space detection
- Spectral resolution
- Prediction of incoming interferences
- Classification of signals

In practice, to identify the presence of the Spectrum Hole (SH), the CR user seamlessly monitor the activities of the PUs who are the licensed users to operate in a particular frequency band. This procedure is known as spectrum sensing (Wang, 2009).

Principle of Spectrum Sensing

The principle of spectrum sensing is illustrated in Figure 1. Let P, Q and S, T be the PU and CR user transmitter-receiver pair respectively. It can be observed that the PU transmitter is sending data to PU receiver in their allotted licensed spectrum band. At the same instant, if a pair of CR user also intends to access the spectrum, then in order to protect the PU transmission, spectrum sensing mechanism should be performed by the CR user transmitter, in which the presence of PU receiver that is present within the coverage of the SU transmitter is determined (Subhedar & Birajdar, 2011).

SPECTRUM HOLE

The unused portion of the authorized spectrum band that can be utilized by CR users is known as spectrum hole or white space. It is illustrated in Figure 2 (Bountouris, 2013). There are two types of Spectrum Holes, namely temporal and spatial.

Temporal

The spectrum hole is said to be temporal, when the PU does not access the authorized spectrum band over a specific time. If both PU and CR user are located in the same geographical area, then they may interfere with each other, when they attempt to transmit over the same spectrum band. This can be alleviated by allowing the CR user to exploit spectrum holes before transmitting data over the spectrum.

Figure 1. Principle of Spectrum Sensing

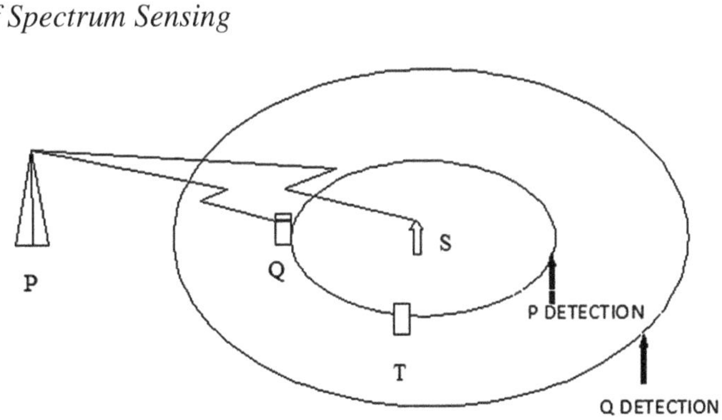

Figure 2. Spectrum Hole

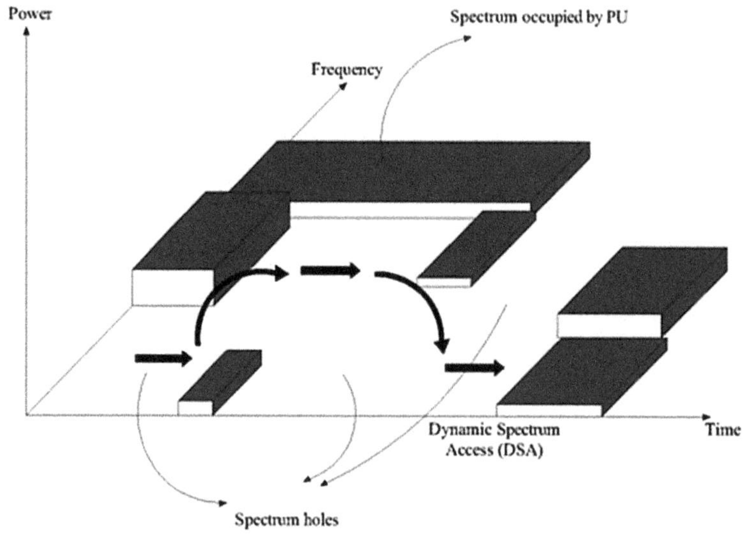

Spatial

The spectrum hole is said to be spatial when the access of PU transmission is restricted to specific geographical area. The region outside this area can be accessed by the CR user. This is possible in relay assisted CRN.

CHALLENGES OF SPECTRUM SENSING

In this section, the challenges in spectrum sensing, namely hardware equipment prerequisites, hidden PU issue, sensing periodicity and noise uncertainty are discussed (Sahai, Hoven, Mishra, & Tandra, 2013).

Hardware Equipment Prerequisites

The major prerequisites needed by the CR users to perform spectrum sensing are increased sampling rate, high precision Analog-to-Digital converters (ADC) with extensive range and high-speed signal processors. Further to support operation in wideband, the radio frequency transceiver modules need to be enhanced. Also high speed processors that can compute algorithms with high computational complexity is needed. They could be either DSP based or FPGA based.

Hidden PU

The hidden PU can induce major issue to the CR user. Due to the path loss and shadowing effect experienced by the CR user, the presence of the PU may be unnoticed which may cause interference to the SU, thereby degrade the QoS of SU.

Sensing Periodicity

The availability of the white space utilized by the CR user need to be sensed periodically in order to overcome the interference caused by the licensed PU. Thus sensing period plays a vital role in determining the QoS of the CR user. This impose additional hardware requirement to enable simultaneously transmission and sensing mechanism.

Noise Uncertainty

Due to the stochastic nature of the channel, it is cumbersome for the CR users to have prior knowledge about the noise power. Hence the CR user receiver needs to estimate the noise power instantly. However, the estimation accuracy is limited due to the errors caused by temperature variations. Hence the CR receivers need to be equipped with more sensitive detectors which can adhere to hostile wireless environment, thereby enhancing the detection accuracy.

Channel Uncertainty

In wireless communication networks, uncertainties in received signal strength are mainly due to channel fading or shadowing. In spectrum sensing, the deep fade experienced by the PU signal may be misinterpreted by the CR user as white space. Then the CR user transmission can become interference to PU. This necessitates development of intelligence/adaptive algorithms that can be differentiate a faded or shadowed primary signal from a white space. Cooperative Sensing is the mechanism that is devised to handle this issue. CR users are grouped into cluster and share their local measurements. A collective decision is then made about the availability of licensed band.

Aggregate Interference

In near future, the possibility of CR users operating on the licensed band of PU may increase exponentially, resulting in uncertainty known as aggregate interference. The detection of white space by the CR user may become faultier due to the impact of aggregate interference even though the PU remains outside the interference range of CR user. This demands more sensitive detectors to enhance the detection accuracy which may induce additional cost on the hardware.

Sensing Interference

As mentioned above, the uncertainties existing in the channel as well as aggregate interference make the spectrum sensing task more challenging. This may induce estimation errors for CR users in detecting the licensed band PU. Context awareness and environment may be the alternative solution that can overcome the estimation of sensing interference.

SENSING APPROACH

The spectrum exploitation techniques of CR users along with the coexistence of PUs can be classified into three techniques namely interweave, underlay and overlay (Zhang, Simula Res. Lab, & Lysaker, 2008).

Interweave

In this technique, the CR user is allowed to utilize the spectrum hole only in the absence of PU. The spectrum hole can be either in time or frequency or space domain. On the arrival of PU, the ongoing communication of CR user is abruptly terminated, since the QoS of PU is given major attention. Along with traditional measurement methods, both polarization and angular measurements are considered to exploit spectrum holes.

Underlay

In this paradigm, the CR users are allowed to transmit along with PUs, only when the interference caused by CR user is within acceptable limit. The allowable interference level at the PUs is governed by a metric namely interference temperature. It is defined as the average level of interference power received at primary receiver. Underlay communication can be realized with the assistance of cognitive beam forming, resource allocation and spread spectrum approaches.

Overlay

In this technique both PU and CR user can access the same spectrum. The interference between them can be avoided by means of advanced modulation and coding systems. Further transparency among PU and SU can be maintained to ensure QoS.

Hybrid

Hybrid paradigm overcomes the drawbacks of the above approaches. In this method, a CR user is allowed to access the unused channel of PU without any restrictions. Moreover on occupancy, it is allowed to transmit with power restrictions. This method executes transmit mode selection/switching to utilize the unused spectrum efficiently. Further, it also maintains the trade-off between sensing and throughput to improve the throughput of CR user.

PERFORMANCE VERSUS CONSTRAINT

The performance of the spectrum sensing technique is influenced by various constraints, namely hidden and exposed terminal caused due to primary user. However, the techniques aim to increase the throughput of CR user by decreasing the probability of collision with PU. It is achieved by designing the spectrum opportunity detector which ensures optimal trade-off between false alarms and missed event detection. It dynamically changes the intensity of the factor considered for detecting the unused band (Mehta, Kumar & Saini, 2013). In this section, various design constraints, namely MAC Layer performance measures, Global Interference Model, Local Interference Model and Sensing Accuracy versus Sensing Period are detailed.

An interference constraint needs to be specified by at least two parameters, namely μ and ρ where μ is the maximum acceptable interference power observed by an active PU and ρ is the maximum outage probability that the interference at an active PU exceeds the noise floor.

In a CRN, the PU activity can be represented by two distinct events namely, presence or absence of PU. Spectrum sensing aims to detect the presence or absence of PUs. The spectrum detection mechanism provides a binary hypothetical solution. They are: H_0 represents the absence and H_1 indicate the presence of PU (Wang, 2009).

Table 1 illustrates the spectrum detection problem in matrix form. Column specifies the PU activity and Row specifies the CR user detection. Two types of error may occur in spectrum detection. They are false alarm and missed detection. When the idle channel is detected as busy it is known as false alarm. When the busy channel is detected as idle then it is known as missed detection. The probabilities of false alarm (P_{FA}) and missed detection (P_{MD}) are the two important parameters that determine the performance of detection.

$$P_{FA} = \Pr\{\text{decides } H_1 \mid H_0\} \tag{1}$$

$$P_{MD} = \Pr\{\text{decides } H_0 \mid H_1\} \tag{2}$$

P_{FA}: Probability of false alarm: Probability of deciding H_1 when H_0 is true.
P_{MD}: Probability of missed detection(1-P_D): Probability of deciding H_0 when H_1 is true.
$P_{D:}$ Probability of detection: Probability of deciding H_1 when H_1 is true.

The trade-off between false alarm and missed detection is given by the Receiver Operating Characteristic (ROC), which gives $P_D = 1 - P_{MD}$ as a function of $P_{FA.}$

MAC Layer Performance Measures

The MAC layer performance is analyzed based on the throughput experienced by both PU and CR user. It depends on the interference influenced by each other. The design goal of the proposed model is to maximize the throughput under maximum outage probability condition.

Consider the cognitive environment represented in Figure 3 (Wyglinski, Nekovee & Hou, 2010). Let I_R be the interference range of PU, I_r be the interference range of CR user, D_r be the detection range of CR user, S and T be the secondary transmitter and receiver, X_1 be the hidden primary transmitter, X_2 be the

Table 1. PU Vs CR Detection

PU Vs CR Detection	H_0	H_1
H_0	0,0	0,1
H_1	1,0	1,1

Figure 3. Spectrum Detection Opportunity

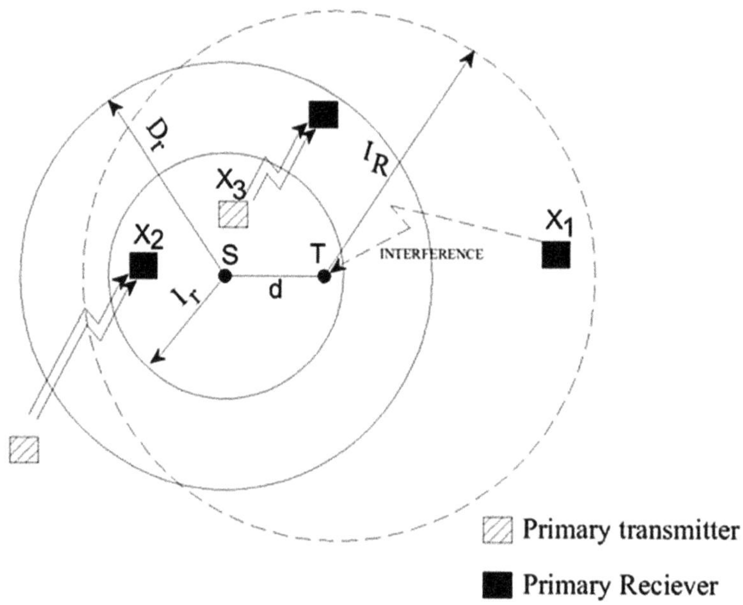

hidden primary receiver, and X_3 be the exposed primary transmitter. There are three susceptible sources of detection errors that may occur in the considered cognitive scenario. They are hidden transmitters, hidden receivers, and exposed transmitters. The hidden transmitter is a primary transmitter (X_1) placed in between I_R and T and outside the communication range of S.

The hidden receiver (X_2) is a primary receiver placed in between the interference range I_r and S, and its corresponding primary transmitter is located outside the detection range of S. The exposed transmitter is a primary transmitter (X_3) placed within the detection range of S, but transmits to a primary receiver located outside the interference range of S. The probability of success (P_s) and the probability of collision (P_C) decide the figure of merit at the MAC layer. It dynamically changes the intensity of the factor considered for detecting the unused band. For the scenario, consider the P_S that depends on three events:

Event 1: When Secondary transmitter (S) detects the absence of primary transmitter (P) within its coverage area and can transmit data to secondary receiver (T) successfully.

Event 2: Secondary receiver (T) receives data successfully, if it is not within the interference range (I_R) of PU.

Event 3: Secondary transmitter (S) receives an acknowledgement from the secondary receiver (T) successfully, if S is not within the interference range (I_R) of PU.

Let C (S, $D_{r,}$ Tx) denote the absence of primary transmitter in the detection range of S. C (T, $R_{I,}$ Tx) denotes the absence of T the interference range of primary transmitter. C (S, $R_{I,}$ Tx) denotes that absence of S within the interference range of primary transmitter. Then the probability of successful transmission (P_S) can be given as:

$$P_{S=} P_r\{ S, D_{r,} Tx \} \& P_r\{ C(T, R_{I,} Tx)\} \& P_r\{ C(S, R_{I,} Tx) \} \quad (3)$$

The probability of collision (P_C) can be given as:

$$P_C = P_r \{ C(S, D_{r,} Tx) \mid C(S, I_{r,} Rx) \} \quad (4)$$

Global Interference Model

In global interference model spectral opportunities are considered as a symmetric one. Here the CR users are allowed to transmit only during the absence of PU. Therefore, the problem of the hidden and the exposed terminals does not have any impact on the global interference model.

- Successful transmissions can happen if there exists a spectral opportunity (H_0).
- Missed detection can result in collision with PU.

The relationship between P_{FA}, P_{MD}, P_S, and P_C are given as:

$$P_S = (1 - P_{FA}) \Pr [H_0] \quad (5)$$

$$P_C = P_{MD} \quad (6)$$

Local Interference Model

In local interference model, the local problems like hidden and exposed terminal problems are considered. It addresses two types of services namely, guaranteed and best effort. Guaranteed services need acknowledgement for successful data transmission, whereas best effort services does not need such acknowledgement.

For both types of models, asymmetric nature of channel is considered and P_C is not equal to P_{MD}.

- For guaranteed services, due to hidden and exposed terminal problems exactly detected opportunities may results in failed data transmission.
- For best-effort services, exactly detected opportunities will always result in successful data transmission eventhough there is hidden and exposed terminal problems.

The basic problem of spectrum opportunities is how to screen primary signals from hidden and exposed terminal problem. Using RTS/CTS handshaking, detection performance can be enhanced. However, it results in decreased throughput (Hattab & Ibnkahla, 2014).

SENSING ACCURACY VERSUS SENSING PERIOD

To alleviate the occurrence of P_{FA} & P_{MD}, it becomes mandatory to improve sensing accuracy. By increasing the sensing period, it is possible to enhance accuracy. However, the prolonged sensing period may greatly reduce the transmission opportunity. Hence it is essential to maintain the trade-off between the sensing period and the sensing accuracy (Biglieri, Goldsmith, Greenstein, & Mandayam, 2012).

Let N be the duration of availability of unused spectrum such that:

$$N = \sum_{t=1}^{k} n_t \quad (7)$$

It consists of k instances. Let 't_s' be the sensing period, t_d be the transition period & L be the threshold period. When t_s = k, then t_d becomes zero. Hence to increase t_d, t_s should be kept less than k i.e.,

If

t_s = k, then t_d=0

If

t_s < L where L < k, then t_d = 0.

BASIC SPECTRUM SENSING TECHNIQUES

Spectrum sensing methods can be characterized into three general classes (Subhedar & Birajdar, 2011):

1. Strategies requiring both PU signal and noise power information;
2. Strategies requiring only information about noise (semi blind detection) and;
3. Strategies requiring no information on PU signal or noise power (blind detection).

MATCHED FILTER DETECTION

A Matched Filter (MF) increases the output signal to noise ratio (SNR) for a given input signal. Matched Filtering correlates the received signal with a reference signal to determine the presence of PU signal in that received signal. The reference signal is the known sequence of the PU signal. The operation of the matched filter based sensing is illustrated in Figure 5 (Jaiswal, Kumar Sharma & Singh, 2013). The impulse response of the matched filter is convolved with received signal which is the time shifted version of a reference signal. Let s(n) be the PU signal, x(n) be the signal by CR user, and w(n) be the AWGN with zero mean and variance σ^2. The binary hypothesis for spectrum sensing can be represented as given in Equations (8) and (9) .

$$H_0: x(n) = w(n) \quad (8)$$

$$H_1: x(n) = s(n) + w(n) \quad (9)$$

Figure 4. Classification of Basic Spectrum Sensing Techniques

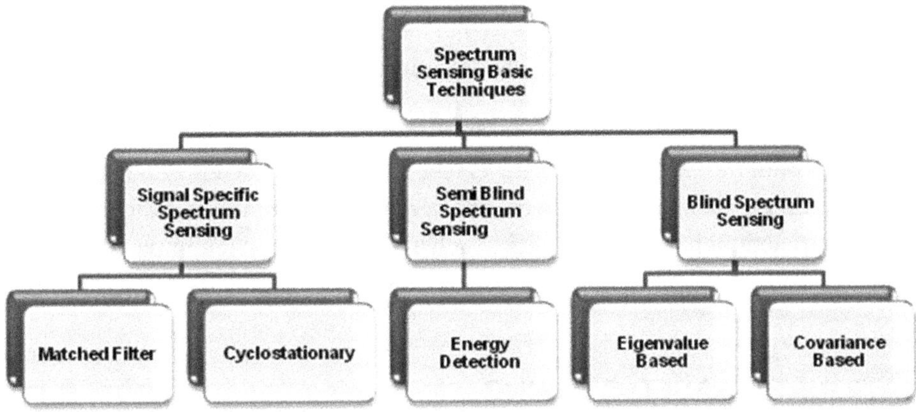

Figure 5. Matched Filtering

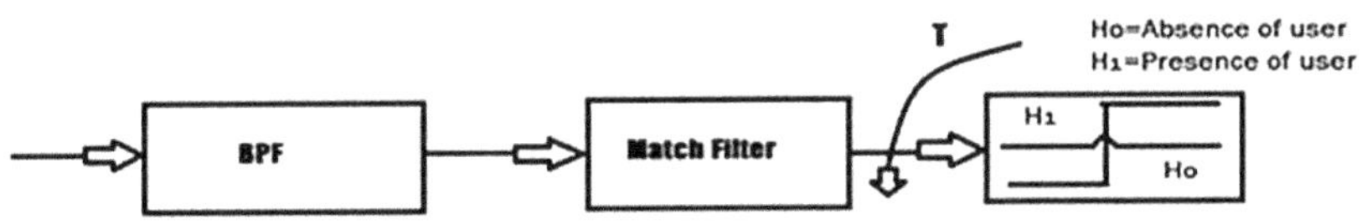

The test statistic taken for matched filter is given by:

$$T(\bar{x}) = \sum_{n=0}^{N-1} x(n)s(n) \tag{10}$$

For H_0

$$T(\bar{x}) = \sum_{n=0}^{N-1} x(n)s(n) \tag{11}$$

$$T(\bar{x}) = \sum_{n=0}^{N-1} x(n)s(n) \tag{12}$$

For H_1

$$T(x) = \sum_{n=0}^{N-1} x(n)s(n) \tag{13}$$

$$T(x) = \sum_{n=0}^{N-1} (s(n) + w(n))s(n) \tag{14}$$

The mean of (12) will be zero and the mean of (14) will be the energy of the signal. These results are compared with a threshold (γ) to detect the presence of PU. For matched filtering the false alarm probability is given as $P_{FA} = Q\left(\frac{\gamma - E}{\sigma\sqrt{E}}\right)$ and the detection probability is given by $P_D = Q\left(\frac{\gamma}{\sigma\sqrt{E}}\right)$.

Matched filtering is a coherent detection technique. Prior knowledge of PU is mandatory for a CR user to detect spectral opportunities. Matched filtering can be used to find PU power which helps the CR user to adjust their transmit parameters according to the interference constraints. Even though matched filtering is computationally efficient, it has few drawbacks such as CR user should completely be aware of every PU and dedicated receiver to acquire signal from different type of primary user (Yucek & Arslan, 2009).

CYCLOSTATIONARY BASED SENSING

Digitally modulated signals have built in periodicity. All periodic signals are cyclostationary in nature. Cyclic nature of the signal properties like preambles, pilots, beacon frames, cyclic prefix, hopping sequence etc can be used for spectrum sensing. The received signal is observed for a particular time interval and the auto correlation of any one of the above-mentioned features is estimated. This result is further compared with a threshold to detect the presence and the absence of PU (Sutton, Lotze, Nolan, & Doyle, 2007). For a given cycle frequency (α), time lag (τ) and cyclic autocorrelation function ($R_\chi^\alpha(\tau)$), Cyclic spectral density function (SDF) is obtained by:

$$S_\chi^\alpha(f) = \int_{-\infty}^{\infty} R_\chi^\alpha(\tau) e^{-2\pi f\tau} d\tau \tag{15}$$

The cyclic spectral coherence function is given by:

$$\rho_\chi^\alpha = \frac{S_\chi^\alpha(f)}{S_\chi^\alpha\left(f + \frac{\alpha}{z}\right) S_\chi^\alpha\left(f - \frac{\alpha}{z}\right)} \tag{16}$$

SDF can be estimated by using frequency or time averaging method. The good estimate of the cyclic SDF is achieved by accomplished by utilizing long perception length and small spectral resolution.

Figure 6. Cyclostationary Feature Detection

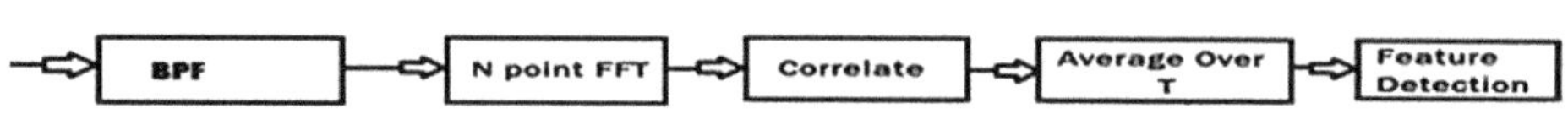

$$\frac{p\left(\bar{X} / H_1\right)}{p\left(\bar{X} / H_0\right)} \geq \gamma \tag{17}$$

Cyclostationary detector performs well in low SNR region. Partial information of PU is necessary and it requires long sensing interval for a better performance which is the drawback of this detector (Bkassiny & Jayaweera, 2014).

ENERGY DETECTION

Energy Detection is simple spectrum sensing technique and is widely used when the PU signal is unknown. In energy detection based sensing, the received signal energy is computed and averaged over a period of time, and compared with a threshold to find the PU presence and absence (Digham, Alouini & Simon, 2007). The test statistics taken for energy detector is given by:

$$T(\bar{X}) = \sum_{n=0}^{n-1} x^2(n) \tag{18}$$

The Likelihood Ratio Test for energy detection is formulated as:

$$LLR = \log\left(\frac{p\left(\bar{X} / H_1\right)}{p\left(\bar{X} / H_0\right)}\right) \geq \gamma \tag{19}$$

The detection probability is given as $P_{FA} = Q_{\chi N}\left(\frac{\gamma}{\sigma^z}\right)$ and false alarm probability is given by $P_D = Q_{\chi N}\left(\frac{\gamma}{\sigma_x^z + \sigma_y^z}\right)$,where γ is the threshold. Multipath fading and shadowing may result in power fluctuations of received PU signals. The performance of energy detection relies on noise power and SNR. Energy detector is a simple non-coherent detector. The major drawbacks of energy detector are these: its performance is poor under low SNR regions and it cannot differentiate between PU and CR user signals.

Figure 7. Energy Detection

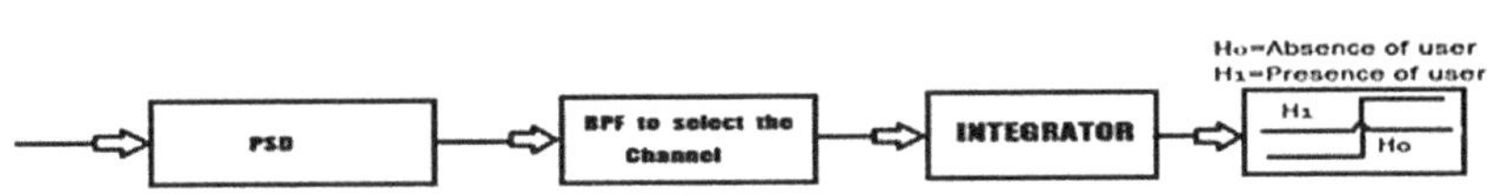

EIGEN VALUE BASED SENSING

This technique uses the Eigen value Decomposition (EVD) of the received signals covariance matrix. CR user has to aggregate received samples in M x N matrix, where M is the receiver dimension. Different Eigen value properties such as Maximum Eigen value (ME), Maximum to Minimum Eigen value (MME) or Signal Condition Number (SCN) can be used for spectrum sensing (Pillay, & Jun Xu, 2013). The advantages of this method are these: non-coherent, sensing performance is highly reliable, and robust to noise uncertainty. The major drawback is high computational complexity (Zeng & Liang, 2009).

COVARIANCE-BASED DETECTION

The covariance-based detection utilizes the dispersive nature of PU signals at the CR receiver… Covariance-based detector predicts the presence and absence of the primary signal based on the received signal covariance matrix. The ratio of maximum to minimum Eigen value of the covariance matrix of the received signal is estimated. By differentiating this difference, desired signal component can be obtained (Zeng & Liang, 2009).

CONCLUSION

Due to the increase in wireless devices, there is a great demand for spectrum. CRN seems to be a promising technique for spectrum scarcity problem. Sensing accuracy has been considered as an critical parameter in determining the performance of CRN. In this chapter, the various challenges and performance measures of CRN in sensing aspect has been discussed. The basis spectrum sensing techniques like matched filter detection, Cyclostationary detection, Eigen value-based detection and covariance based detection has been discussed. Matched filtering method is a coherent detection method. The major limitation of matched filtering is it requires synchronization and PU prior knowledge. Energy detection does not need any prior knowledge of PU. It cannot differentiate whether the received signal is a PU signal or CR user signal. It also has poor performance in low SNR regions. Cyclostationary detector performs well in low SNR regions and it is robust to noise uncertainty. Due to channel fading the Cyclostationary features may be completely lost and this technique consumes more sensing time too. Eigen value-based detector is a non-coherent detector. The major drawback of this method is high computational complexity.

Even though the Cyclostationary method is computationally complex its sensing accuracy is quite good when compared to other sensing techniques. Environment scenario-based spectrum sensing detection selection may increase the overall performance of the CRN. Moreover advance spectrum sensing techniques provide more spectral opportunities than basic spectrum techniques which have been addressed in the upcoming chapter.

REFERENCES

Alexander, M. (2010). *Cognitive Radio Communications and Networks - Principles and Practice*. Elsevier Inc.

Ali, A., & Hamouda, W. (2017). Advances on Spectrum Sensing for Cognitive Radio Networks: Theory and Applications. *IEEE Communications Surveys and Tutorials*, *19*(2), 1277–1304. doi:10.1109/COMST.2016.2631080

Biglieri, E., Goldsmith, A. J., Greenstein, L. J., & Mandayam, N. B. (2012). *Principles of Cognitive Radio*. Cambridge University Press. doi:10.1017/CBO9781139236850

Bkassiny & Jayaweera. (2014). Robust, Non-Gaussian Wideband Spectrum Sensing in Cognitive Radios. *IEEE Transactions on Wireless Communications, 13*(11), 6410-6421.

Bountouris. (2013). *Spectrum Sensing Techniques for Cognitive Radios* (Thesis). Technical University of Crete, Greece, Department of Electronic and Computer Engineering.

Digham, F. F., Alouini, M. S., & Simon, M. K. (2007). On the energy detection of unknown signals over fading channels. IEEE Transactions on Communications, 55(1), 21-24.

Hattab & Ibnkahla. (2014). Multiband Spectrum Access: Great Promises for Future Cognitive Radio Networks. *Proceedings of the IEEE*, 102(3), 282-306.

Jaiswal, M., Sharma, A. K., & Singh, V. (2013). A Survey on Spectrum Sensing Techniques for Cognitive Radio. *Conference on Advances in Communication and Control Systems*, 647-660.

Mehta, Kumar, & Saini. (2013). Comparison of Spectrum Sensing Techniques in Cognitive Radio Networks. *International Journal on Electronics & Communication Technology, 4*(3), 33-37.

Pillay, N., & Xu, H. J. (2013). Eigen value-based spectrum sensing using the exact distribution of the maximum eigenvalue of a Wishart matrix. *IET Signal Processing*, *7*(9), 833–842. doi:10.1049/iet-spr.2012.0320

Sahai, Hoven, Mishra, & Tandra. (2006). Fundamental tradeoffs in robust spectrum sensing for opportunistic frequency reuse. *IEEE Journal on Selected Areas in Communications*.

Subhedar & Birajdar. (2011). Spectrum Sensing Techniques in Cognitive Radio Networks: A Survey. *International Journal of Next-Generation Networks*, *3*(2).

Subhedar & Birajdar. (2011). Spectrum Sensing Techniques in Cognitive Radio Networks: A Survey. *International Journal of Next-Generation Networks, 3*(2).

Sutton, P. D., Lotze, J., Nolan, K. E., & Doyle, L. E. (2007). Cyclostationary signature detection in multipath Rayleigh fading environments. *Proceedings of IEEE International conference on Cognitive Radio Oriented Wireless Networks and Communication*. 10.1109/CROWNCOM.2007.4549833

Wang, W. (2009). Spectrum Sensing for Cognitive Radio. *Proceedings of the IEEE Computer Society*, 110-112.

Yucek & Arslan. (2009). A Survey of Spectrum Sensing Algorithms for Cognitive Radio Applications. *Proceedings of the IEEE Communications Surveys & Tutorials, 11*(1), 116-130.

Zeng, Y., & Liang, Y.-C. (2009). Eigen value-based Spectrum Sensing Algorithms for Cognitive Radio. *IEEE Transactions on Communications*, *57*(6), 1784–1793. doi:10.1109/TCOMM.2009.06.070402

Zeng, Y., & Liang, Y. C. (2009). Spectrum-sensing algorithms for cognitive radio based on statistical covariances. *IEEE Transactions on Vehicular Technology*, *58*(4), 1804–1815. doi:10.1109/TVT.2008.2005267

Zhang & Lysaker. Dynamic Spectrum Access in Cognitive Radio Wireless Networks. *Proceeding of IEEE International Conference on Communications,* 4927 – 4932.

Chapter 5
Review of Spectrum Sensing Techniques in Cognitive Radio Networks

Ala Eldin Omer
American University of Sharjah, UAE

ABSTRACT

Most frequency spectrum bands are licensed to certain services to avoid the interference between various networks, but the spectrum occupancy measurements show that few portions of this spectrum are fully efficiently used. Cognitive radio is a future radio technology that is aware of its environment, internal state, and can change its operating behavior (transmitter parameters) accordingly. Through this technology the unlicensed users can use the underutilized spectrum without causing any harmful interference to the licensed users. Its key domains are sensing, cognition, and adaptation. The spectrum sensing problem is one of the most challenging issues in cognitive radio systems to detect the available frequency bands. This chapter introduces the concepts of various transmitter detection techniques, namely energy detection, matched filter detection, and cyclostationary feature detection. The chapter also discusses other sensing techniques that are introduced to enhance the detection performance of the conventional energy detector. Additionally, the introduced sensing techniques are implemented using extensive MATLAB simulations and their performances are evaluated and compared in terms of sensing time, detection sensitivity, and ease of implementation. The implementation is based on BPSK and QPSK modulation schemes under various SNR values for AWGN noisy channel with Rayleigh fading.

INTRODUCTION

Recent advances in wireless communication and the extensive development of new wireless services and applications have led to a transition from the conventional voice-only communications to multimedia-based applications. Consequently, spectrum resources are facing huge demands with needs for higher data rates. However, this interest is impeded by the shortage of radio spectrum, due to the legacy method of static assignment of the radio spectrum to licensed users. With providing each new service with a

DOI: 10.4018/978-1-5225-5354-0.ch005

unique fixed frequency band in a traditional allocation way, most of the primary spectrum has become fully assigned, accordingly, it has become very difficult to allocate the newly expanded services within the existing spectrum. Due to present government polices of static assignment, unlicensed users are restricted from accessing the licensed spectrum instead they turn to share several frequency bands that are interference-prone and heavily-populated. Therefore, the radio spectrum scanning measurements investigate a huge scarcity in certain frequency bands with only partial occupation most of the time, while the rest of the spectrum is being heavily and fully used (Haykin, 2005).

The researches in spectrum utilization have shown that the main cause of spectrum scarcity problem is the underutilization of frequency bands in the fixed spectrum policy in which the wireless spectrum (i.e. frequency carriers) are leased to some operators for a long period of time up to several years. Therefore, an urgent need is emerged for new innovative and developed techniques that can exploit the available spectrum more efficiently. Researchers have been motivated to find new wireless communication technologies to overcome the radio spectrum scarcity and meet the increasing demand for the wide range of wireless service. Cognitive Radio arises to be an enticing solution to the inefficient utilization and spectral congestion problems that allows the secondary users to opportunistically exploit the less occupied frequency bands by the licensed users in the network (Haykin, 2005) ((FCC), November 2002).

Cognitive Radio (CR) networks have been introduced by researches to overcome the spectrum scarcity problem. It provides an efficient usage of the underutilized radio spectrum through dynamic spectrum access methodology. CR has different system models and cognitive functionality levels. The first CR standard using TV bands is IEEE 802.22 Wireless Regional Area Networks (WRAN), in which the medium access is controlled by a base station (BS). The definition adopted by Federal Communications Commission (FCC):

Cognitive Radio: A radio or system that senses its operational electromagnetic environment and can dynamically and autonomously adjust its radio operating parameters to modify the system operation, such as maximize throughput, mitigate interference, facilitate interoperability, access secondary markets. ((FCC), November 2002)

Generally, Cognitive radio architecture consists of: primary user, secondary user, and shared channel. The licensed spectrum is a shared resource between primary and secondary users with higher priority for the primary users in accessing the channel. Therefore, when a primary user is detected, the secondary users have to vacate the channel in a timely manner to avoid any possible interference. In order to reduce secondary users connection loss and primary users interference, the secondary users need to avoid the channels with high occupancy probability within a given time period. Therefore, the secondary users should be able to evaluate the channel(s) availability through predicting the traffic pattern of the primary user(s).

The Cognitive Radio technology will enable the unlicensed user to determine the available portion of the spectrum to be used by detecting the presence of primary user(s) (Spectrum Sensing), select the best available channel based on the user requirements (Spectrum Management), coordinates the access to the channel with other unlicensed users (Spectrum Sharing) and change to some other frequency channel whenever the primary user is detected (Spectrum Mobility) (Akyildiz, Lee, Vuran, & Mohanty, 2006).

Being the focus of this chapter, spectrum sensing is the most important motif for establishing the Cognitive Radio and it is concerned about being aware of the spectrum usage and presence of licensed users in a geographical area in specific frequency band(s). This awareness can be obtained by local spectrum sensing at Cognitive Radios (Akyildiz, Lee, Vuran, & Mohanty, 2006; Shojaeifard, Zarringhalam, & Shikh-Bahaei, 2011). When Cognitive Radio is considered, a general perspective for spectrum sensing is considered rather than its traditional understanding of measuring the energy of spectral content over a frequency band. It can be viewed to determine: characteristics of the spectrum usage across multiple dimensions (time, space, frequency and code) and the types of signals occupying the spectrum (modulation technique, bandwidth and carrier frequency). Hence, robust signal analysis techniques with additional computational complexity are needed to achieve such efficient sensing.

Several methods based on single Cognitive Radio user have been introduced to obtain good spectrum sensing performance (Haykin, 2005; Mitola, 2000). Those techniques can be classified into three categories as shown in Figure 1: primary transmitter detection, primary receiver co-operative detection and interference-based detection. The goal of this chapter is to implement different transmitter detection techniques for sensing the frequency spectrum and compare them in terms of detection performance under different SNR conditions, sensing time required, detection sensitivity and the ease of implementation.

BACKGROUND

With spectrum sensing, the secondary users become able to adapt to the radio environment by exploiting spectrum opportunities for transmission without causing an interference to the primary licensed users. This can be done through a real-time wideband spectrum sensing capability to detect weak primary signals in a wide spectrum range. After the spectrum holes have been identified then the secondary users can utilize them for their transmission. All spectrum sensing approaches aim to decide between the two hypotheses, defined as:

$$y(n) = \begin{cases} v(n), & H_0 \\ s(n)h(n) + v(n), & H_1 \end{cases} \tag{1}$$

Figure 1. Classification of Spectrum Sensing Techniques

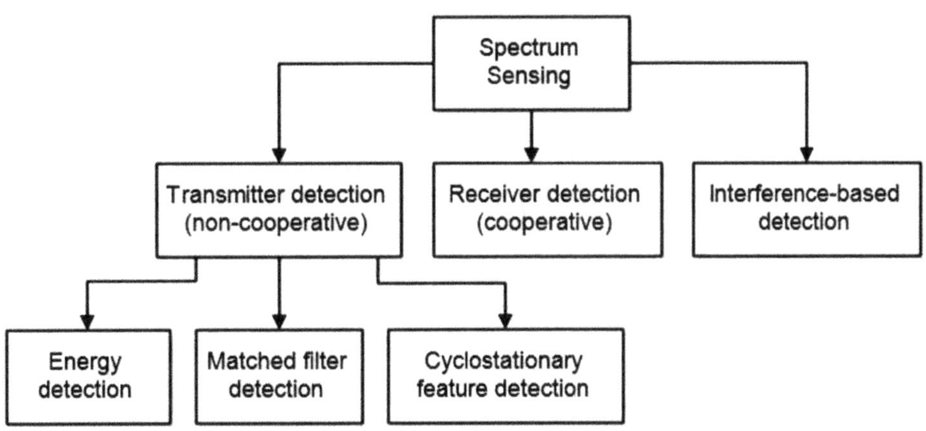

where $y(n)$ is the signal received by the secondary user, $s(n)$ is the transmitted signal of the primary licensed user, $v(n) \sim N(0, \sigma_v^2)$ is the additive white Gaussian noise (AWGN) with zero mean and variance σ_v^2 σw2, $h(n)$ is the amplitude gain of the channel. H_0 is the null hypothesis, which indicates that the channel is free to use with no primary user existence over the spectrum, while, H_1 is the hypothesis that indicates the primary user is online utilizing the channel.

Errors occur in spectrum sensing algorithms due to the false alarms and missed detections. The first occurs when the sensed spectrum is available but H_1 is decided. Consequently, it leads to low spectrum utilization as the secondary users miss the available transmission opportunities. The second error occurs when the sensed spectrum is allocated by primary user(s) but H_0 is decided, which leads to undesirable harmful interference to the licensed users. Thus the performance evaluation of any spectrum sensing is conducted through the probability of false alarm P_F and the probability of detection P_D which are related over the Receiver Operating Characteristic (ROC) curves.

In this chapter we will discuss the implementation of different transmitter detection techniques for sensing the frequency spectrum and compare between them in terms of detection performance under different SNR conditions, the sensing time required, the detection sensitivity and the ease of implementation. Those detection techniques are: Single-threshold Energy detection (ED), Matched filter detection (MFD), Cyclostationary feature detection (CFD). Moreover, an adaptive double threshold Energy detection based and an improved Energy detection approaches are introduced to improve the performance accuracy of the conventional Energy detection.

It is worth mentioning that, the primary user signal on which the different spectrum sensing approaches can be applied needs to be generated in the transmitter side with different parameters such as: the operating frequency, the modulation scheme and the sampling frequency. Then this signal is passed through the channel that is also characterized with specific attributes such as: fading, noise … etc. Therefore, the parameters for our transmitter model are set as follows: the operating frequency on which the primary signal is transmitted is 200 Hz, the signal was sampled with 4 KHz (i.e. 20 times the operating frequency). We have applied two modulation schemes BPSK (binary phase shift keying) and QPSK (quadrature phase shift keying). While for the channel modelling, we assume an AWGN behaviour under various SNR constraints with fading impulse response that follows the Rayleigh distribution.

Single-Threshold Energy Detection

The Energy detector (also known as Periodogram) is a non-coherent detection approach that is commonly used for monitoring and detecting the primary user activity over the spectrum for its low computational and implementation complexities. Additionally, in comparison with other sensing methods the conventional Energy detection is found to be more generic as receivers do not need any previous knowledge about the primary users signal(s). It is the optimal detector when the secondary user(s) cannot gather sufficient information about the primary user signal(s) (Doyle, 2009). The Energy detector has several methods of implementation, for example in the block diagram shown in Figure 2 the received signal is transformed from the time domain to the frequency domain using the Fast Fourier Transform (FFT), then the energy content in each frequency band of the received signal is measured resulting in the Power Spectral Density (PSD).

Figure 2. The Periodogram Energy Detector block diagram

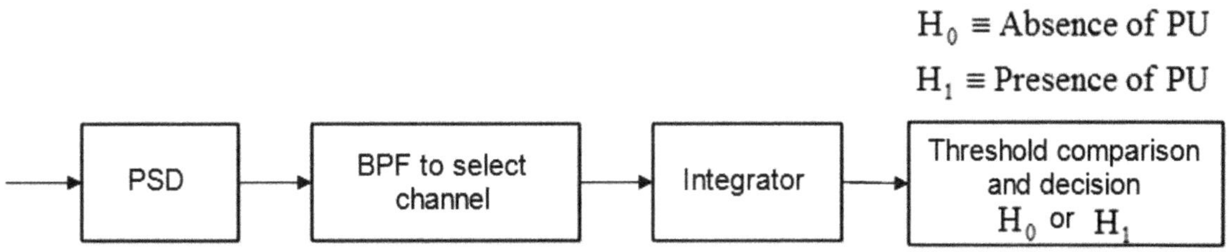

After obtaining the PSD the signal is applied to band pass (BP) filter in order to reduce the noise and select the bandwidth/channel of interest resulted in a noise of a band-limited flat spectral density at the filter output. After that the signal is integrated over the time interval. In the last stage of the Energy detector, the output signal of the integrator is compared against a pre-set threshold to determine whether the primary user is present in the frequency spectrum or not. The threshold value could be set as fixed or variable based on the channel conditions and the noise floor.

One of the drawbacks of the energy detection sensing approach is that a threshold will be easily influenced by unknown or changing noise levels and channel conditions, as a result the detector will be confused by the presence of any in-band interference. Additionally, this detector will have a poor performance under low SNR conditions.

Another simple way of implementing the Energy detector is to determine the signal energy using Equation (4). To elaborate, assume that the signal detected at the receiver side has the following expression:

$$y(n) = s(n)h(n) + v(n), \tag{2}$$

where, $y(n)$ is the received signal under test, $s(n)$ is the transmitted primary user signal, $v(n)$ is the AWGN sample, $h(n)$ is the amplitude gain of the channel and n is the sample index. Note that $s(n) = 0$ when there is no transmission by the primary user (offline). The decision metric for the energy detector can be written as:

$$T = \sum_{n=0}^{N} \left| y(n) \right|^2, \tag{3}$$

where N is the size of the observation vector used by the detector (number of samples). The decision on the occupancy of a frequency band can be obtained by comparing the decision metric T against a pre-set threshold λ. This is equivalent to distinguish between the two hypotheses given in Equations (1) and (2).

The sensing performance of the Energy detection algorithms is summarized in two probabilities: The probability of detection P_D and the probability of false alarm P_F. The detection probability is the probability of detecting a primary signal on a certain frequency band when it is truly present. Thus, a large detection probability is desired for efficient sensing. It has the following formulation:

$$P_D = P_r(T > \lambda / H_1), \tag{4}$$

The probability of false alarm is the probability that the sensing test incorrectly decides that the tested frequency band is occupied by the primary user signal when actually it is not. In order to efficiently utilize the available transmission opportunities by the secondary users, P_F should be kept as small as possible as follows:

$$P_F = P_r(T > \lambda / H_0), \tag{5}$$

The threshold value is selected in such a way to obtain an optimum balance between P_D and P_F. However, this will require a prior knowledge of the noise and detected signal powers. The estimation of noise power is quite possible compared to the signal power estimation that could be difficult with the changes in the characteristics of the on-going transmission and the distance between the licensed user and Cognitive Radio. In practice, the threshold is chosen to meet a target probability of false alarm P_F according to the noise power (i.e. variance). Therefore, the threshold value can be calculated through the following equation:

$$\lambda = Q^{-1}\left(P_F\right) x \sqrt{2N\sigma_v^4} + N\sigma_v^2, \tag{6}$$

where Q^{-1} represents the inverse Gaussian tail probability Q-function, P_F is the targeted probability of false alarm, N is the number of samples of the received primary signal used by the detector, σ_v^2 is the variance of the Gaussian noise added to the primary user signal.

Match Filter Detection

The second transmitter detection approach for sensing the spectrum is the Matched filter which is a coherent detector and is known as the optimum method for detecting the primary user signal(s) when the transmitted signal is known for the receivers. The MFD is considered as an optimal detector in the radio environment with Gaussian noise as it maximizes the received SNR. The operation of the MFD can be expressed as (Garhwal & Bhattacharya, 2012):

$$x\left(n\right) = \sum_{k=-\infty}^{\infty} h\left(n-k\right) y\left(k\right), \tag{7}$$

where $y\left(n\right)$ is the unknown received signal given either by (1) or (2) convolved with $h\left(n\right)$ which is the impulse response of the Matched filter that is matched to the reference primary signal to maximize the SNR as follows:

$$h\left(n\right) = s\left(T-n\right). \tag{8}$$

The Match filter technique is useful for checking and detecting the spectrum availability only if the secondary users receivers have a previous information about the primary users signals. Therefore, there is a need to generate local carriers on the receiver side based on that prior knowledge such as: type of modulation, operation frequency, sampling frequency … etc.

In the Match filter detector block diagram shown in Figure 3, the band pass filter is applied to select the channel then the presence of the primary user is determined by convolving the received unknown signal with time shifted version of the locally generated carrier used for PU signal. Finally, the matched filter output is compared against a pre-set threshold to decide between H_0 and H_1 regarding the spectrum availability.

The advantage of the Matched filter sensing technique is that, less time is needed to get high detection performance. However, it suffers from some drawbacks such as: a devoted receiver with a prior knowledge is required for each primary user signal. Moreover, the implementation complexity of the sensing unit is large where the local carriers need to be generated with large amount of consumed power while various receiver algorithms are simultaneously executed for sensing the spectrum (Garhwal & Bhattacharya, 2012).

Cyclostationary Feature Detection (CFD)

Another transmitter detection technique is known as Cyclostationary feature detection (CFD) which is based on the introduction of periodic redundancy into a signal and exploiting the built in features in the primary user signal such as: modulation, sampling and operating frequency. The CFD block diagram shown in Figure 4 exploits the periodicity in the received primary user signal to identify the presence or absence of primary users in the scanned spectrum band (Mehta, Kumar, & Saini, 2013).

One of the powerful features of this detector its ability to recognize the modulated information signal and differentiate it from the additive noise. Thus, it can easily identify the noise from the primary user signal because most of the man-made signals used in communication systems such as BPSK, QPSK, AM and OFDM exhibit Cyclostationary behaviour, while the noise signals (typically white Gaussian noise) do not experience this behaviour. The CFD exploits the unique information embedded on the primary user signal such as: modulation scheme and the operating frequency that do not exist in noisy signals. Moreover, the CFD can classify the primary users signals using this built-in information.

The CFD measures a property of the received signal known as Spectral Correlation Function (SCF) or cyclic spectral density (CSD) that is given by:

Figure 3. The Matched Filter Detector block diagram

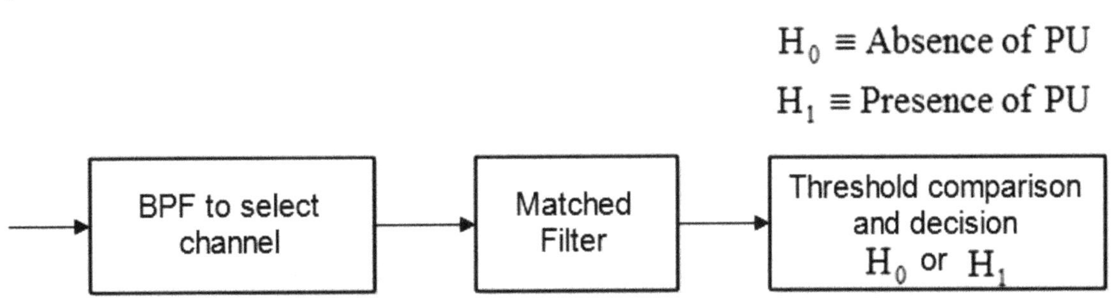

Figure 4. The Cyclostationary Feature Detector block diagram

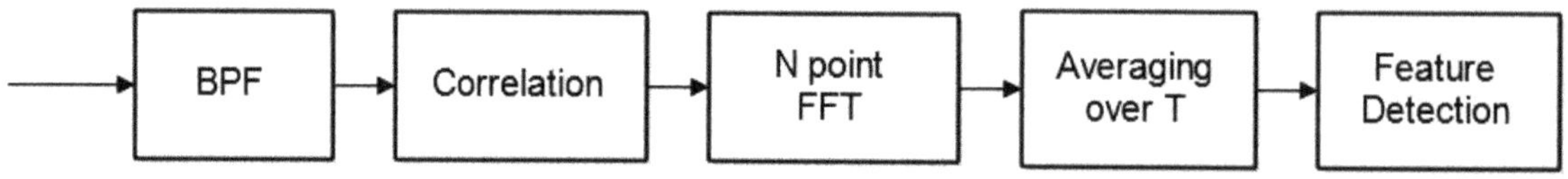

$$S_y^{\alpha} = \int_{-\infty}^{\infty} R_y^{\alpha}\left(\tau\right) e^{-j2\pi f\tau} d\tau, \tag{9}$$

where $R_x^{\alpha}\left(\tau\right)$ is the cyclic autocorrelation function (CAF) of the transmitted primary user signal y(n) and α is the cyclic frequency. The CSD function outputs peak values when the cyclic frequency α is equal to the fundamental frequencies of transmitted signal y(n). Also it determines the type of modulation used for the primary signal transmission by the shape of peaks appear at that frequency. Generally, the receivers with CFD capability extract the cyclic frequencies to be used as features for recognizing the transmitted primary users signals or otherwise are assumed to be known (Doyle, 2009; Shojaeifard, Zarringhalam, & Shikh-Bahaei, 2011).

This detection technique is robust to noise discrimination and performs better than the conventional Energy detector and Match filter under low SNR conditions. However, it has a higher computational complexity and longer time for sensing the spectrum is needed compared with other sensing techniques.

Adaptive Double-Threshold Energy Detection based

The choice of an optimal single threshold for the Cognitive Radio in the traditional energy detection is difficult because of the noise uncertainty in the transmission systems. Moreover, under low SNR conditions the performance may ultimately differ from the targeted metrics (Tandra & Sahai, 2008). Therefore, an adaptive double threshold (ADT) sensing method has been proposed to improve the detection performance at fixed probability of false alarm P_F under low SNR conditions. This technique can be used as a second stage when the first conventional Energy detection fails to detect the primary user signal.

As shown in Figure 5, the region bounded between λ_1 and λ_2 is known as the confusion region where it is difficult to differentiate the primary user signal from the noise using a single threshold. Therefore, the ADT technique is introduced with two applied thresholds, the upper bound threshold λ_1 and the lower bound threshold λ_2 that are determined based on the maximum and minimum levels of the noise variance, respectively (Song et al., 2010) (Ejaz, ul Hasan, Lee, & Kim, 2013).

The noise uncertainty in wireless systems can be described as $\left[1/\rho\sigma_w^2, \rho\sigma_w^2\right]$, where $\rho > 1$ is a parameter that determines the noise uncertainty. Now, the two thresholds can be formulated as follows to obtain a specific false alarm probability P_F (Song, Alemseged, Tran, Villardi, Sun, Filin, & Harada, 2010; Bagwari, Tomar, & Verma, 2013):

Figure 5. The primary signal and noise energy distribution

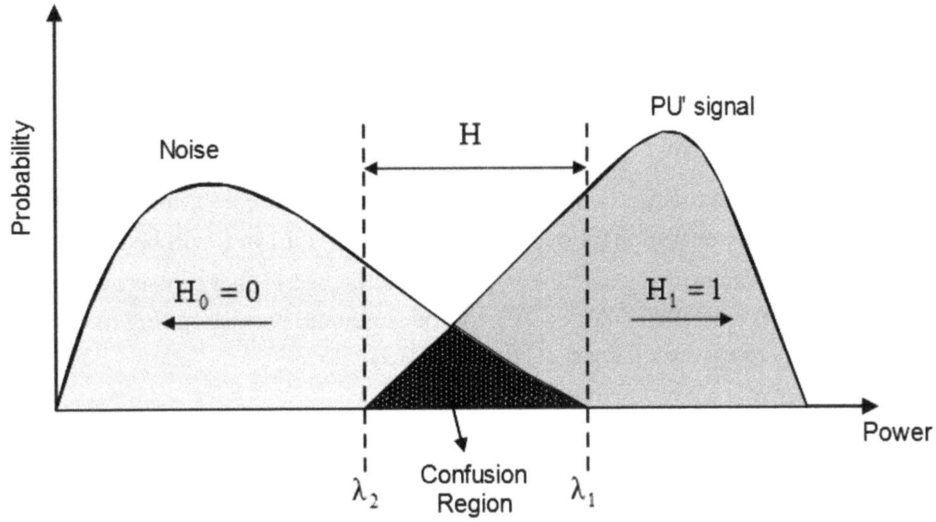

$$\lambda_1 = Q^{-1}\left(P_F\right) x \sqrt{2N\rho\sigma_v^4} + N\rho\sigma_v^2, \tag{10}$$

$$\lambda_2 = Q^{-1}\left(P_F\right) x \sqrt{2N / \left(\rho\sigma_v^4\right)} + N / (\rho\sigma_v^2), \tag{11}$$

where P_F is the targeted false alarm probability (assumed to be 0.1 in our implementation).

The confusion region shown in Figure 5, is divided into four equal sub-regions as shown in Figure. 6. (λ_1 A-AB-BC-C λ_2), where D is the equal gap between the sub-regions. Therefore, the new sub-threshold will be:

Figure 6. The divisions of the confusion region into four equal sub-regions

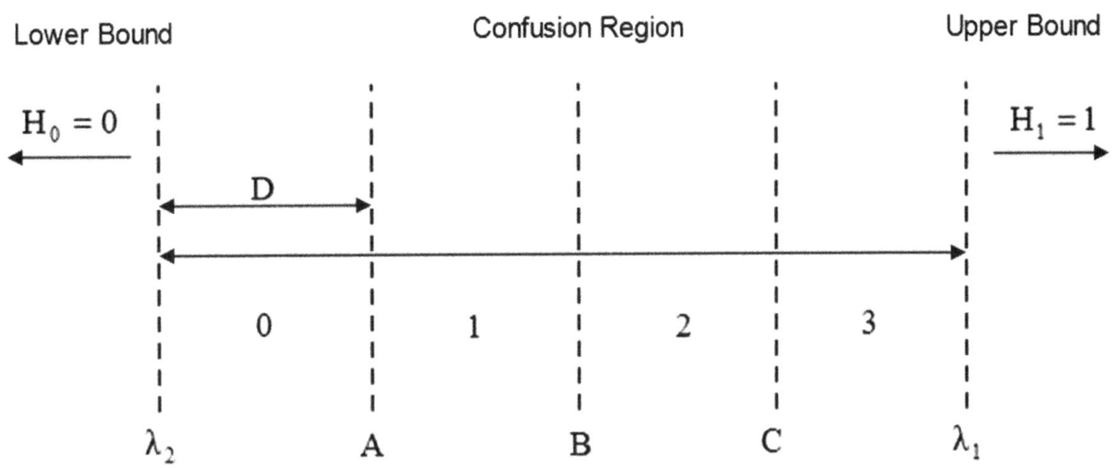

$$ST = \begin{cases} A = \lambda_2 + D \\ B = A + D \\ C = B + D \\ \lambda_1 = C + D \end{cases}, where\, D = \frac{(\lambda_1 - \lambda_2)}{4}. \tag{12}$$

To elaborate, if the decision metric T in Equation (4) that is calculated for the received primary user signal falls outside the confusion region (above λ_1 or below λ_2) then 1 or 0 will be generated to indicate the presence or absence of the primary user, respectively. While, if the energy value lies inside the confusion region then respective decimal values (DV) will be generated corresponding to which sub-region that T falls in as follows (Bagwari, Tomar, & Verma, 2013):

$$DV = \begin{cases} 0, & \lambda_2 < T \leq A \\ 1, & A < T \leq B \\ 2, & B < T \leq C \\ 3, & C < T \leq \lambda_2 \end{cases} \tag{13}$$

Now, the decision at the CR end user on the detection of the primary user signal can be expressed with the following logic rule (LR):

$$LR = \begin{cases} H_0 = 0, & T \leq \lambda_2 (PU\, absent) \\ H = DV, & \lambda_2 < T < \lambda_1 (NOT\, sure), \\ H_1 = 1, & T \geq \lambda_1 (PU\, present) \end{cases} \tag{14}$$

A decision on the detection of primary users is made when the metric T falls in the confusion region based on the value of n which is compared against a specific threshold according to the targeted false alarm probability:

$$n = \{DV, \lambda_2 < T < \lambda_1 . \tag{15}$$

Improved Energy Detection

Due to the poor performance of the traditional Energy detector under low SNR conditions and its dependency on SNR and number of samples, an improved Energy-based detection technique (K & S, 2014) has been introduced to improve the detection probability shown in Figure 7.

The working principle of this improved technique begins with the traditional mechanism of comparing the decision metric T of a received primary signal given by Equation (4) against a predefined threshold over an observation course according to the noise floor in Equation (6) (Ejaz, ul Hasan, Lee,

Figure 7. The improved technique for Energy Detection

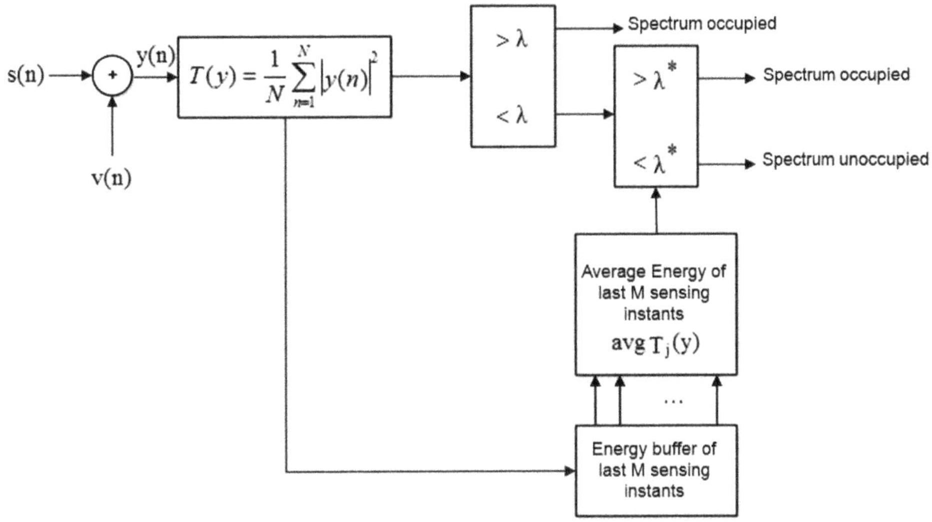

& Kim, 2013). If the decision is H_0 (i.e. spectrum unoccupied) then the improved detection technique takes place to verify the first decision by comparing the averaged energy of the previous stored M sensing samples with a threshold. If the average is larger, then the decision threshold is compared against the immediate past sensing instant energy value to avoid the possibility if the primary user has just leave the spectrum. Then, the decision will be either H_1 (i.e. spectrum occupied) or H_0 based on the last comparison.

In this algorithm s(n) and v(n) are assumed to be real signals of Gaussian distribution. They are also assumed to be independent of each other as well as the tested statistic values of the past instants.

$$avgT_j(y) = mean\left(T_i(y), T_{i-1}(y), \ldots T_{i-M}(y)\right) = \frac{1}{M}\sum_{j=i-M}^{i} T_j(y). \tag{16}$$

Then the new average metric in (16) is compared against the following estimated threshold:

$$\lambda^* = \sigma_v^2\left(\sqrt{\frac{2}{N}}Q^{-1}\left(\hat{P}_F\right) + 1\right). \tag{17}$$

The threshold above takes three different values according to the targeted false alarm probability in the following situations: first, using threshold corresponding to $\hat{P}_F = P_F$ to protect the primary user from being interfered but will result in losing the available transmissions for the secondary users. Second, using threshold with $\hat{P}_F = P_F / 3$ to maximize the secondary user spectrum utilization. The last situation is to compromise between the previous two cases and using a threshold that utilizes $\hat{P}_F = P_F / 2$. Using the improved technique under any of the above thresholds will yield a better performance than the traditional energy detector as we can see later in the implementation (K & S, 2014).

Performance Evaluation

In this section we implemented the spectrum sensing techniques introduced in this chapter and evaluate their performance using extensive MATLAB simulation. First in the transmitter side we generate the primary user signal on which the different spectrum sensing approaches can be applied and we set the different parameters such as: the operating frequency, the modulation scheme, and the sampling frequency. Then, the signal is passed through the channel that also has specific characteristics such as: fading, noise … etc.

Towards this end, we have set the parameters for our transmitter model as follows: the operating frequency on which the primary signal is transmitted is 200 Hz, the signal was sampled with 4 KHz (i.e. 20 times the operating frequency). We have applied two modulation schemes BPSK (binary phase shift keying) and QPSK (quadrature phase shift keying). The channel model was assumed to be AWGN with zero mean and varied variance according to a range of SNR values under test from (-20 dB up to 20 dB) along with fading impulse response that follows the Rayleigh distribution. The constellation diagram of the primary user signal at 20 dB for both BPSK and QPSK are shown in Figure 8.

Figure 9, shows the power spectral density of the detected primary user signal by the Periodogram Energy detector under different SNR conditions using BPSK modulation scheme. Figures 9(a) and 9(b) show how the detector clearly identifies the 20 KHz primary user signal under good SNR (0 up to 20 dB), while it can not easily identify the primary user signal under low SNR (-20 dB) as shown in Figure 9(c). Same results are demonstrated for the QPSK case of modulation.

In the Matched filter detection technique, the received primary user signal was correlated at good SNR for both BPSK and QPSK with the locally generated carriers in the receiver side as shown in Figure 10(a). Figure 10(b) clearly shows that, the red signal gives the maximum cross correlation value at zero lag which indicates that the primary user signal was transmitted on this specific carrier and the primary user is present on the spectrum. This observation will be clear under good SNR conditions and identifiable up to low SNR of about (- 5 dB).

Figure 8. The BPSK and QPSK constellation of the primary user signal

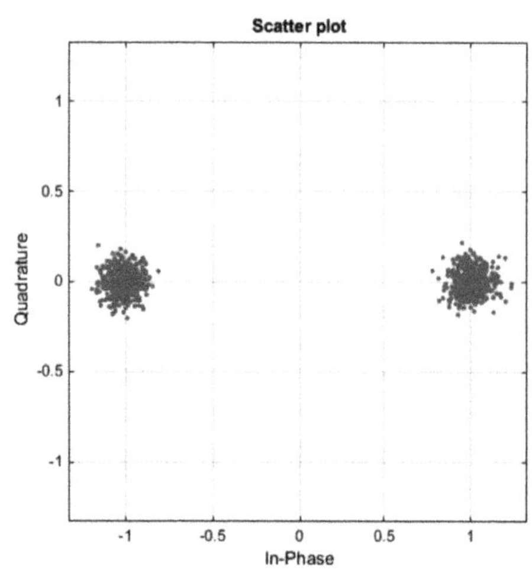

(a)

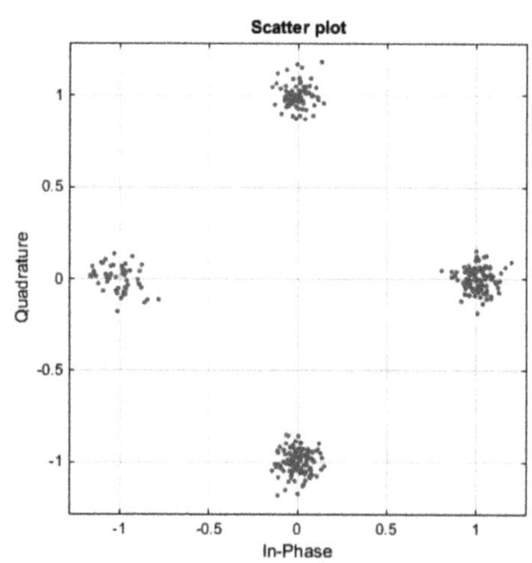

(b)

Figure 9. The Periodogram ED output under high and low SNR: (a) Under high SNR (20 dB); (b) Under zero SNR; (c)Under low SNR (-20 dB)

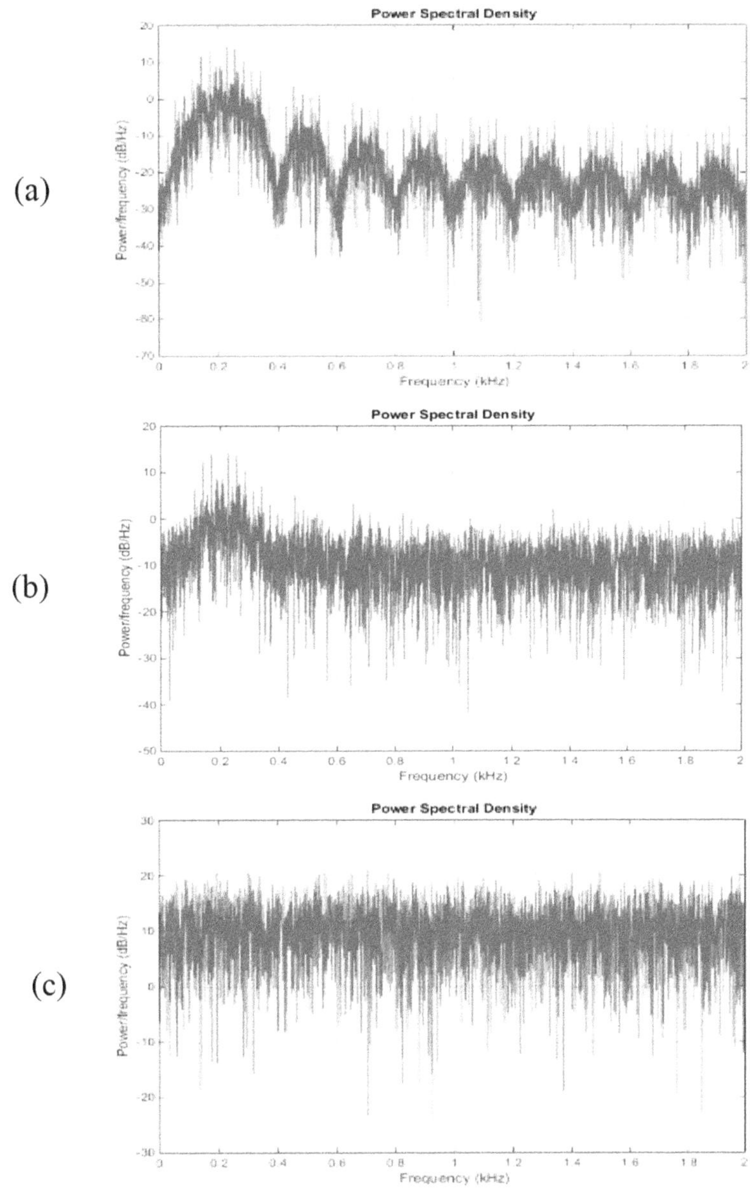

The third implemented sensing technique is the Cyclostationary detector which resulted into peak outputs at the centre frequency as well as double the frequency of the primary user signal at 400 KHz under good SNR of about 20 dB for the BPSK. While two peaks appear when the QPSK modulation scheme is used as shown in Figure 11.

However, under low SNR conditions the Cyclostationary detector can not differentiate the primary user signal as tested at -20 dB in Figure 12 for both BPSK and QPSK.

Figure 10. The Matched Filter Detector output at high SNR: (a) Carriers used in modulation (b) Cross correlation between the RX and carriers

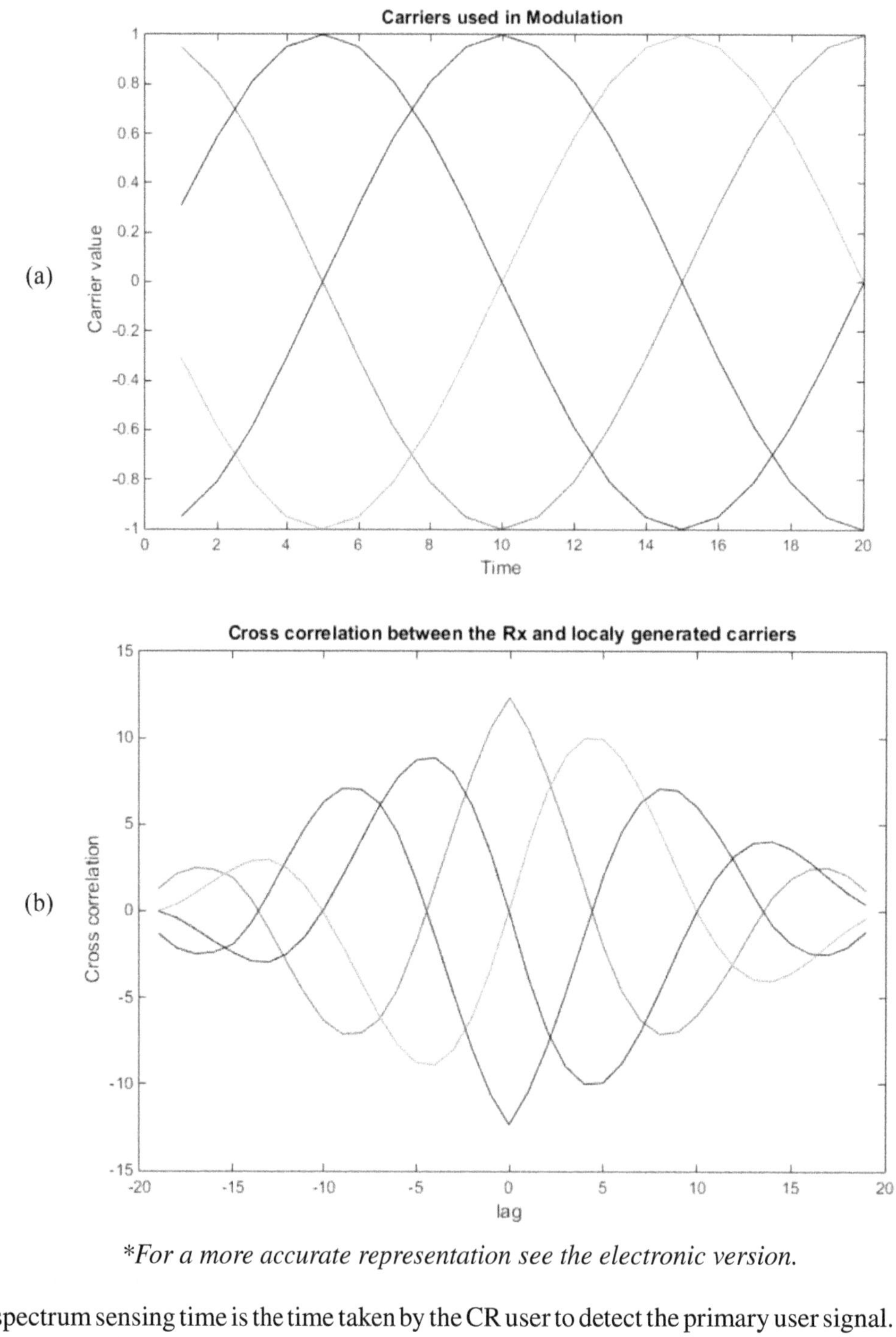

**For a more accurate representation see the electronic version.*

The spectrum sensing time is the time taken by the CR user to detect the primary user signal. Generally, this time is proportional to the number of samples used by the detector, less sensing time is preferable to increase the transmission time for the CR user. In what follows we obtain the computational time for the implemented sensing techniques using the testing machine: Intel(R) - Core(TM) i7-4510U, 2-2.6

Figure 11. Cyclostationary Detector outputs for BPSK and QPSK at high SNR: (a) BPSK (b) QPSK

(a)

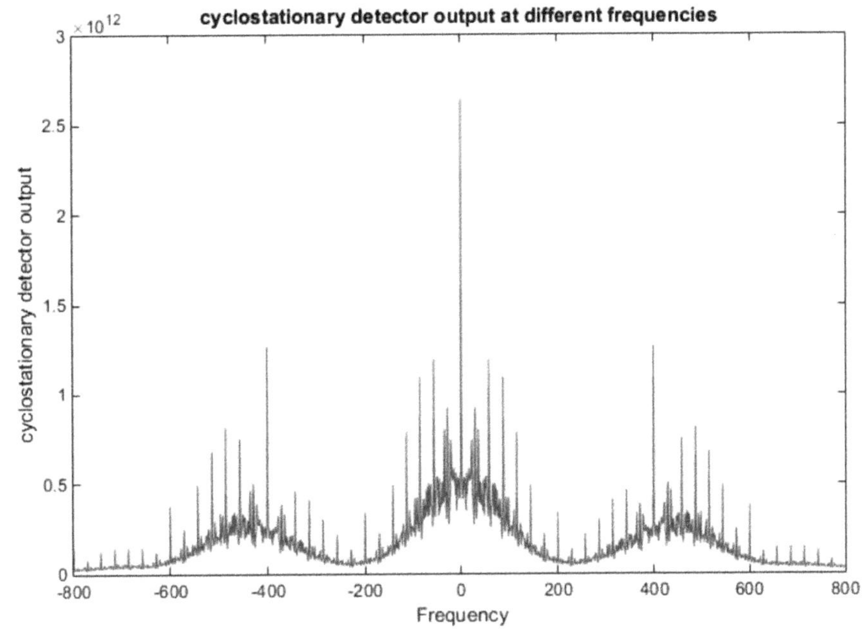

(b)

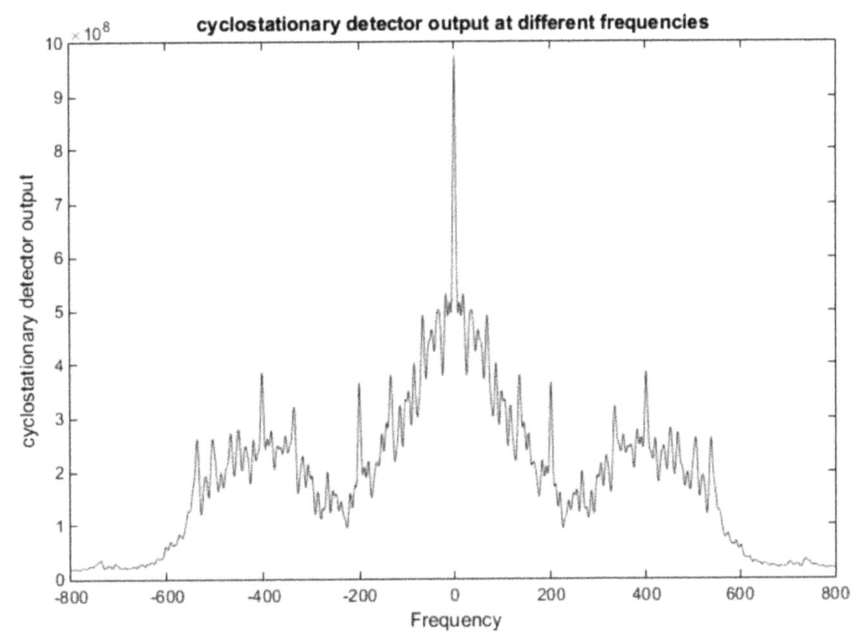

GHz processor and 8 GB RAM using MATLAB platform. Table 1, shows the sensing time required by each sensing technique under the same channel conditions (AWGN channel with Rayleigh fading) for both modulation techniques BPSK and QPSK.

Figure 13 shows the detection probability for the traditional Energy detector under the SNR range from -20 to 10 dB for different number of samples *N*. The same plot was obtained for the improved Energy detector. In both approaches we can clearly depict that increasing the number of samples *N* will

Figure 12. Cyclostationary Detector outputs for BPSK and QPSK at low SNR: (a) BPSK (b) QPSK

(a)

(b)

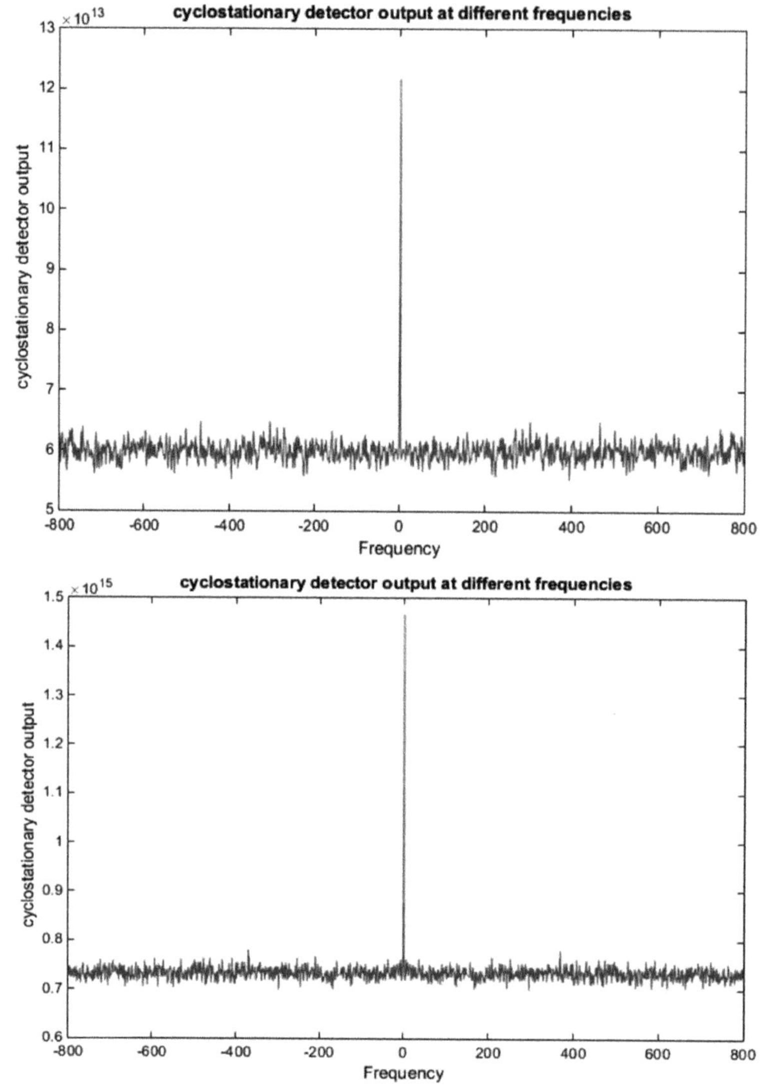

Table 1. Sensing time comparisons for the ED, Matched Filter and Cyclostationary Detectors

Modulation Techniques	Sensing Technique		
	Energy Detection	**Matched Filter Detection**	**Cyclostationary Feature Detection**
BPSK (N=1260)	0.375 (s)	0.078 (s)	1.25 (s)
QPSK (N=540)	0.188 (s)	0.062 (s)	0.73 (s)

[a] *Number of samples (N)*

Figure 13. Detection probability vs. SNR for the traditional and improved EDs: (a) Traditional Energy Detector; (b) Improved Energy Detector

(a)

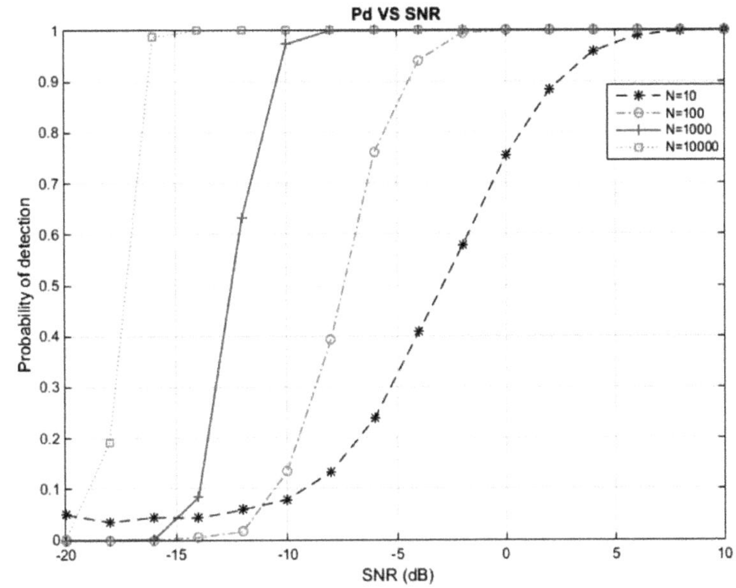

(b)

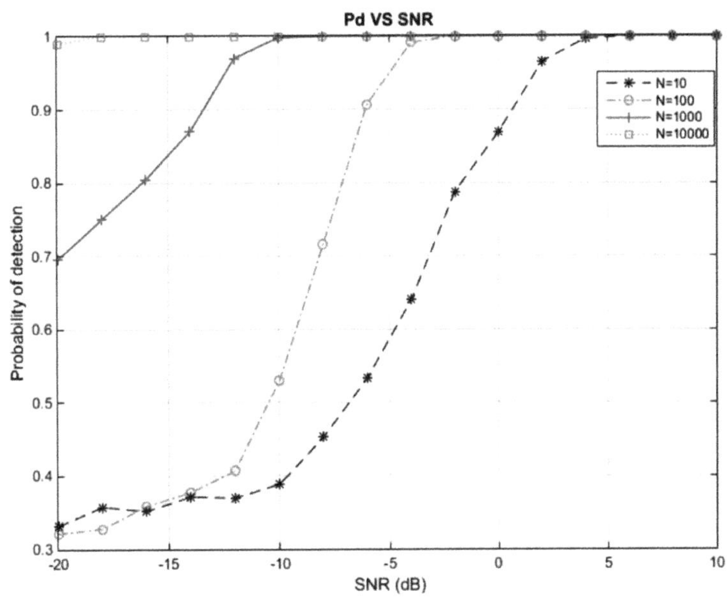

**For a more accurate representation see the electronic version.*

result in an improved detection performance. However, the improved Energy detector can achieve the same performance level of the traditional Energy detector at lower number of samples thus decreasing the sensing time required for detection. For example the traditional Energy detector achieves 0.75 detection probability at -6 dB using $N = 100$ samples, while the improved Energy detector can achieve up to 0.54 at -6 dB using only $N = 10$ samples.

Figure 14 shows the detection performance under the SNR range of values from -20 to 10 dB and the ROC curves at -20 dB for the improved Energy detector under three different thresholds that correspond to the targeted false alarm probabilities discussed previously using $N = 1000$ samples and $M = 10$ previous stored sensing instants. The figure clearly demonstrates that the best performance is obtained when the threshold is corresponding to 0.1 false alarm probability a situation that protects the primary user against the possible interference (i.e. situation (i)).

The receiver operating characteristics (ROC) curves at -10 dB SNR for both the traditional and improved Energy detector are plotted in Figure 15. We used $N = 100$ samples and $M = 5$ previous stored sensing instants for the improved Energy detector. Clearly the improved Energy detector provides better detection performance for the given probability of false alarm.

Figure 14. Detection probability versus SNR and the ROC for the improved ED under different thresholds: (a) ROC curves under different thresholds (b) Probability of Detection versus SNR

(a)

(b)

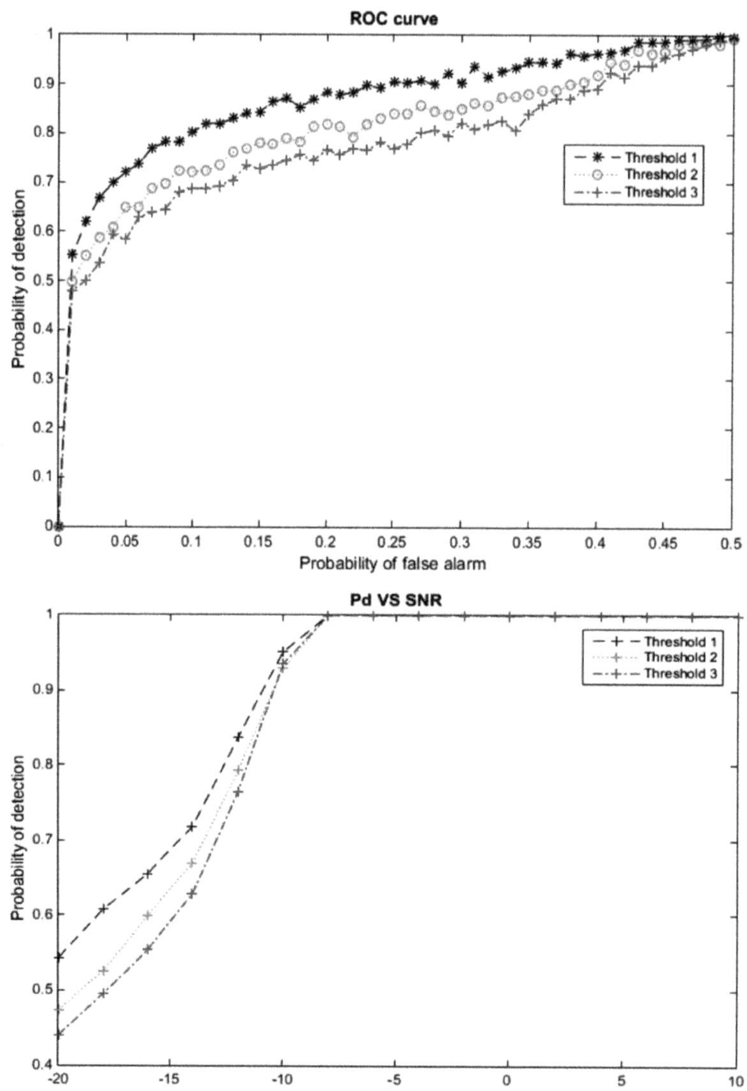

**For a more accurate representation see the electronic version.*

Figure 15. ROC curves for the traditional and improved ED at -10 dB SNR

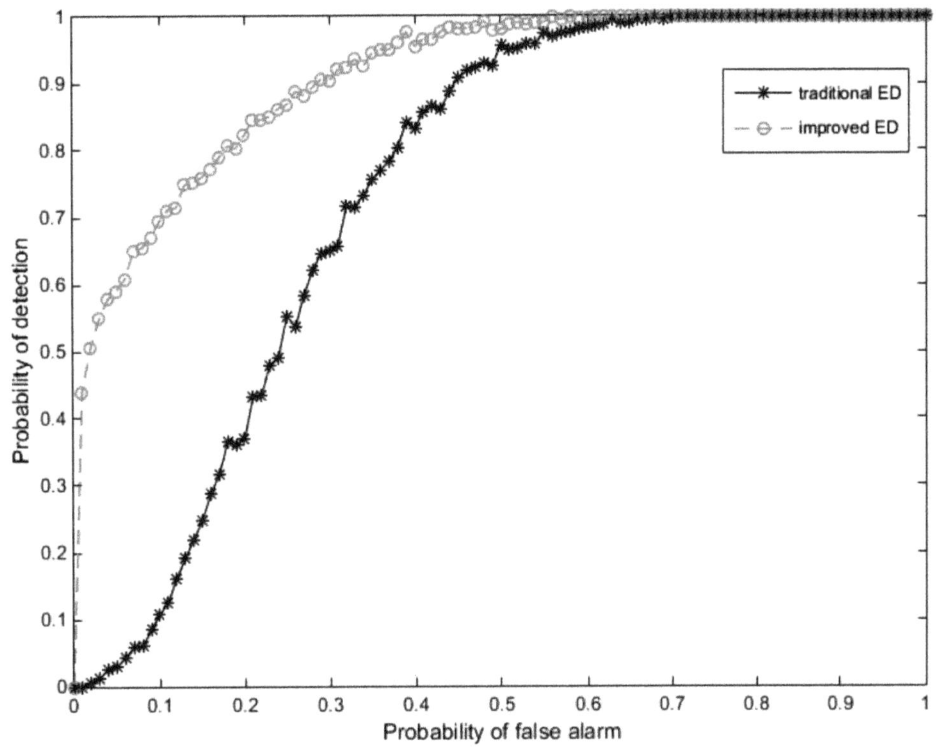

The previous performance observations are confirmed in Figure 16, in which the detection probability is plotted against the SNR using $N = 100$ samples, targeted false alarm probability of 0.1 and $M = 5$ previous stored sensing instants for the improved Energy detector. In all previous observations the signal and noise are assumed to be real Gaussian. The improved ED shows better detection performance than the traditional Energy detector under low SNR conditions.

Another proposed technique for improving the traditional Energy detector performance is the adaptive double threshold Energy detector. Figure 17, shows the detection performance under a range of SNR values from -20 dB to 20 dB for different number of samples. The targeted probability of false alarm was set to 0.1, the test was done using QPSK modulation assuming an AWGN noisy channel with Rayleigh fading. Clearly we can notice that as the number of samples increases the overall performance for detecting the primary user signal improves accordingly.

All techniques of spectrum sensing introduced in this chapter have been implemented and tested against a generated primary user signal with operating frequency 200 Hz and sampling frequency 4 KHz using two types of modulation BPSK and QPSK that resulted into different number of samples. The channel model was assumed to be single path Rayleigh fading AWGN with zero mean and variable variance according to the SNR range under test from -20 dB up to 20 dB.

Figure 18 shows the detection performance against the SNR for the Periodogram Energy detector, Matched filter detector, CFD, adaptive double threshold and the improved Energy detector. We can clearly depict that, the Periodogram Energy detector provides the lowest detection performance under low SNR values and its threshold depends on the provided number of samples. The Matched filter de-

Figure 16. Detection performance comparison between the traditional and improved ED

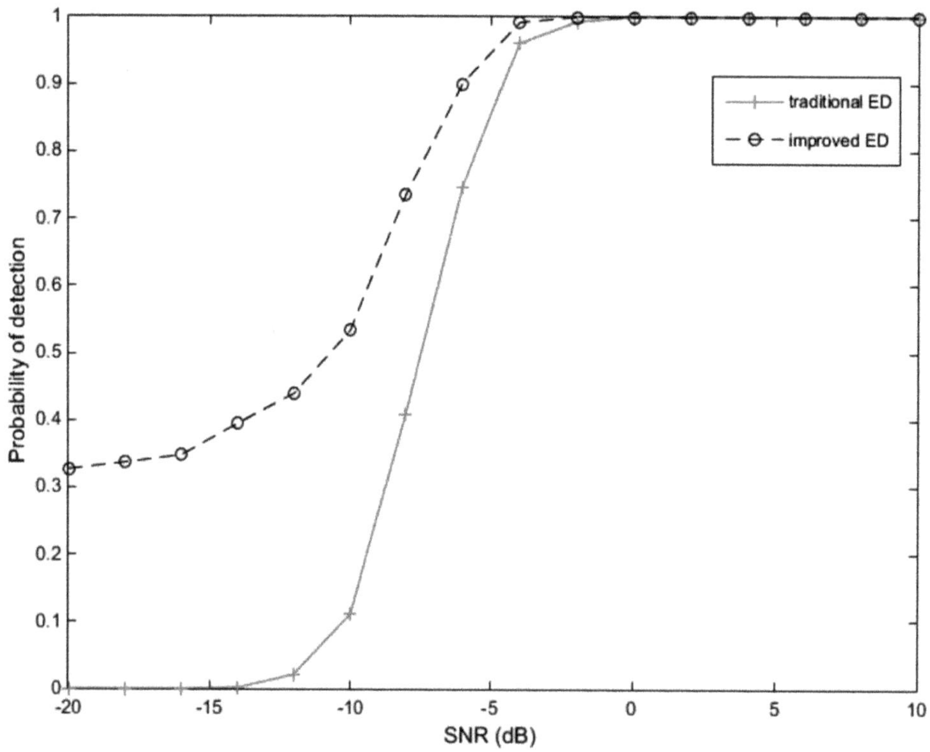

tector provides better detection performance even under low SNR values comparing to the traditional Energy detector. Additionally, it needs less sensing time which in turn maximizes the transmission time for the CR user. However, this detector requires prior knowledge about the primary user signal and that is in most cases difficult to achieve with some complex implementation of generating the local carriers using local oscillators in the receiver side. The CFD on the other hand, achieves very good detection performance and accuracy compared to the previous two sensing techniques even under low SNR values for example 0.7 detection probability at -10 dB and 0.33 at -12 dB compared to 0.1 dB and 0.13 dB for both Periodogram ED and MFD respectively. The CFD figures out the built in features in the primary user signal and uses them for detection like, the operating frequency and the modulation type used in the transmitter side as discussed earlier. This detector is computationally complex and requires more time for sensing the spectrum as previously shown in Table 1.

Figure 18 also shows that the proposed techniques for improving the performance of the conventional Energy detector provide an enhanced detection performance compared to the other techniques. The adaptive double threshold technique achieves about 0.24 detection probability at -14 dB SNR, while the improved Energy detector outperforms by achieving 0.33 detection probability at -16 dB SNR.

Figure 17. Detection probability versus SNR for the adaptive double threshold ED

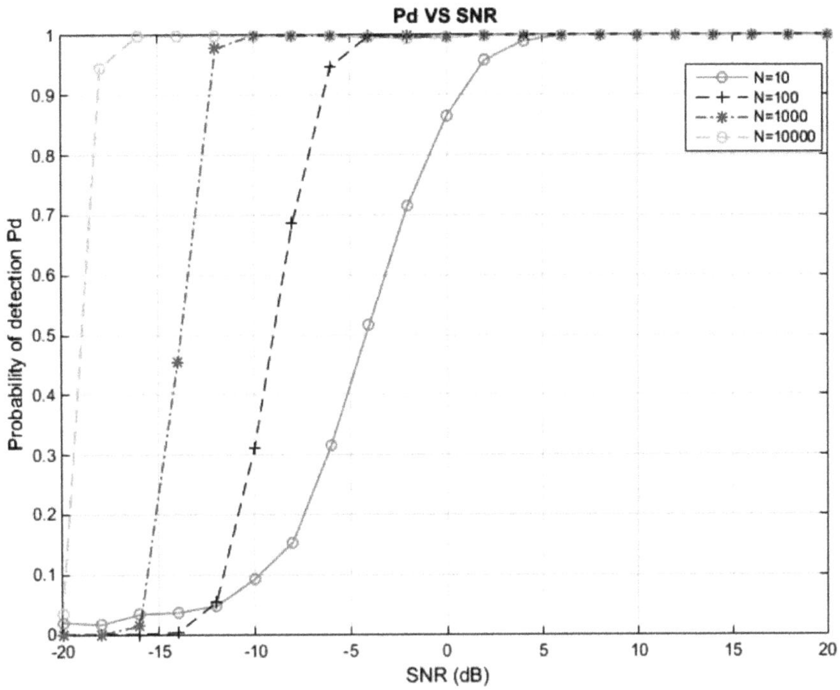

**For a more accurate representation see the electronic version.*

Figure 18. Detection performance comparison between all sensing methods

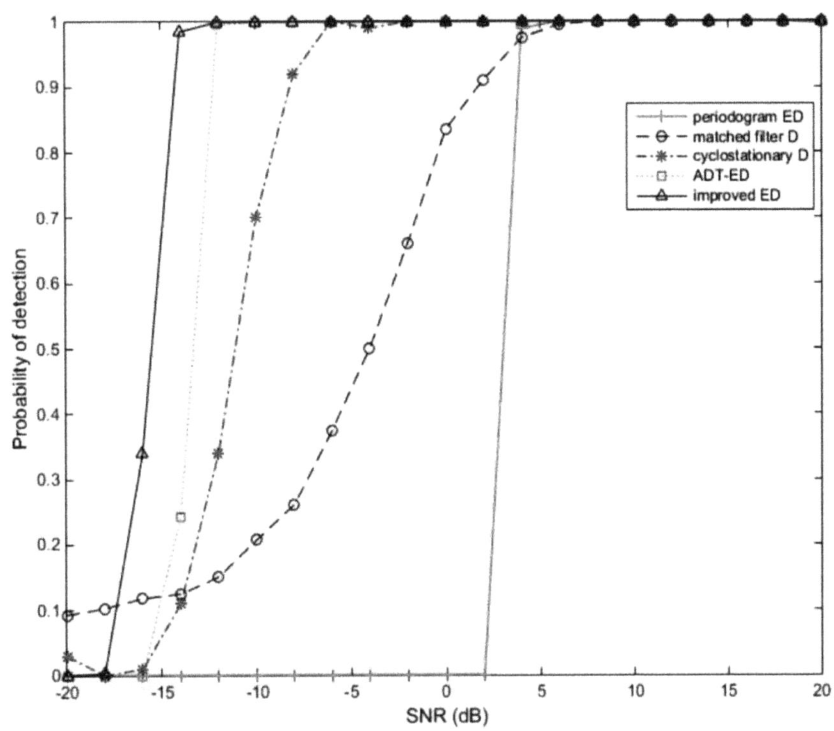

**For a more accurate representation see the electronic version.*

CONCLUSION

In this chapter different spectrum sensing techniques have been implemented in MATLAB software platform and their detection performance was evaluated under different SNR conditions. The well-known Energy detection technique requires no knowledge about the primary user signal with low level of complexity for both computation and implementation. However, its accuracy is sensitive to the SNR and the number of samples used which in turn affect the sensing time required. The performance drawbacks of the traditional Energy detector under low SNR conditions could be handled with the adaptive double threshold ED and the improved ED, these methods improves the detection accuracy at low SNR conditions and can achieve similar performance to the traditional ED with less number of samples, thus, minimizing the sensing time and maximizing the secondary user transmission time. Such optimized techniques can be used as a second stage after the conventional Energy detector if it fails to detect the primary user signal in the scanned spectrum or to verify its decision. The coherent Matched filter is optimum if the primary user signal is known to the receiver side, it provides high detection accuracy with less sensing time but has high implementation complexity for generating the local carriers in the receiver side. Lastly, the Cyclostationary feature detection is computationally complex with more sensing time but it performs well under low SNR conditions with high probability of failing under strong channel fading conditions.

REFERENCES

Akyildiz, I. F., Lee, W.-Y., Vuran, M. C., & Mohanty, S. (2006). NeXt generation/dynamic spectrum access/cognitive radio wireless networks: A survey. *Computer Networks*, *50*(13), 2127–2159. doi:10.1016/j.comnet.2006.05.001

Bagwari, A., Tomar, G. S., & Verma, S. (2013). Cooperative spectrum sensing based on two-stage detectors with multiple energy detectors and adaptive double threshold in cognitive radio networks. *Canadian Journal of Electrical and Computer Engineering*, *36*(4), 172–180. doi:10.1109/CJECE.2014.2303519

Doyle, L. (2009). *Essentials of cognitive radio*. Cambridge University Press. doi:10.1017/CBO9780511576577

Ejaz, W. (2013). I3S: Intelligent spectrum sensing scheme for cognitive radio networks. *EURASIP Journal on Wireless Communications and Networking*, (1), 1–10. doi:10.1186/1687-1499-2013-26

FCC. (2002). *Et docket no 02-135 spectrum policy task force report*. FCC.

Garhwal, A., & Bhattacharya, P. P. (2012). *A Survey on Dynamic Spectrum Access Techniques for Cognitive Radio*. Academic Press.

Haykin, S. (2005). Cognitive radio: Brain-empowered wireless communications. *Selected Areas in Communications. IEEE Journal on*, *23*(2), 201–220. doi:10.1109/JSAC.2004.839380

K, M., & S, R. (2014). *Improved Sensing Accuracy using Enhanced Energy Detection Algorithm with Secondary User Cooperation in Cognitive Radios*. Retrieved from http://www.ijcnis.org/index.php/ijcnis/article/view/533

Mehta, T., Kumar, N., & Saini, S. S. (2013). Comparison of Spectrum Sensing Techniques in Cognitive Radio Networks. *IJECT, 4*.

Mitola, J. (2000). *Cognitive Radio: An Integrated Agent Architecture for Software Defined Radio* (PhD). KTH. Retrieved from http://www.citeulike.org/group/17835/article/9342875

Shojaeifard, A., Zarringhalam, F., & Shikh-Bahaei, M. (2011). Joint Physical Layer and Data Link Layer Optimization of CDMA-Based Networks. *IEEE Transactions on Wireless Communications*, *10*(10), 3278–3287. doi:10.1109/TWC.2011.080311.101301

Song, C., Alemseged, Y. D., Tran, H. N., Villardi, G., Sun, C., Filin, S., & Harada, H. (2010). *Adaptive Two Thresholds Based Energy Detection For Cooperative Spectrum Sensing*. Paper presented at the 2010 7th IEEE Consumer Communications and Networking Conference.

Tandra, R., & Sahai, A. (2008). SNR Walls for Signal Detection. *IEEE Journal of Selected Topics in Signal Processing*, *2*(1), 4–17. doi:10.1109/JSTSP.2007.914879

Chapter 6
Sensing Techniques for Next Generation Cognitive Radio Networks:
Spectrum Sensing in Cognitive Radio Networks

Dhaya R.
Rajalakshmi Engineering College, India

Rajeswari A.
Velammal Engineering College, India

Kanthavel R.
V V College of Engineering, India

ABSTRACT

Cognitive radio is the technology used to solve the problem of spectrum underutilization by performing spectrum sensing, spectrum management, spectrum sharing, and spectrum mobility. The primary goal of cognitive radio is open spectrum sharing. Spectrum is a scarce and valuable natural resource that has to be used very effectively. The static allocation of spectrum to the licensed users will lead to wastage of resources when the spectrum is unused by the licensed user. Spectrum sensing methodology helps in detecting the spectrum holes and enables the unlicensed users to access the unused bands in the licensed spectrum effectively without interfering the licensed users. Cognitive thinking takes wireless communication to the next level by sensing the electromagnetic environment and dynamically adjusts its operating parameters in order to achieve maximum throughput, mitigate interference, facilitate interoperability, etc. The chapter presents the basics of cognitive radio networks, its architecture, its application, and advantages of cognitive radio networks.

DOI: 10.4018/978-1-5225-5354-0.ch006

INTRODUCTION

Cognitive Radio is:

an intelligent wireless communication system that is aware of its environment and makes use of the method of understanding just by building to learn from the environment and adapt to statistical variations in the input given.

The Cognitive environment is designed to meet –highly reliable communication whenever and wherever needed and efficient utilization of the radio spectrum. Dynamic Spectrum Management is the order of the day whereas Static Spectrum Management was followed since 1900s. Initially, cognitive radio concept was first proposed by *Joseph Mitola III* in the year 1998 and later was published in an article by Mitola and Gerald Q.Maguire in the year 1999. The first cognitive radio wireless regional area network standard was developed by IEEE 802 LAN/MAN standard committee in the year 2011.

Conventional radio assumes that it operates in an interference free spectrum band. So, there is no necessity to dynamically change the parameters. The view of unlicensed spectrum in the conventional radio band is as the "Wall of Interference" whereas the view of unlicensed spectrum in the cognitive radio band is as the "Windows of Opportunity". The "Opportunistic Use" of shared frequency as like unlicensed spectrum requires continuous sensing of the environment. The users who have been given legal rights to access the portion of the spectrum is called primary user. They have highest priority in accessing the spectrum. The users who can don't have legal rights but can access the portion of the spectrum allotted to the primary user without causing interference to the primary user is called as secondary user. The unused frequency bands in the spectrum are termed as "whitespaces" or "spectrum holes". The secondary user can access the licensed spectrum when the primary user is not using the band and when it senses the arrival of the primary user it either shifts to the next available spectrum hole or the secondary user reduces the percentage of the spectrum usage and avails it to the primary user giving them the first priority. The detection and usage of the spectrum hole by the cognitive user in shown in the Figure 1.

The cognitive radio not only makes use of the spectrum hole in the licensed spectrum but also the unlicensed spectrum when it gets an opportunity. This is shown in Figure 2. The abilities of a cognitive user are:

- **Spectrum Sensing:** Senses and detects for the unused spaces by the primary users in the spectrum band.
- **Spectrum Sharing:** Makes use of the unused white spaces of the primary users shares the spectrum hole information with the cognitive users.
- **Prediction:** Predicts the arrival of the primary user into the spectrum.
- **Fairness:** There may be more than one cognitive user on the spectrum. So, there should be fairness in the distribution of spectrum utilization opportunity among the cognitive users.
- **Routing:** The packets are transmitted to the destination effectively considering the parameters such as network life span, load on the network, shortest path, delay etc.
- **Reconfiguration Capability:** Make adjustments or reconfigure based on the environmental outcomes.
- **Environment Sensing:** Sense the environment for adaptable changes.
- **Trust and Security:** Build a trustworthy environment and a secure network.

Figure 1. Detection and usage of spectrum hole

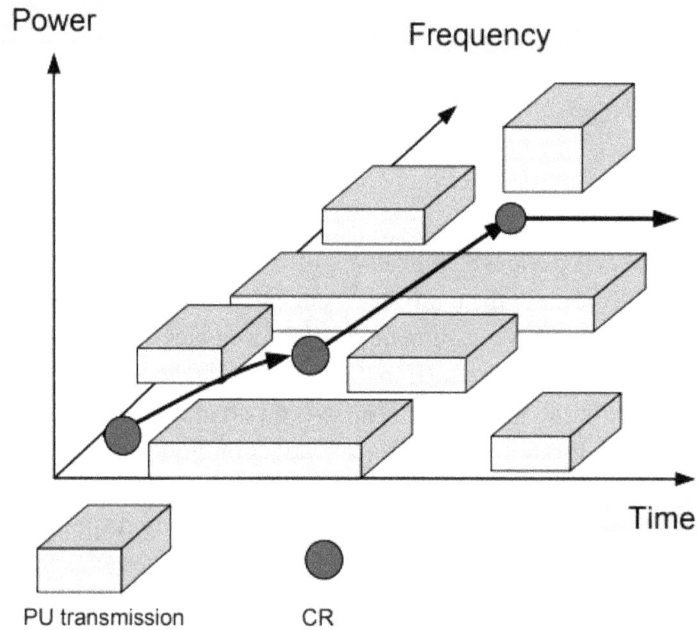

- **Power Control:** Control over the transmitted power considering the boundaries.

The cognitive radio is characterized by the cognitive ability to sense the information of the radio environment, the reconfigurable nature of the spectrum which includes the operating frequency of the band based on the transmission power and modulation, the cognitive cycle, the communication technology and finally the need of cognitive radio.

The Cognitive Cycle in Figure 3 shows the cycle involved in complete utilization of the spectrum. The cycle has four processes: Spectrum Sensing, Spectrum Management, Spectrum Mobility and Spectrum Sharing.

1. **Spectrum Sensing:** The process of identifying the unused spectrum and using it without causing interference to other users especially the primary user. It is an important requirement for the cognitive radio to sense the spectrum hole effectively.
2. **Spectrum Management:** The process of capturing the best available spectrum to meet the user communication needs.
3. **Spectrum Mobility:** It is the process of changing the frequency of operation by the cognitive user.
4. **Spectrum Sharing:** Sharing is the major issue as the available spectrum has to be shared fairly among the available users.

Figure 2. Spectrum Usage in Cognitive Radio Wireless Networks

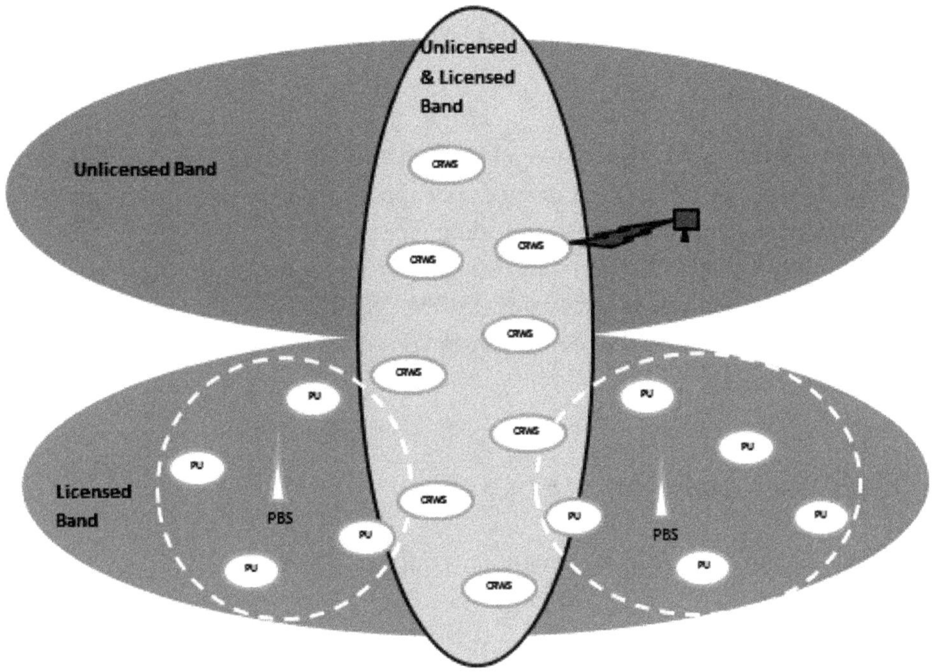

Figure 3. Cognitive Cycle

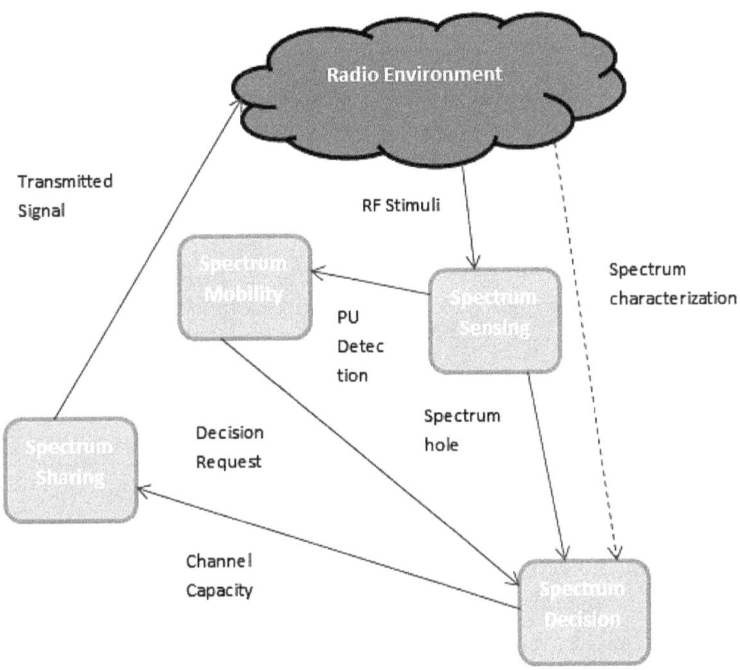

TYPES OF COGNITIVE RADIO

Basically, there are two types of cognitive radio based on the transmitting and receiving parameters considered. They are:

1. **Full Cognitive Radio:** In this type of Cognitive radio, all the possible parameters are considered. The possible parameters include the channel occupancy, type of data transmitted, modulation, frequency band, duration of transmission, coding etc.
2. **Spectrum Sensing Cognitive Radio:** In this type of cognitive radio, only the radio frequency spectrum is the only parameter to be considered. The cognitive user senses the available spectrum, selects the best available channel in coordination with other users and releases it when a licensed user is detected.

APPLICATIONS ON COGNITIVE RADIO

The cognitive radio networks can be used in various applications. They are explained as follows.

Military Services

Cognitive Radio is applied to the autonomous wireless networks in the battlefield to avoid jamming of certain frequencies, to provide immediate and effective communication at the time of emergency or attacks, to maintain the secrecy of communication by detecting the intruders in the network.

Public Safety

Wireless communication is widely used at the time of emergency. The safety workers are equipped with wireless devices to prevent or respond to the incidents by quickly accessing the emergency services. The safety network can be designed as shown in the Figure 4.

The technology has extended from voice to video streaming, picture transfer, messaging, email, database access etc. The reliability, data rate, delay requirements vary from one service to the other service. Obviously, at the time of emergency, the spectrum will be utilized to the maximum resulting in jamming. Cognitive Sensing helps to solve these issues by enabling intelligent thinking.

Mobile Content Providers

The mobile users expect the information to be quickly available in their smart phones as they search for something. The challenge for the content providers is to increase the average revenue per user due to added data services, to provide the information at the fastest rate, to provide the more relevant information with respect to the searched key terms. The mobile content provider incorporates cognitive thinking to meet all its challenges and facilitate the user with all they need.

Figure 4. CR in public safety

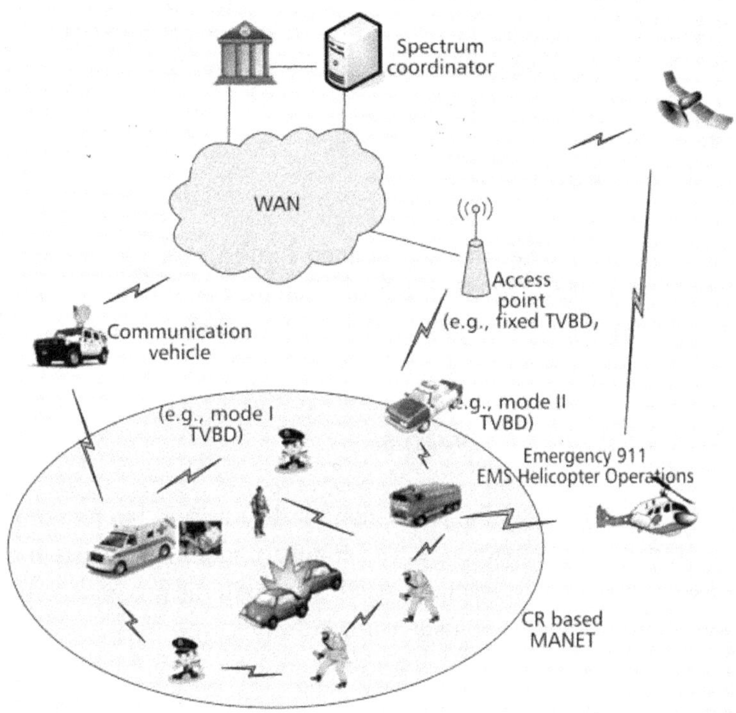

Smart Grids

Smart grid consists of three layers namely- the physical power layer, the communication networking layer and the application layer. The smart grid transforms the way power is generated, consumed, delivered and billed. Adding cognitive intelligence to the grid increases the grid reliability, improves demand handling and responsiveness, increases efficiency, better harness and integrated renewable energy sources there by reducing the cost of provider and consumer.

There can be three segments in the smart grid network – the home area network, the field area network and the wide area network. An example network is designed as shown in Figure 5.

Figure 5. Smart grid network

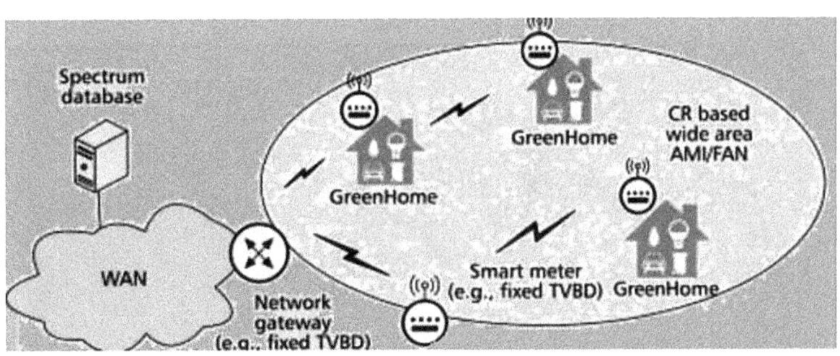

Indoor Application

The cognitive monitoring sensor system can be used for indoor applications such as detecting the contaminant leakage, fire possibility, sensor faults etc. This is highly useful in large industries and buildings for efficient thinking and communication to take place.

Real Time Surveillance Application

Cognitive thinking helps in many real time surveillance applications. Underwater surveillance is one of the important real time surveillance applications that can make use of the cognitive radio network to build a highly secure, efficient monitoring system that enables both detection for intruders and routing the data to the system off shore. The architecture is shown in Figure 6.

Transportation and Vehicular Networks

The wireless sensors are fixed up in the vehicles to monitor the location and for communication between the vehicles as well as at the time of emergency. The cognitive monitoring system also enables the passenger to select the best shortest path to reach the destination. All the possible paths are checked for the shortest distance using the most famous Dijkstra's algorithm.

Figure 6. Underwater Surveillance Cognitive Radio Wireless Sensor Network Architecture

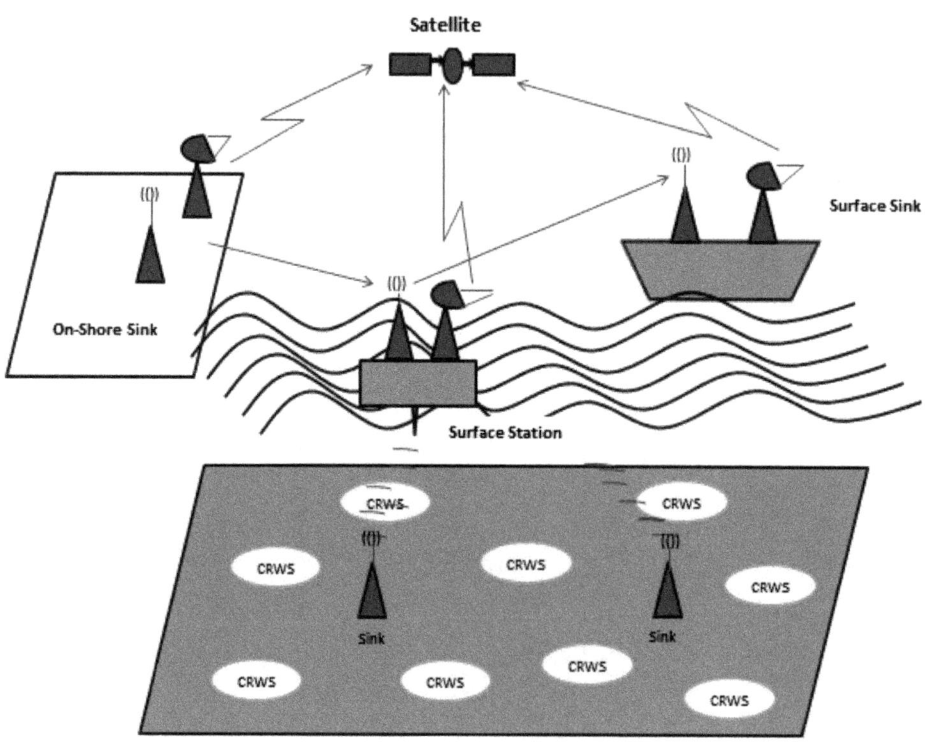

Health Care

The tremendous growth in technology has produced a vital impact on the medical field. The licensed 2.4 GHz industrial, medical and scientific (ISM) band is not suitable for life critical medical applications due to interference and congestion. A medical band can be set with a MBAN coordinator to communication if the hospital is registered within the protection band. The body sensor is fixed into the patient through which temperature, pressure, blood oxygen, and electrocardiogram (ECG) are monitored. If the sensor loses its connectivity with the hub, it stops transmitting in the MBAND frequency and makes use of the restriction free spectrum. The cognitive radio helps in this critical medical data communication.

The various other applications of cognitive radio are bandwidth intensive applications, rural broadband telecoms, incumbent wireless carriers, cable markets and so on.

CHALLENGES IN SPECTRUM SENSING

- **Hardware Requirement Issues:** The spectrum sensing application requires cost consuming hardware such as high resolution analog to digital convertor, bandpass filter, high speed signal processor, low power processor, antennas, and power amplifiers and so on. This automatically increases the overall cost of the system.
- **Hidden Primary User Problem:** Hidden primary user problem is the one that is similar to the hidden terminal problem in CSMA/CD. The figure 7 below shows the coverage area of the primary user and cognitive user. When the primary user is out of the coverage area of the cognitive user, the CU will not be able to sense the arrival of the PU. Now PU is said to be hidden to the CU. Cooperative sensing is the technique proposed for handling the hidden primary user problem.
- **Detecting Spread Spectrum Users:** The wireless devices can operate on fixed frequency or spread spectrum. There are two types of Spread spectrum-FHSS and DSSS. It is very difficult to detect a primary user using a spread spectrum technique because even if the bandwidth of the primary user is narrow, the power of the user is spread over much wider bandwidth.

Figure 7. Hidden PU problem

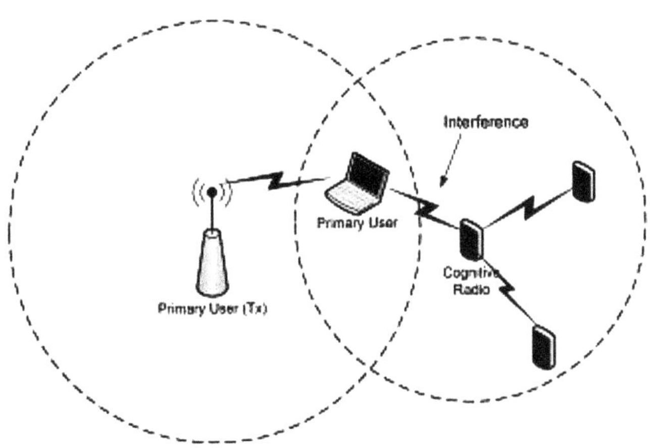

- **Decision Fusion:** The decisions made by the cognitive user in utilizing the spectrum can be either hard or soft. Soft decisions are much preferred compared to the hard decisions even though hard decisions are considered complete. There may be more than one cognitive user in the spectrum. The decisions of all the cognitive users have to be fused using various diversity schemes.
- **Security:** The CU has to detect the arrival of the PU in the spectrum. At times, any intruder can interpret themselves as the primary user in the network. This false interpretation is called the Primary User Emulation Attack (PUEA). The attacker in the network may stop transmitting the information to other node, they can cause jamming or they can allow take overall control of the network.
- **Sensing Frequency Involved and Duration Taken for Sensing:** The Sensing frequency involved is nothing but how frequent the spectrum should be sensed for the arrival of primary user. The primary user can claim for its operating frequency band anytime. The cognitive user has to sense the PU arrival and leave the spectrum at the same time as fast as possible. So, the CU can have some fixed time to sense the spectrum and at the same time it is enough to sense the spectrum that has a varying pattern instead of sensing the entire spectrum. There are several other sensing parameters that affect the sensing methods.

SPECTRUM SENSING ALGORITHMS

Matched Filtering Algorithm

If the transmitted signal of the primary user is known in advance, then Matched Filtering algorithm is the optimum solution for detecting the presence of the primary user. Matched Filter Algorithm is a non-blind detection algorithm. The algorithm works as follows as shown in Figure 8.

Step 1: The received signal is initially passed into the band pass filter.
Step 2: The measured signal around the band is then convolved with the match filter's impulse response which is the time shifted version of the primary user signal.
Step 3: Finally the threshold comparison is done to find the presence of the PU.

Advantages:

1. Optimal performance.
2. Low Computational cost.
3. Takes short time to detect certain probability of misdetection.

Figure 8. Matched Filtering Algorithm

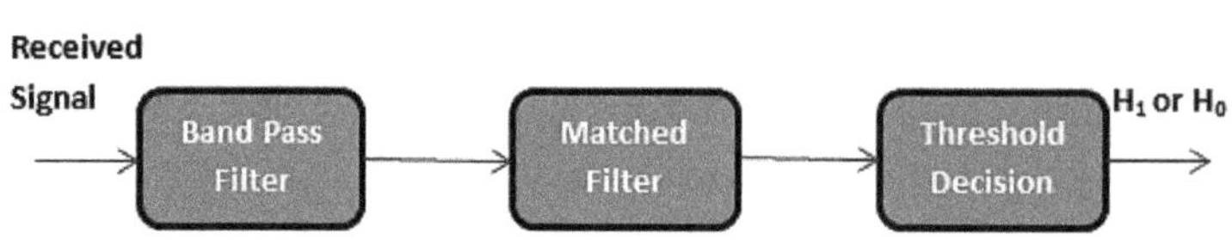

Disadvantages:

1. Requires prior information about every primary user signal.
2. Power consumption is high.
3. Requires separate receiver for every type of primary user.

Energy Detector Algorithm

Energy Detection algorithm is blind detection algorithm which does not require the prior knowledge about the primary user. As the name suggests, it detects the energy of the received signal and compares it with the noise signal which process there by involves the following steps as shown in Figure 9.

Step 1: The received signal is passed initially through the band pass filter.
Step 2: The band filtered signal is then passed into the squaring device to measure the signal.
Step 3: The energy measured signal is then passed into the integrator to integrate over time.
Step 4: The time integrated signal is then compared with the predefined threshold value to determine the presence of the primary user. The threshold value depends on the noise variance.
Step 5: If the energy detected is greater than the energy of the noise threshold then there is the presence of primary user otherwise it is absent.

Advantages:

1. Low complexity.
2. Implementation is easier.
3. Faster decision making probability.
4. Low computational cost.
5. Does not require prior information about the primary user.

Disadvantages:

1. The threshold value depends on the noise variance.
2. Unable to differentiate cognitive user from the primary user.
3. Poor performance in the low signal to noise(SNR) condition.
4. This detection method does not work for spread spectrum techniques like direct sequence spread spectrum and frequency hopping spread spectrum.

Figure 9. Energy detector algorithm

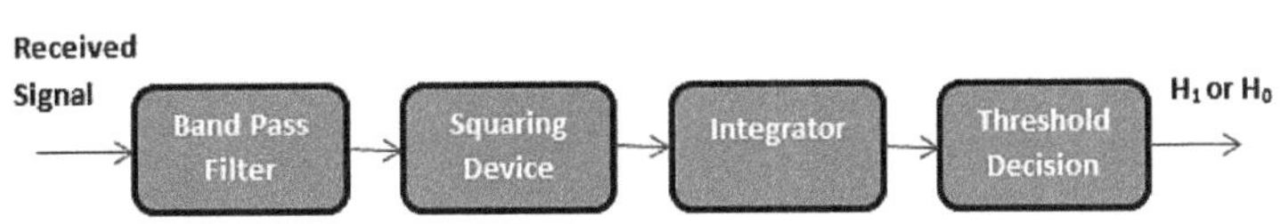

Spectral Correlation Algorithm (Cyclostationarity Algorithm)

The presence of PU is determined by the cyclostationary features. The cyclostationary feature such as mean and autocorrelation are due to the periodicity in the signal. Cyclostationary feature detection requires prior knowledge about the primary user characteristics in the spectrum. The block diagram of the detection is shown in Figure 10.

Step 1: The received signal is filtered around a selected band.
Step 2: The Fourier series expansion is done which is called as Cyclic Auto-Correlation Function (CAF) and its related frequency is called Cyclic Frequency.
Step 3: The Cyclic Spectral Density (CSD) is calculated by taking the Fourier transform of the CAF. CSD represents the density of correlation between two components.
Step 4: Spectral Correlation is done and average is found over time T.
Step 5: Finally the Cyclic Frequency Detector(CFD) arrives at the decision of the presence of the PU.

Advantages:

1. Can work better than Energy Detector algorithm in low SNR circumstances.
2. Easy to differentiate noise and PU signal as noise is a Wide Spread Spectrum (WSS) with no correlation and PU signal has cyclostationary feature.
3. Robust against interference.
4. Increases overall CR throughput i.e., while one CR is detecting it doesn't mean that the other CR should be silent.

Disadvantages:

1. Requires prior information about the PU characteristics in the spectrum.
2. High Computational cost.
3. Long sensing time and high computational complexity due to this it is used less common compared to energy detector algorithm.

Radio Identification Based Sensing Algorithm

Radio identification involves getting complete knowledge about the spectrum characteristics by identifying the transmission channel used by the PUs and communicating to those technologies at times. This identification and communication enables the cognitive radio with a higher dimension knowledge and

Figure 10. Cyclostationarity algorithm

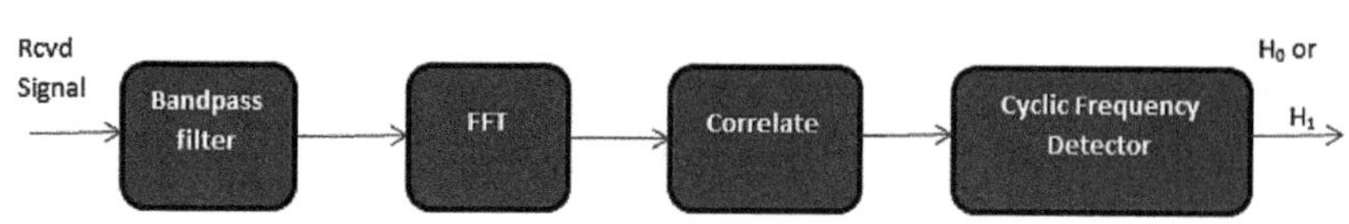

higher accuracy. There are two main steps in identifying the available technology and enabling communication through them. They are:

1. **Initial Mode Identification (IMI):** The cognitive device searches for the available technology by following the power on.
2. **Alternative Mode Monitoring (AMM):** The cognitive device keeps monitoring the other modes while it is transmitting currently in one mode.

Radio identification based sensing decides on the presence of PU by gaining several features from the received signal, finding the technologies used and by communicating to them.

Advantages:

1. Cognitive radio communication leads to high accuracy.
2. Communicating with the technology leads to higher dimension knowledge.

Disadvantages:

1. Classification methods used are little complex.
2. Mode identification consumes time.

Waveform Based Sensing Algorithm

Waveform based sensing method can be used only for known, fixed pattern primary users. The fixed patterns include preamble, midambles, regularly transmitted pilot patterns, spreading sequences etc. The preamble is the known sequence that is transmitted before every burst or slot. The midamble is the known sequence transmitted in the middle of the burst. When the pattern is known, the received pattern can be compared with the known copy of itself and decided on the presence of PU.

Advantages:

1. Outperforms energy detector algorithm in convergence time and reliability.
2. Performance increases as the length of the known pattern increases.
3. Simple method of detection.

Disadvantages:

1. Can be used only if the pattern is known.
2. Error prone if the known pattern is short.

The comparison of all the five sensing algorithms with respect to their complexity and accuracy is shown in Figure 11.

Figure 11. Sensing algorithms comparison

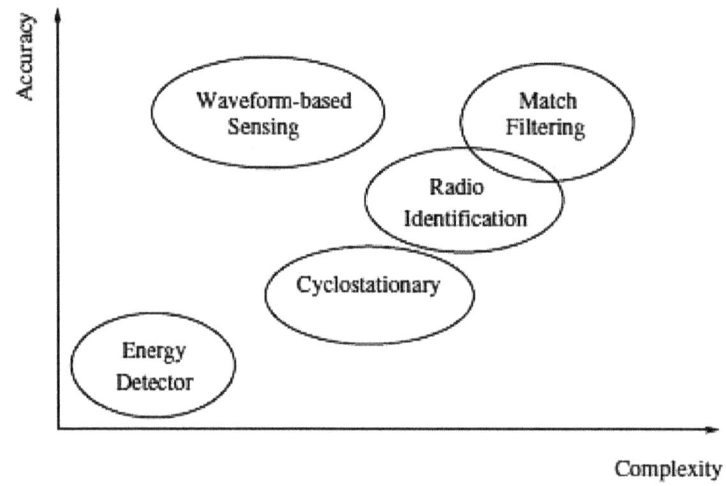

TYPES OF SPECTRUM SENSING

Parallel and Sequential Sensing

When the CU senses for the vacant spectrum, if the sensing of all the frequency spectrum happens at the same time i.e., 1st spectrum to Nth spectrum in parallel then it is called as parallel sensing. It requires N sensing device to sense N frequency spectrum.

In sequential sensing, the frequency spectrum is sensed one by one. It may take too long to find a vacant spectrum. The sensing is shown in Figure 12.

Synchronous and Asynchronous Sensing

There can be more than one cognitive user in a spectrum to sense the channel for a spectrum hole. If all the CRs have the same sensing schedule to sense the channel then it is synchronous sensing. It can be done when the primary user's behavior is known and the CR is in sync with the PU. All the CRs stops transmission and senses the channel at the same time.

Figure 12. (a) Parallel sensing (b) Sequential sensing

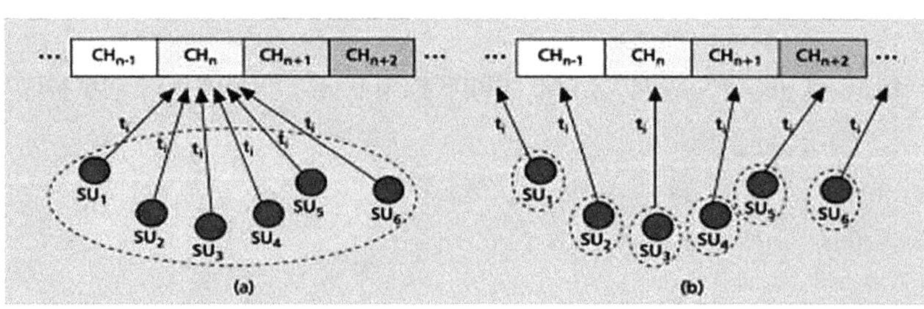

The PUs behavior will not be known to the CR always. So, in those kinds of channels, each CR can have sensing schedule to sense that channel. This way of sensing is called as asynchronous sensing. Since the sensing schedule is different, the CR cannot differentiate another CR and a PU because when a CR senses there can be another CR transmitting or even the PU itself may be transmitting. This is the drawback of asynchronous sensing. The asynchronous sensing is shown in Figure 13.

Proactive and Reactive Sensing

In Proactive sensing, Cognitive Radio senses the channel even if the collected information about the channels will not be transmitted immediately. The channel is sensed periodically and hence the sensing cost is high.

In Reactive sensing, Cognitive Radio senses the channel only if it will transmit or receive the channel information. It will take longer time to find the idle channel but whereas it is energy efficient as it does not sense the spectrum periodically.

Local and Cooperative Sensing

The sensing method is based on the contribution made to the decision taken. When the CR senses itself locally and uses that sensed information to give a decision on channel state i.e. whether the channel is idle or busy, then it is called as local sensing. Local sensing are otherwise known as Non-Cooperative Sensing.

As we all know there may be more than one CR in the channel, so each CR shares its sensing information with other CRs and utilizes outcome of other CRs to make a decision on the channel state. This method is called Cooperative Sensing. The major drawback of Cooperative sensing is that it is highly prone to error and faces hidden Primary user problem and channel fading problem.

Distributed and Centralized Sensing

This sensing technique also deals with decision making. If every CR makes its own decision without any central point of coordination, then it belongs to distributed sensing.

If the decision is made by a central manager such as a base station or an access point, then it belongs to centralized sensing. The central manager collects all the information from each CR before making

Figure 13. Asynchronous sensing

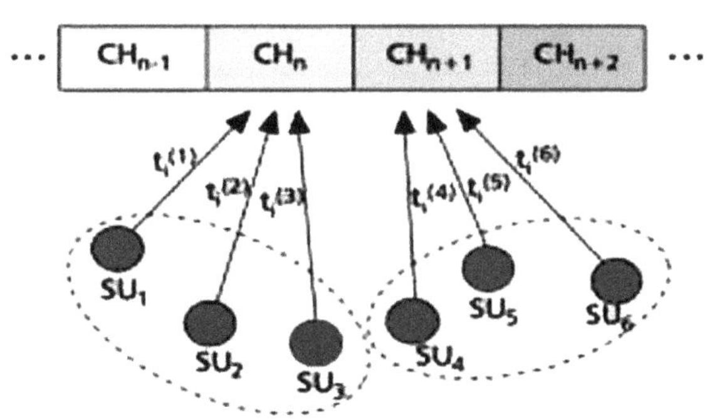

a decision and this involves an extra cost for transmitting the sensed data. One more drawback is the problem due to failure of central manager. The distributed sensing is shown in Figure 15 and the centralized sensing is shown in Figure 14.

Out-of-Band and In-Band Sensing

Based on the range of channels sensed there are two kinds of sensing – in-band and out-band sensing. If the CR senses the channel in which it is already transmitting for the presence of PU, then it comes under in-band sensing.

Figure 14. Centralized Sensing

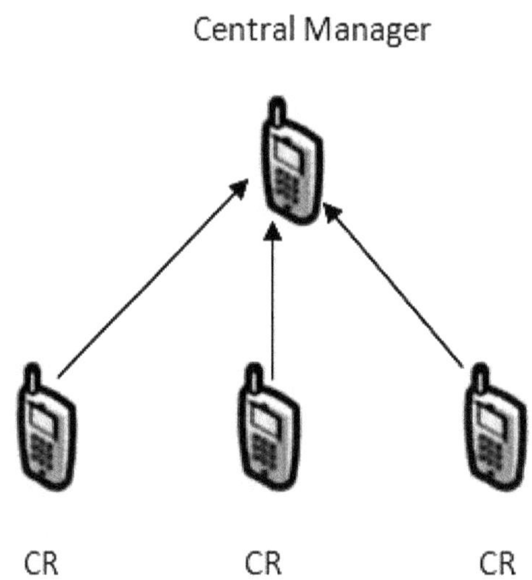

Figure 15. Distributed sensing

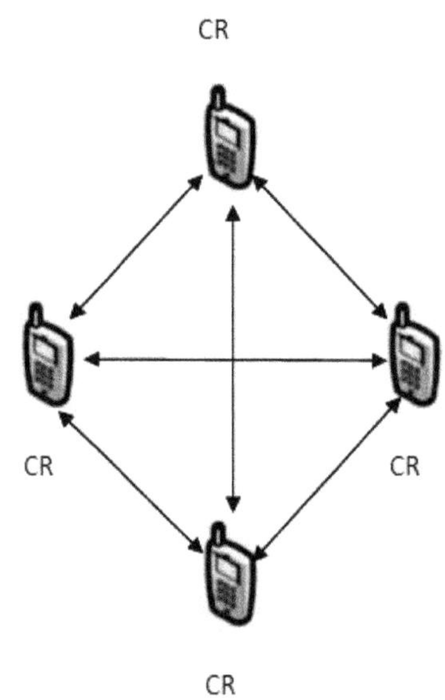

If the CR senses the channels other than the channel in which it is present, then it comes under out-band sensing. It aims to find the other spectrum holes or to switch to another channel as a PU has already appeared in the channel it is present.

CONCLUSION

Cognitive radio is a powerful concept which is discussed in this chapter right from the basic definition, advantages, application, cognitive cycle, types of CR and finally elaborating the concept of spectrum sensing in detail along with the types and algorithms of spectrum sensing. The various sensing algorithms are compared. The future study can be done by implementing the cooperative communication concept with the cognitive radio concept i.e., cooperative cognitive radio communication to improve the utilization of the available spectrum resource to the best without any interference.

REFERENCES

Bayhan. (2013). *Overview of Cognitive Radio Basics and Spectrum Sensing*. CN-S2013.

Garhwal, A., & Bhattacharya, P. P. (2011). A Survey on Spectrum Sensing Techniques in Cognitive Radio. *International Journal of Computer Science & Communication Networks, 1*(2), 196-206.

Iacobucc. (2013). *Reconfigurable Radio Systems: Network Architectures and Standards*. Wiley Publications.

Kokare & Kamble. (2014). Spectrum Sensing Techniques in Cognitive Radio Cycle. *International Journal of Engineering Trends and Technology, 9*(1).

Kumar, R. (2014). Analysis of Spectrum Sensing Techniques in Cognitive Radio. *International Journal of Information and Computation Technology, 4*(4), 437-444.

Kumari. (2016). Spectrum Sensing Techniques for Cognitive Radio Networks: A Review. *International Journal of Innovative Research in Computer and Communication Engineering, 4*(3).

Mehta, T., Kumar, N., & Saini, S. S. (2013). Comparison of Spectrum Sensing Techniques in Cognitive Radio Networks. *International Journal of Electronics & Communication Technology, 4*(3), 33-37.

Sridhar, B., & Srinivasulu, D. T. (2013). A Novel High Resolution Spectrum Sensing Algorithm for Cognitive Radio Applications. *IOSR Journal of Electronics and Communication Engineering, 8*(4), 30-38. Retrieved from http://www.radio-electronics.com/info/rf-technology-design/cognitive-radio-cr/cooperative-spectrum-sensing.php

Subhedar & Birajdar. (2011). Spectrum sensing techniques in cognitive radio networks: A survey. *International Journal of Next-Generation Networks, 3*(2).

Wang, Ghosh & Challapali. (2011). Emerging Cognitive Radio Applications: A Survey. *IEEE Communications Magazine*, 74-81.

Y¨ucek, T., & Arslan, H¨. (2009). *A Survey of Spectrum Sensing Algorithms for Cognitive Radio Applications. IEEE Communications Surveys & Tutorials, 11(1).*

KEY TERMS AND DEFINITIONS

Cognitive Radio: A radio or system that senses its operational electromagnetic environment and can dynamically and autonomously adjust its radio operating parameters to modify system operation, such as maximize throughput, mitigate interference, facilitate interoperability, access secondary markets.

Licensed Band: The individual companies pay a licensing fee for the exclusive right to transmit on assigned channels within that band in a given geographic area. The industrial, scientific and medical (ISM) radio bands are radio bands (portions of the radio spectrum) reserved internationally for the use of radio frequency (RF) energy for industrial, scientific and medical purposes other than telecommunications. Licensing is a way of ensuring that wireless operators do not interfere with each other's transmissions.

Primary User: A user who has higher priority or legacy rights on the usage of a specific part of the spectrum.

Secondary User: A user who has a lower priority and therefore exploits the spectrum in such a way that it does not cause interference to primary users.

Spectrum: The wireless spectrum consists of electromagnetic radiation and frequency bands. Spectrum relates to the radio frequencies allocated to the mobile industry and other sectors for communication over the airwaves. Spectrum includes both coverage and capacity bands, meaning mobile operators can connect more people and offer faster speeds.

Spectrum Sensing: The task of obtaining awareness about the spectrum usage and existence of primary users in a geographical area.

Unlicensed Band: Unlicensed wireless technologies don't require any permission, so long as products and users comply with the rules associated with that unlicensed band. Unlicensed wireless technologies are vulnerable to interference.

Chapter 7
Spectrum Sensing Techniques:
An Overview

Rajib Biswas
Tezpur University, India

ABSTRACT

Cognitive radio has come a long way in the recent years with the advent of improved algorithms and instrumentation. However, for ease and efficient working of cognitive radio, there is a need to have effective detection of spectrum sensing. The objective of spectrum sensing techniques is to find spectrum holes which can be accessible by the users of cognitive radio. The deployment of suitable sensing techniques reduces undesirable congestion in traffic and enhancement of spectrum usage. All these require sensing techniques whose main goal is oriented towards efficient identification and subsequent deployment of spectrum. This chapter is aimed to give a brief overview of some spectrum sensing techniques. An attempt is made to give the characteristics of the highly deployable sensing schemes. Accordingly, the merits and demerits are comprehensively highlighted. Further, emphasis has been given to relevant future challenges.

INTRODUCTION

Cognitive radio has come a long a long way since its inception. In wireless communication, there is constant demand of either extra bandwidth or efficient bandwidth. In such cases, cognitive radio helps in identification of unused spectrum without causing any deterrence to the functional environment. It complements the existing wireless communication networks by optimizing the licensed frequency bands with a view to improving quality of services. This necessitates the effective deployment of spectrum utilization. Again, in order to effectively utilize the spectrum, there must be a way to find out holes in spectrum. Here, spectrum holes refer to unutilized portion which remain unutilized by primary user. These holes are to be provided access to be used by secondary user in right time and location. As a whole, there is a need of a reliable technique that's main objective is to detect the idle spectrum. In this direction, cognitive radio has come a long way in implementing effective monitoring of spectrum usage. Being endowed with adaptability with a varying radio environment, the CR users avail the liberty

DOI: 10.4018/978-1-5225-5354-0.ch007

of using spectrum which is licensed in an opportunistic way, albeit not causing any interference with licensed users. This reduces undesirable congestion in traffic and enables enhancement of spectrum usage. All these require sensing techniques whose main goal is oriented towards efficient identification and subsequent deployment of spectrum. This chapter is aimed to give a brief overview of some spectrum sensing techniques, outlining modus operandi as well as relevant future challenges.

CHARACTERISITCS OF SPECTRUM SENSING

There are two types of users, one is primary user and the other is secondary user. Primary users are categorized as licensed users whose frequency bands are already allocated. On the other hand, secondary users fall in the category of opportunistic users. Since they basically utilize the unallocated frequency bands or unlicensed portion, hence there is always greater demand to find the spaces. The function of exploring holes in frequency bands which is solely licensed to primary user, for opportunistic user by networks in secondary category is referred to as spectrum sensing. This is indeed a vital point which directly influences the cognitive radio. With this, cognitive radio becomes capable of adapting itself to neighboring environment, thereby tracking holes in spectrum. Whatever the unused potion exists after monitoring the available spectrum bands, the CR user also ensures that the communication is within their reach.

SPECTRUM SENSING METHODS

There are several ways of sensing spectrum. Chiefly, it is categorized into two-namely co-operative and non-co-operative. However, there arises another classification known as Interference based sensing. It is depicted in Figure 1. Again, the co-operative sensing is split into three ways, viz., matched filtering, energy detection and cyclostationary detection. In the following section, the sensing techniques are elaborated.

Matched Filtering

One of the effective ways of spectrum sensing is conceived by Matched Filtering. As it requires very less time for detection with considerable accuracy, there is much enhancement in the signal to noise ratio. Generally, it triggers probability of false alarm or probability of missed detection within a very

Figure 1. General Classification of Spectrum Sensing Techniques

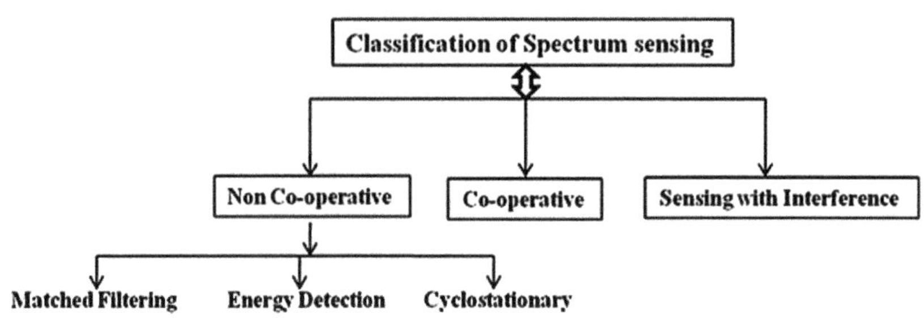

short span of time. This works as a general linear filter by acting as a detector of primary signal in a coherent manner. One can extract desired signal which appears as output. Whatever be the noise, it is highly attenuated. Previous data corresponding to each primary user is essential in order to establish filtering of required signal. This again necessitates a receiving section which is entirely committed for each primary user. The signal at the receiver undergoes convolution with filter response which experiences time shifting. Suppose, the secondary user attains information about the primary user. In that case, there is a practice of using detector which is optimized as well as eliminates stationary Gaussian noise. Mathematically, it can be expressed in the following way:

$$S(s) = \sum_{k=-\infty}^{k=+\infty} I\left(s-k\right)Q\left(k\right)$$

In this expression, *Q (k)* represents the unknown signal and impulse response of the matched filter is implied by *I*. This impulse response is actually correlated with the reference signal in order to increase the signal to noise ratio. Basically, it reduces the noise amplitude present in the signal, thereby boosting the signal content. This will lead to a sharp contrast in peak amplitude of signal to noise. As shown in the block representation, the input is allowed to pass through the band pass filter. Subsequently, it proceeds through the matched filter where the desired content is filtered out.

Although Matched Filter works efficiently in terms of SNR improvement, it suffers from some drawbacks. For example, prior knowledge of primary user is of utmost importance. Without that, there is a drop of efficiency. Similarly, it also increases the hardware cost as there is need of a dedicated receiver system for each primary user, thereby consuming more power. Apart from that, one has to adopt distinct algorithms in order to identify various primary users.

Energy Detection

As the aforementioned technique suffers from some severe drawbacks, in that case, energy detection technique comes as a savior. Energy detection can be used even though secondary user is quite ignorant of primary users. It primarily depends on a threshold level which is going to identify the idle spectrum. Let's say that the received content is above a threshold level. The cognitive users will then perceive it to be busy line where there is no access of idle spectrum. Similarly, threshold level incase becomes greater than the input, then CR user immediately identifies it to be idle spectrum with sufficient access.

Figure 2. Block Diagram of Matched Filter

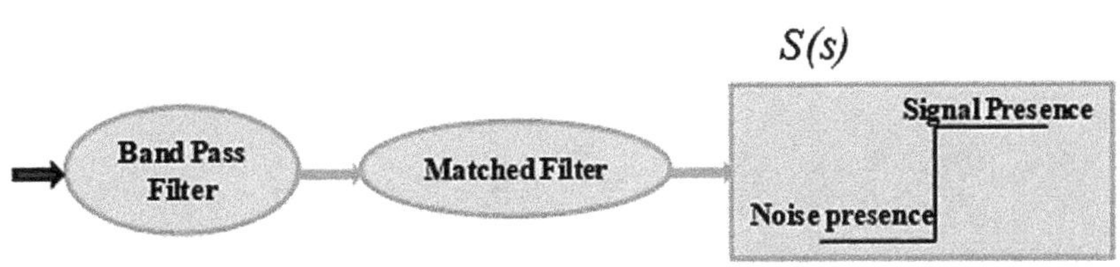

WORKING PRINCIPLE OF ENERGY DETECTION

Energy detection is an optimal way to detect primary signals when priori information of the primary signal is unknown to secondary users. It measures the energy of the received waveform over a specified observation time. *PD* (probability of detection) and *PF* (probability of false alarm) are the decisive parameters towards energy based detection. Based on these probabilities, one can ascertain whether the spectrum is available or not.

As shown in Figure 3, the input is initially allowed to pass through a pre-noise filter. One can also utilize a band pass filter. The main idea is to suppress the noise or in other words select a certain range of frequencies. The filtered content then proceeds through a squaring device. In between, there is a analog to digital converter which remain hidden in the above figure. The squaring device actually accounts for squaring the amplitudes which eventually enter the averaging unit. The output finally is obtained by comparing with the threshold. As a matter of fact, this is mainly accomplished by periodogram or radiometry. One known procedure is Welch Periodogram. Eventually, irrespective of frequency or time domain, the threshold detector is going to yield information about available spectrum.

Merits and Demerits

Energy detection method can be referred to as the easily implementable scheme in terms of sensing elements. It is preferable to matched filter because of its low computational complexities. Another edge over its peers is that it can work without any information of primary users. As the main modus operandi is based on threshold-dependence on the noise floor, hence the signal can be easily identified.

In similitude to other methods, it is not also fully efficient. Some challenges associated with energy detection are mention worthy. As for instance, the setting of the threshold is one of the crucial factors as it is the back bone of energy detection. In case threshold is set in a wrong way, it will lead to interference with other primary users in the loop. Going in the same flow, this detection scheme suffers another flaw. It is very much susceptible to interference caused by primary users and noise. Additionally, it becomes inefficient in low signal to noise ratio values.

Figure 3. Block diagram of Energy Detection Scheme

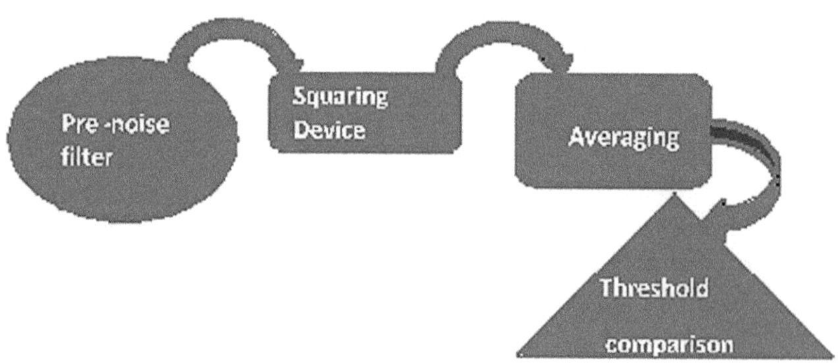

CYCLOSTATIONARY FEATURE DETECTION

Another robust technique of spectrum sensing is cyclostatinary feature detection. One can easily sense from the name that the detection is entirely based upon finding the cyclic stationary characteristics. In general, when we look into a modulated signal, the constituents that come into our mind are sine wave carriers, pulse trains, cyclic prefixes and so on. These, as a matter of fact add periodic features in their statistical regime in the context of mean, correlation etc. Generally, these periodic features are observed in frequency as compared to time domain. Exploiting the cyclic correlation function (CAF), the presence of signal in the given spectrum can be sensed out. Thus, capitalizing on statistical information entailing correlation feature of primary user, the secondary user can enhance the sensing accuracy. Without resorting to power spectrum density, the cyclic correlation function facilitates faster sensing of available signal.

Basics

As we modulate a signal with a sinusoidal carrier, cyclic prefixes, code or hopping sequences, the statistical properties such as mean, autocorrelation reveal periodicity. In other words, we can cite it as induced cyclostationarity in the modulated spectrum. In this detection scheme, we capitalize on this cyclostationary feature whose main objective is to identify a signal property called Spectral Correlation Function/ cyclic correlation function.

Drawbacks and Advantages

Cyclostationary detection suffers from some drawbacks. Although it works better than energy detection, we cannot totally ignore the lacunas it possesses. As for example, this detection can be deployable only for few primary signals. For a large number of primary signals, the main characterizing function becomes erroneous which severely affects the sensing capability. Apart from that, we also emphasize about frequency domain which makes the sensing time lengthy. This also increases the cost and complexity for the time shift correlation.

Yet, we can't rule out use of this detection scheme. One of the figures of merit of this detection is that it has got a significant capacity to work even under low SNR values. It is very much immune to low level of noise making it quite robust in that regime.

VIA MULTIPLE ANTENNA METHOD

Multiple Antenna Techniques are implemented in conjunction with sensing techniques such as energy detection and cyclostationary feature detection. This is performed with the sole objective of lessening the shadowing effects caused by the channel which leads to considerable augmentation in the detection probability, thereby enhancing the performance. There are several ways of implementing this scheme. Generally, there are three models. They are SIMO (Single input multiple output), MISO (Multiple input and single output) and MIMO (multiple input multiple output). In addition to these, this technique again entails three deployment schemes in order to achieve the desired signal. However, we stick to only selection combining method; schematic of which is provided below.

Figure 4. Schematic of a multiple antenna detection scheme. S1, S2… Sr represents the channel gain for each representative antenna.

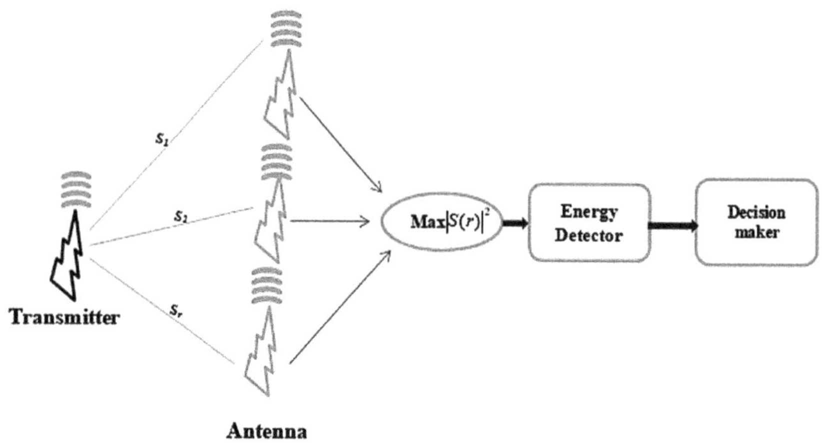

In this technique, it is assumed to have knowledge of channel state information. This enables one to select the branch endowed with optimal gain. Once selected, the data is retrieved from that branch or antenna. Afterwards, the retrieved data is passed to detector to choose the desired one on the basis of threshold level. Since, cognitive radio elects the optimal one; hence it operates like a single input single output. This results in constancy in the threshold level.

Multiple antenna technique possesses some advantageous features. One of them is that one can attain higher sensitivity without worrying about fading to a considerable extent. This is achieved because of existence of spatial diversity yielded by various spatial paths which is inherent in multiple antenna method. Again, the Bit Error Rate improvement can be raised to certain level by resorting to this detection procedure

CHALLENGES WITH SPECTRUM SENSING

As per reports of Fullmer and Garcia-Luna-Aceves 1997, spectrum sensing suffers from a hidden node problem which leads to fading effects of the wireless channel. Usually, this node refers to the primary user which is hidden in the background. With respect to the transmitting node, there arises considerable fading in the spectrum sensing despite having a good channel in the receiving side. Then the node after detecting a free medium begins transmitting, thereby causing interference in the primary mode of transmission. This issue can be resolved by incorporating sensing in a co-operative manner, as suggested by Akyıldız, et al. (2006).

Additionally, there are other challenges related with cognitive radio sensing. As for example, a blind man's sensitivity is mostly determined by his sense of hearing. When he attempts to walk from one side of the road to other, he has to utilize his intellect of sense of hearing. H¨oty¨a, et al. (2007) highlighted that spectrum sensing is also impaired by its basic sense of hearing. As a result, swift detection of spectrum holes is badly influenced which necessitates more advanced techniques providing ample bandwidth rapidly.

Another setback to spectrum sensing is the wide band sensing. While doing, one gets stuck so as to define the boundaries of the spectrum to sense. One solution to that is use of narrow spectrum. However, it comes up with its own demerits. When worked on limited spectrum, there is every possibility of oversampling or under sampling of the received signal in accordance with Nyquist Criterion. Again, wide band sensing also requires costlier components such as analog front-end.

Apart from this, the existence of multiple environments also obstructs the well-functioning of the spectrum sensing. In a more detailed way, we can say there may be multiple numbers of licensed and unlicensed users. The unlicensed users which are categorized as secondary users provide additional restraint in the form of interference which leads to other kind of complicacies.

In addition to the aforementioned demerits, there exist other parameters such as sensing time which also affects the spectrum sensing.

CONCLUSION

In this chapter, we have outlined the importance of spectrum sensing in cognitive radio. We have documented some robust method of spectrum sensing techniques. Each of them is highlighted with their demerits and merits. The challenges that one can encounter in spectrum sensing are considerably discussed. Although, with advent of advanced technologies, there is tremendous growth observed in spectrum allocations and their subsequent identifications. If we alter the signal to noise ratio in nearby environment, it may lead to direct adverse impact on spectrum sensing. Given to its' multifaceted characteristics, there should be synergistic approach and co-ordination between regulatory and technical bodies in order to fully utilize spectrum sensing.

REFERENCES

Akyildiz, I. F., Lee, W. Y., Vuran, M. C., & Mohanty, S. (2006). Next generation dynamic spectrum access/cognitive radio wireless networks: A survey. *Computer Networks*, *50*(13), 2127–2159. doi:10.1016/j.comnet.2006.05.001

Fanan, A. M., Riley, N. G., Mehdawi, M., Ammar, M., & Zolfaghari, M. (2014). Survey: A Comparison of Spectrum Sensing Techniques in Cognitive Radio. *Int'l Conference Image Processing, Computers and Industrial Engineering (ICICIE'2014).*

Fullmer, C. L., & Garcia-Luna-Aceves, J. (1997). Solutions to hidden terminal problems in wireless networks. *Proceedings of the ACM SIGCOMM97 Conf. on Applications, Technologies, Architectures, and Protocols for Computer Communication, 97*, 39-49. 10.1145/263105.263137

Höyhtyä, Hekkala, Katz, & Mämmelä. (2007). Spectrum Awareness: Techniques and Challenges for Active Spectrum Sensing. *Cognitive Wireless Networks,* 353-372.

Lee, J.-H., Baek, J.-H., & Hwang, S.-H. (2008). Collaborative Spectrum Sensing using Energy Detector in Multiple Antenna System. *10th International Conference on Advanced Communication Technology ICACT, 1*, 427 - 430.

Loyka, S. L. (2001). *Channel Capacity of MIMO Architecture Using the Exponential Correlation Matrix. IEEE Communications Letters*, *5*(9), 369–371.

Lpez-Bentez, M., & Casadevall, F. (2012). Improved energy detection spectrum sensing for cognitive radio. *IET Communications, 6*(8), 785.

Mounika, Chandra, & Kumar. (2013). Spectrum Sensing Techniques and Issues in Cognitive Radio. *International Journal of Engineering Trends and Technology, 4*(4).

Subhedar, M., & Birajdar, G. (2011, June). Spectrum sensing techniques in cognitive radio networks: A survey. *International Journal of Next-Generation Networks*, *3*(2), 37–51. doi:10.5121/ijngn.2011.3203

Yucek, T., & Arslan, H. (2009). A survey of spectrum sensing algorithms for cognitive radio applications. *IEEE Communications Surveys and Tutorials*, *11*(1), 116–130. doi:10.1109/SURV.2009.090109

Chapter 8
Advanced Spectrum Sensing Techniques

P. T. V. Bhuvaneswari Mohan
Anna University, India

Bino Jesu Stephen
Anna University, India

ABSTRACT

Spectrum sensing plays an important role in cognitive radio networks. Based on the bandwidth considered for spectrum sensing, it can be classified into narrowband and wideband spectrum sensing. Traditional spectrum sensing methods are devised for narrow band as it focuses on narrow frequency a range that is the channel bandwidth is lesser than the coherence bandwidth of the channel. To provide more spectral opportunity to cognitive user and to increase the throughput, cognitive radio network needs techniques that exploits spectral opportunities over a wide frequency range. Wideband spectrum sensing techniques aim to sense a channel bandwidth that exceeds the coherence bandwidth of the channel. Narrowband sensing techniques cannot be directly employed to perform wideband spectrum sensing as they make a single binary decision. In this chapter, the advanced spectrum sensing techniques and their taxonomy are discussed in detail.

INTRODUCTION

The main aim of spectrum sensing is to provide more spectrum access to Cognitive Radio (CR) users. Based on the bandwidth sensed, spectrum sensing can be classified into Narrowband and Wideband (Abdelmohsen Ali, & Walaa Hamouda, 2017). Narrowband spectrum sensing algorithms focuses on narrow frequency range i.e. the channel bandwidth taken for spectrum sensing is lesser than the coherence bandwidth (Bountouris, 2013). Due to rapid fluctuating nature of wireless environment, the spectral opportunities provided by narrowband sensing techniques that enable continuous spectrum access to CR users undergo difficulty to meet out the demand of CR users. Hence, there arises a need for advanced spectrum sensing techniques. Wideband spectrum sensing techniques focuses on wide frequency range i.e. the channel bandwidth sensed is greater than the coherence bandwidth (Sun, Nallanathan, Wang, & Chen, 2013).

DOI: 10.4018/978-1-5225-5354-0.ch008

LIMITATIONS IN NARROW BAND SENSING

Time consumption involved in sensing is considered as one of the major limitation of Narrowband sensing. It limits the local observations which lead to incomplete sensing information. The traditional narrowband spectrum sensing techniques are Matched filtering, energy detection, Cyclostationary detector and covariance based detector (Sun, Nallanathan, Wang, & Chen, 2013). Matched filtering technique (Jaiswal, Kumar Sharma & Singh, 2013) correlates the known reference signal with the received signal to detect the presence of Primary User (PU). However, it requires prior knowledge of PU signal and need to maintain synchronisation. In energy detection technique (Digham, Alouini, & Simon, 2007), the received signal energy is compared with a threshold to detect the presence of PU. When compared to the previous technique it does not require any prior information about PU. However due to the stochastic nature of wireless channel, it is cumbersome to differentiate PU from CR user. Cyclostationary detector (Sutton, Lotze, Nolan, & Doyle, 2007) utilizes the cyclic correlation property of modulated signals. It performs well in low SNR regions. It consumes more time for spectrum sensing and it is computationally complex. In covariance based detector (Zeng & Liang, 2009), covariance matrix of the received signal is used to detect the signals. It utilizes the dispersive nature of channels. It is also affected by noise uncertainty.

Significance of Wideband Sensing

Wideband spectrum sensing techniques aim to sense a channel bandwidth that exceeds the coherence bandwidth of the channel. The excessive bandwidth increases the hardware complexity. Wideband sensing can detect more amount of unused spectrum of PU and provides more spectral opportunities to CR users. This not only increases the throughput of CR user but also allows them to switch between various spectrum holes. This provides seamless communication and thereby enhances the QoS of CR users.

Challenges and Requirements of Wideband Sensing

In additional to the challenges faced by traditional sensing techniques such as hardware requirements, hidden PU problem, sensing periodicity and noise uncertainty (discussed in Chapter 3) wideband spectrum sensing also has the following challenges:

1. **Wideband Signal Acquisition:** The fundamental challenge of wideband sensing is to devise mechanism to capture the entire wideband signal. Advanced reception techniques such as MIMO can be used to receive signals which make the RF front end hardware more complex (Tian, Leus, & Lottici, 2010). Depending on the hardware the RF circuitry processes the acquired signals as multiple RF chains or single RF chain.
2. **User Hierarchy:** When applying cognitive concept to heterogeneous network, the PUs distribution can be of different hierarchy. This kind of modulation schemes followed may be varied based on the spectrum used. Hence to protect the PU performance, it is essential that the CR user need to sense the PU with very low SNR.
3. **Wireless Fading:** The detection performance is limited by the received signal strength. The PU signal may be affected by deep fading, shadowing and local interference. This decreases the sensing accuracy and increases the possibility of missed detection. Cooperative spectrum sensing can be

a solution to this problem but the tradeoff between cooperative sensing gain and sensing overhead has to be considered.

4. **Dynamic Range:** Dynamic Range (DR) describes the range of the input signal levels that can be reliably measured. It is the ability to accurately measure small signals in the presence of the large signals. The DR is mainly limited by ADC and DR in (dBFS/. $\sqrt{\text{Hz}}$)an be given by:

$$DR = SNR + 10 * \log\left(f_s / 2\right) \tag{1}$$

where fs is the sampling frequency. DR is limited by noise and spurious signals (Treichler, Davenport, Laska, & Baraniuk, 2011).

5. **Sampling Rate and Sparsity Order:** To determine a suitable sampling rate, the sparsity order of the unused spectrum should be known. Sparsity order depends on the spectrum occupancy of PUs which is time varying in nature (Sun, Nallanathan, Wang, & Chen, 2013). Therefore the sampling rate depends on the sparsity level and the tradeoff between sampling rate and dynamic sparsity order should be considered.

CLASSIFICATION OF WIDEBAND TECHNIQUES

Wideband sensing can be further classified into two types namely Nyquist wideband sensing and Sub-Nyquist wideband sensing based on the sampling rates used to acquire signals.

Nyquist Wideband Sensing

Nyquist Wideband sensing acquire signals having sampling rate greater than nyquist rate. Therefore these techniques require hardware that can support high sampling rates and complex digital processing algorithms.

Multiband Joint Detection

Multiband joint detection algorithm sense the PU signal over multiple frequency bands. Initially Wideband signal (x (t)) is captured by using a high sampling rate ADC as shown in Figure 1. A Serial to Parallel (S/P) convertor is used to convert the stream of samples into parallel data streams. Fast Fourier Transform (FFT) is used to convert the wideband signals to frequency domain. The wideband spectrum is divided into 'n' number of narrowband channels. Power Spectral Density (PSD) of individual bands which represents the power present in each band of signal is estimated. Further, Binary hypothesis test is used to detect spectral opportunities in each band. The optimal threshold value is jointly chosen by considering the entire sub-bands. Based on the outcome, it is concluded that performance of Multiband joint detection algorithm out performs narrowband techniques (Quan, 2009).

Figure 1. Multiband Sensing

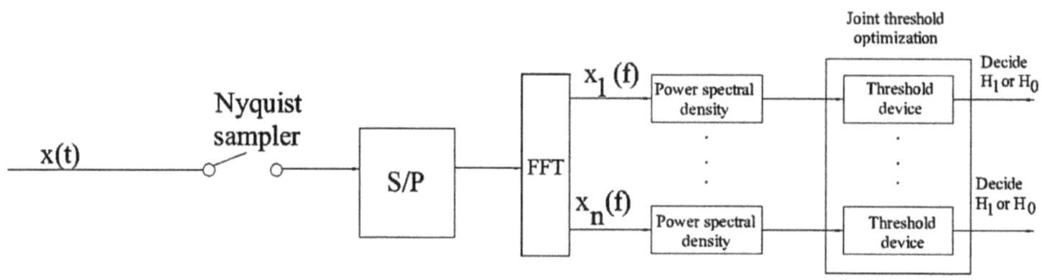

Filter Bank Detection

In filter bank detection, a bank of prototype filters with different shifted central frequencies is used to process the received wideband signal (Farhang-Boroujeny, 2008). To minimize the high sampling rate, frequency mixing techniques is used. A local oscillator (LO) signal is mixed with wideband signals for down converting to lower frequency signal as shown in Figure 2. The down converted baseband signals are then processed by a series of low pass filter whose central frequency can be adjusted to any selected band. Further, narrowband sensing algorithms are applied to detect the spectrum holes. Due to the parallel structure of the filter bank, the implementation of this method requires a large number of RF components which increases the system complexity.

Sub Nyquist Wideband Sensing

Sub-Nyquist wideband sensing utilizes the sparseness property of communication signal (Qin, Gao, & Parini, 2016). It acquires wideband signals using sampling rate lesser than nyquist rate and detect spectral opportunities using this partial measurements. In this section, discuss the sub-nyquist wideband sensing is discussed detail.

Figure 2. Filter bank Sensing

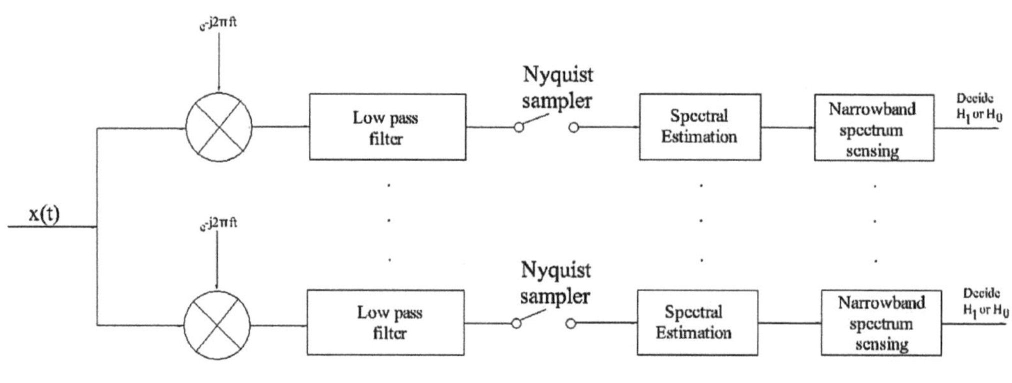

Compressed Sensing

Compressed sensing aims to recover the extracted version of the original sparse signal through measurements (Tian & Giannakis, 2007). In compressed sensing, the received signal y(t) is down converted by mixing the wideband signal with local oscillator signal as shown in Figure 3. Then a series of low pass filter are used to process the signals. To implement compressive sensing in CRN, a low rate sub-Nyquist sampler is required to obtain the few measurements. Then, the spectrum is reconstructed from the measured samples. Further, a localization and identification for the occupied sub-bands are explored to finalize the wideband sensing decisions.

Multi-Coset Sampling Based Sub-Nyquist Wideband Sensing

Multi–coset sampling uses the information of samples taken from multiple subsets or cosets as shown in Figure 4. Each sampling channel has different time offsets. It is shown that any sparse ruler can produce a multicoset sampling design that guarantees the full rank condition for spectrum sensing (Aziz, Traore, Nafkha, & Le Guennec, 2014). Sampling pattern should be carefully chosen to reconstruct the signals. Further narrow band sensing techniques are used to find the spectrum holes (Jayaweera, 2014).

Figure 3. Compressive sensing

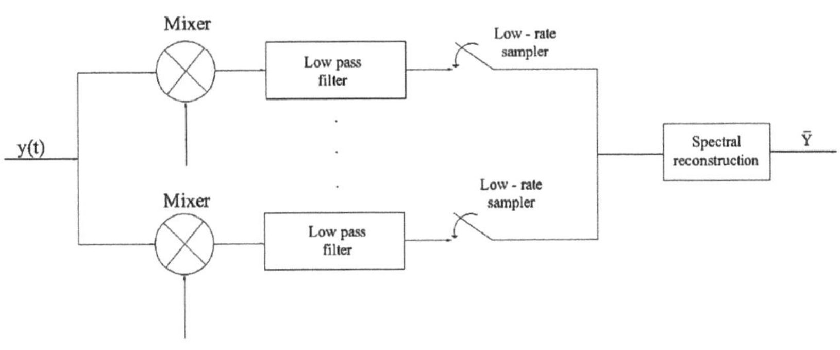

Figure 4. Multi channel sub nyquist

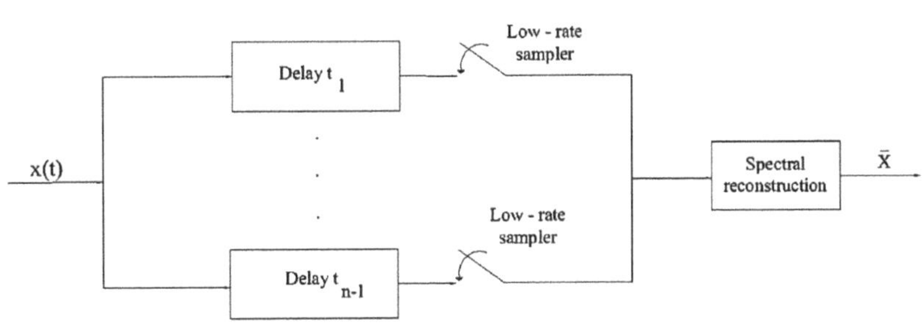

ADVANCED SPECTRUM SENSING TECHNIQUES

Due to the multipath nature of wireless environment, PU signal received by the CR user may be degraded one, which leads to unreliable sensing results. To overcome this, advance spectrum sensing techniques such as cooperative spectrum sensing is applied.

Cooperative Sensing

The main aim of spectrum sensing is to increase the probability of detection. It can be achieved if there exists cooperation among a group of CR users. The CR users have to sense and transmit. Cooperative sensing not only increases the sensing accuracy but also increases the time duration for CR user to transmit. If there exist a proper coordinating scheme among CR users, cooperative wideband sensing may be an optimum one. The cooperative sensing can be classified as centralized, distributed and relay assisted (Sharan & Wankhede, n.d.).

Centralized Approach

In centralized approach, CR users sensing information is collectively given to a fusion centre as shown in Figure 5. Fusion centre collects the sensing information from CR users within its coverage area. Finally, fusion centre determines the availability of spectrum and allocates optimal frequency of operation for CR users (Cabric, Mishra, & Brodersen, 2004).

Figure 5. Centralized Approach

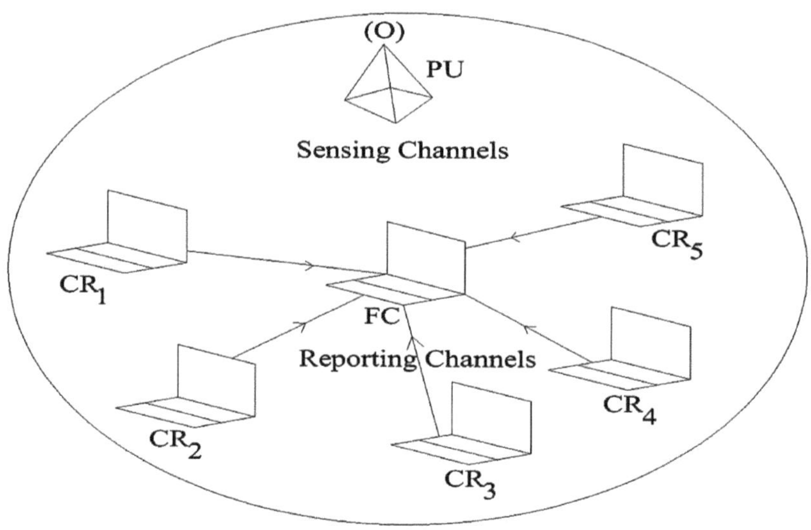

Distributed Approach

In distributed approach there is no centralized coordinator. Each node shares its spectrum sensing information with all other nodes within its coverage area as shown in Figure 6. Spectral decision is taken by each node individually (Liu, Jia, Gu, & Tan, 2013).

Relay-Assisted Cooperative

To avoid the effects of multipath and fading relay nodes are used between CR users and fusion centers as shown in Figure 7. This enhances the accuracy of spectrum sensing.

Even though cooperative spectrum sensing increases the spectral accuracy it is difficult in real time complex to establish cooperation among multiple CR users. IT can be worsted further that if the CR users happen to be malicious nodes, it may spoil the entire network operation. Hence a cooperative spectrum sensing scheme with appropriate security mechanism ensure optimum results in wideband spectrum sensing (Akyildiz, Lo, & Ravikumar, 2011).

CONCLUSION

In this chapter the challenges in the design and implementation of wideband spectrum sensing is addressed. The wideband spectrum sensing is classified into nyquist and sub-nyquist based on the sampling types

Figure 6. Distributed Approach

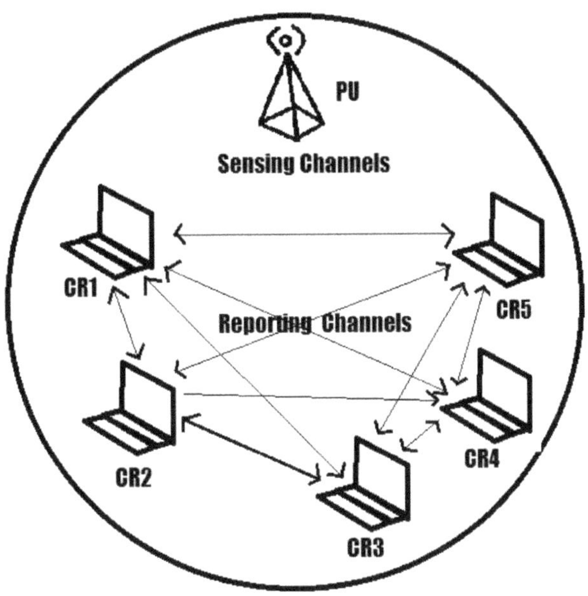

Figure 7. Relay-assisted cooperative approach

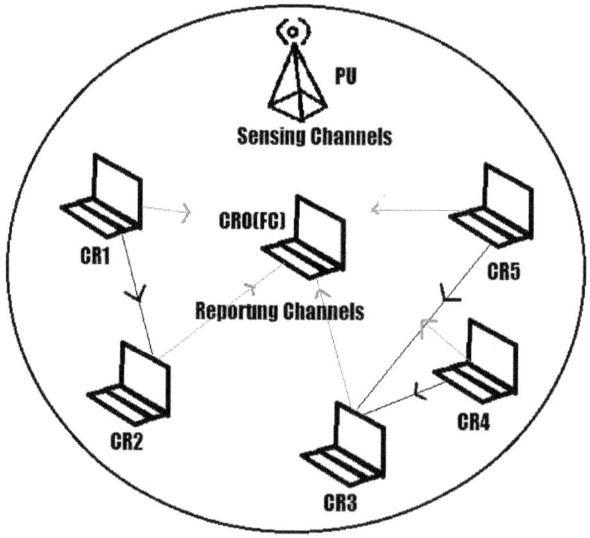

taken to acquire signals. The various nyquist and sub-nyquist wideband sensing has been discussed in detail. Compressive sensing uses the sparse property of wideband signals. Further to enhance the cooperative wideband sensing cooperative wideband sensing has been suggested. The types of cooperative sensing and its practical implementation complexity have also been discussed in detail. Even though advance wideband sensing technique produces good results, its implementation is complex. Reduction of complexity of wideband spectrum sensing can be given more research focus.

REFERENCES

Akyildiz, Lo, & Ravikumar. (2011). Cooperative spectrum sensing in cognitive radio networks: A survey. *Elsevier Journal on Physical Communication*, 40-62.

Ali, A., & Hamouda, W. (2017). Advances on Spectrum Sensing for Cognitive Radio Networks: Theory and Applications. *IEEE Communications Surveys and Tutorials*, *19*(2), 1277–1304. doi:10.1109/COMST.2016.2631080

Aziz, B., Traore, S., Nafkha, A., & Le Guennec, D. (2014). Spectrum sensing for cognitive radio using multicoset sampling. *Proc. IEEE Global Commun. Conf. (GLOBECOM)*, 816–821. 10.1109/GLOCOM.2014.7036909

Bountouris. (2013). *Spectrum Sensing Techniques for Cognitive Radios* (Thesis). Technical University of Crete, Greece, Department of Electronic and Computer Engineering.

Cabric, A. D., Mishra, S., & Brodersen, R. (2004). Implementation issues in spectrum sensing for cognitive radios. Processing of Asilomar conference on signals, systems, and computers, 1, 772–776. doi:10.1109/ACSSC.2004.1399240

Digham, F. F., Alouini, M. S., & Simon, M. K. (2007). On the energy detection of unknown signals over fading channels. IEEE Transactions on Communications, 55(1), 21-24.

Farhang-Boroujeny, B. (2008, May). Filter Bank Spectrum Sensing for Cognitive Radios. *IEEE Transactions on Signal Processing*, *56*(5), 1801–1811. doi:10.1109/TSP.2007.911490

Jaiswal, M., Sharma, A. K., & Singh, V. (2013). A Survey on Spectrum Sensing Techniques for Cognitive Radio. *Conference on Advances in Communication and Control Systems*, 647-660.

Jayaweera. (2014). Robust, Non-Gaussian Wideband Spectrum Sensing in Cognitive Radios. *IEEE Transactions on Wireless Communications, 13*(11), 6410-6421.

Liu, X., Jia, M., Gu, X., & Tan, X. (2013). Optimal Periodic Cooperative Spectrum Sensing Based on Weight Fusion in Cognitive Radio Networks. *Sensors (Basel), 13*(4), 5251–5272. doi:10.3390130405251 PMID:23604027

Qin, Z., Gao, Y., & Parini, C. G. (2016). Data-assisted low complexity compressive spectrum sensing on real-time signals under sub-nyquist rate. *Proc. IEEE Trans. Wireless Commun., 15*(2), 1174–1185. 10.1109/TWC.2015.2485992

Quan, Z. (2009, March). Optimal Multiband Joint Detection for Spectrum Sensing in Cognitive Radio Networks. *IEEE Transactions on Signal Processing*, *57*(3), 1128–1140. doi:10.1109/TSP.2008.2008540

Sun, H., Nallanathan, A., Wang, C.-X., & Chen, Y. (2013, April). Wideband spectrum sensing for cognitive radio networks: A survey. *IEEE Wireless Communications*, *20*(2), 74–81. doi:10.1109/MWC.2013.6507397

Sun, H., Nallanathan, A., Wang, C.-X., & Chen, Y. (2013). Wideband spectrum sensing for cognitive radio networks: A survey. IEEE Wireless Communication, 20(2), 74–81.

Sutton, P. D., Lotze, J., Nolan, K. E., & Doyle, L. E. (2007). Cyclostationary signature detection in multipath Rayleigh fading environments. *Proceedings of IEEE International conference on Cognitive Radio Oriented Wireless Networks and Communincation*. 10.1109/CROWNCOM.2007.4549833

Tian, Z., & Giannakis, G. (2007). Compressive Sensing for Wideband Cognitive Radios. *Proc. IEEE Int'l. Conf. Acoustics, Speech, and Sig.*, 1357–60.

Tian, Z., Leus, G., & Lottici, V. (2010). Compressed Sensing Techniques for Dynamic Resource Allocation in Wideband Cognitive Networks. *Proc. IEEE International Workshop on Signal Processing Advances in Wireless Communications (SPAWC)*. 10.1109/SPAWC.2010.5670984

Treichler, Davenport, Laska, & Baraniuk. (2011). Dynamic range and compressive sensing acquisition receivers. *Proceedings Defense Applications Signal Processing (DASP),* 1–7.

Zeng, Y., & Liang, Y. C. (2009). Spectrum-sensing algorithms for cognitive radio based on statistical covariances. *IEEE Transactions on Vehicular Technology*, *58*(4), 1804–1815. doi:10.1109/TVT.2008.2005267

Section 3
Improved Sensing Techniques

Chapter 9
Multiple Antennas-Based Improved Sensing Detector

Ashish Bagwari
Uttrakhand Technical University, India

Jyotshana Kanti
Uttrakhand Technical University, India

Geetam Singh Tomar
THDC Institute of Hydropower Engineering and Technology, India

ABSTRACT

5G is one of the newest and most acclaimed engineering technology of wireless era. This technology is also known as CRNs, in which spectrum detection plays a key role. Through this chapter, the authors explore multiple antennas-based improved sensing detector (MA_ISD). In the said scheme by using adaptive threshold and multiple antennas patterns, its easy to mitigate sensing failure issue enriches reliability. The presented scheme uses two detectors (TD) concept. The concept of two detectors (i.e., TD scheme has been applied in which multiple antennas are used for electing the quality signals). The considered model upgrades the detection operation and acquires limited or inferior detection time. The first energy detector uses a single adaptive threshold (ED-SAT) while another energy detector employs two adaptive thresholds (ED-TAT). The threshold value is accommodative as it relays on noise variance (σ_ω^2), the behavior of noise variance transforms in accordance with noise signal. Both the detectors run collaboratively and their gain is then supply to a decision device which operates OR functions. In this research work, results shows that with cooperation of two antennas (Nr=2) in multiple antennas-based improved sensing detector (MA_ISD) technique delivers enhancement in detection performance by 24.6%, 53.4%, 37.9%, and 49.6%, in contrast to existing schemes (i.e., EDT-ASS-2015 scheme, ED and cyclo-2010, adaptive SS-2012, and conventional-ED) scheme at -12dB SNR, respectively. During the time, proposed technique also reduces the sensing time in the order of 47.0 ms, 49.0 ms, and 53.2 ms as compared to existing schemes (EDT-ASS-2015, Adaptive SS-2012, and ED and Cyclo-2010) scheme at - 20 dB SNR, respectively.

DOI: 10.4018/978-1-5225-5354-0.ch009

INTRODUCTION

There is several essential sensing techniques have been proposed by the investigators to sense licensed signal . In (Maleki, Pandharipande, & Leus, 2010), authors presented two-stage detectors, energy detector and cyclostationary detector respectively. Generally detector also has some limitation like complex computations and additional sensing time. Moreover, in (Ejaz, Hasan, & Kim, 2012), authors suggested adaptive spectrum sensing scheme to mitigates the sensing period. These two stage detectors can't operate at the same time. So it can be stated that system complexity sustains while having minimum sensing time period. Furthermore, in (Sobron, Diniz, Martins, & Velez, 2015), authors presented adaptive sensing technique using energy detector (EDT-ASS). In which, cost-function and the presence/absence of PU's signal is concluded.

In this Chapter, the two detectors i.e. ED_SAT and ED_TAT performs sensing task concurrently in order to improve detection performance using multiple antennas. Adaptive Thresholds value leads less sensing failure problem (Liu, Hu, & Wang, 2012). The results of detectors inset to decision device (DD) who takes ultimate decision using OR-rule. Decision device with value 1 shows frequency band is occupied i.e (H_1), else it is free (H_0). The proposed model is much more concerned about spectrum sensing failure (Liu, Hu, & Wang, 2012), and fading problem while the other (Maleki, Pandharipande, &Leus, 2010; Ejaz, Hasan, & Kim, 2012; Sobron, Diniz, Martins, & Velez, 2015) techniques overlooked at the same. Adaptive threshold scheme diminishes the problem of sensing failure and the use of multiple antenna deals the fading problem. Meanwhile, (Pandharipande, & Linnartz, 2007; Taherpour, Nasiri-Kenari, & Gazor, 2010), remarks that the use of multiple antennas gives the reliability for spectrum sensing. The main feature of this technique to tackle out sensing failure and fading problem. Simulation results approve that the proposed model advances detection performance at $P_f = 0.1$, operates well at low SNRs, and decreases sensing period also.

BACKGROUND

Energy Detector and Cyclostationary Detector (ED and Cyclo-2010) (Maleki, Pandharipande, & Leus, 2010)

In this section, authors proposed two-stage spectrum sensing detector, who carried energy detector and cyclostationary detector. Further, authors assumed that the total numbers of channels are L, and channels are sensed serially by secondary users. First stage is known as coarse sensing stage, carries energy detector to take spectrum sensing decision, if the energy of detected signal is greater than or equal to a certain threshold, the channel is announced to be busy. Else, second stage detector known as fine sensing stage, carries cyclostationary detector will come on picture and try to identify licensed frequency band. If the decision metric in second stage is above a certain threshold, the channel is announced to be busy. Else, it is declared to be free and available for cognitive radio users for their use. Authors are examined the performance of proposed two-stage detectors in terms of the probabilities of detection alarm, probability of false alarm, and sensing time.

Limitations

The two stage detection scheme (Maleki, Pandharipande, & Leus, 2010) enhances the detection performance. Two stage detector, in which first stage belongs to energy detector scheme and second stage dedicated to cyclostationary scheme both delivers good sensing performance, but the maine issue is regarding their computing complexity and the requirement of more sensing time period.

Adaptive Spectrum Sensing (Adaptive SS-2012) (Ejaz, Hasan, & Kim, 2012)

Cognitive Radio is a key technology based on dynamic spectrum allocation scheme that helps to mitigate scarcity of frequency band or spectrum. The technique of sensing licensed signal in radio environment is called spectrum sensing. In this section, the adaptive spectrum sensing detector carries estimated signal to noise ratio device, energy detector, and one-order cyclostationary detector for spectrum sensing. This approach carries two detectors and assuming N numbers of channel to be sensed by CR users. Once cognitive radio will sense the channel, it acquires signal & passes to estimated SNR device to estimates the SNR of the sensed channel. Therefore, based on estimated SNR value and desired threshold, system will choose one of the appropriate detector out of two for signal detection i.e. either energy detector or one-order cyclostationary detector. After studied most of the research papers we can conclude that energy detector does not perform well at very low SNR. That's why this approach go with one-order cyclostationary detector under low SNR and same for high SNR it go with energy detector to detect the licensed channels.

Limitations

The adaptive spectrum sensing technique (Ejaz, Hasan, & Kim, 2012) is a new system, which support or run only one of the detection technique at a particular instance of time based on the calculated SNR. Despite the fact that it reduces the mean sensing time, it brush-off spectrum sensing failure problem (Liu, Hu, & Wang, 2012).

Adaptive Sensing Technique Using Energy Detector (EDT-ASS-2015) (Sobron, Diniz, Martins, & Velez, 2015)

In this segment, authors suggested a competent and firm ED-based technique which conducts the sensing process using an adaptive parameter rather than soft-mixer of energy estimation. To do that, least mean squares (LMS) algorithm has been applied to an MSE minimization problem of a new cost-function. The cost-function is used to execute both single-node and weighted cooperative SS with the help of an adaptive test statistic, depends on pre-processed intakes and anticipated signals is selected adequately to correct the well-known deflection coefficient of the test statistic. In this way, the proposed method substantially improves the deflection coefficient compared with that achieved in conventional ED methods.

Limitations

In (Sobron, Diniz, Martins, & Velez, 2015), authors introduce adaptive sensing technique using energy detector (EDT-ASS-2015). Here, authors only focused to cost-function and consider the availability and inadequacy of primary signal. There are no remarks regarding the sensing failure problem (Liu, Hu, & Wang, 2012).

System Model and Main Assumptions

Two Hypothesis tests, H_1 (alternate hypothesis) and H_0 (null hypothesis) has been studied to detect the presence and absence of primary user signal. H_1 declares that PU signal present and received signal *r(n)* can be expressed as(Ejaz, Hasan, & Kim, 2012)

$$r\left(n\right) = x\left(n\right) \times h\left(n\right) + \omega\left(n\right), H_1 \tag{1}$$

H_0 (null hypothesis) declares that PU signal absent and received signal *r(n)* can be declared as

$$r\left(n\right) = \omega\left(n\right), H_0 \tag{2}$$

In Equation (1) and (2), *r(n)* is that signal which is sensed by the CR consumers. *x(n)* is licensed users signal, *w(n)* is additive white gaussian noise having zero mean and σ_ω^2 denotes noise variance, *h(n)* is the channel gain, and number of samples denotes by *n, where the range of n from 1 to N.*

PROPOSED SYSTEM MODEL

Multiple Antennas Based Improved Sensing Detector (MA_ISD)

Figure 1 illustrates the unique schemed sensing model having multiple antennas based improved sensing detector (MA_ISD). Here we presume N_r numbers of antennas recepting N number of cases, and CR uses the selection combiner (SC) scheme that randers maximum SNR value signal out of N_r decision statistics. Therefore, the upper stream and lower stream receives the signal $r = max\ (r_1, r_2, r_3, \ldots, r_{Nr})$.

In Figure 1, upper stream consists ED with a single adaptive threshold, similar to conventional-ED, except adaptive threshold that's why it is an improved form of conventional-ED. ED using single adaptive threshold measures the energy (*E*) of received signal (Bagwari, & Singh, 2012) and balance it with the adaptive threshold ($\lambda 1$),to generate output (*L1*) and surpass it to decision device (DD) in the form of binary bits. If the observed energy (E >= $\lambda 1$) then detector (L1) delivers bit 1 else 0. Similarly, the lower stream consists ED with two adaptive thresholds (ED_TAT), this detector is diverse from the upper stream detector as it includes two adaptive thresholds. Acquiring two adaptive thresholds is more profitable and relevant to eliminate sensing failure problem (Liu, Hu, & Wang, 2012). Now, the energy calculated by ED_TAT is compared to thresholds (γ) and gives output (L2). If calculated energy is greater and equal to γ, then the output L2 delivers bit 1 else bit 0. Decision device (DD) receives the input from ED_SAT and ED_TAT detectors. DD sum-up the output of L1&L2 by substituting OR rule operation. If the sum of L1&L2 is greater or equal to 1, shows H1 (the channel is occupy), else shows H0 (the channel is vacant) as shown in Figure 1.

Let consider r$j_(k)$ as a received PU signal at the jt^h antenna for the kt^h data stream, N counts as total number of samples, the total number of antennas are Nr and the channel between PU and CR is Rayleigh fading channel. Therefore, the output of the SC can be written as:

$$SC_{o/p} = \max \sum_{j=1}^{N_r} \left[\sum_{k=1}^{N} \left| r_j(k) \right|^2 \right] \tag{3}$$

Figure 1 exemplifies that the CR claims multiple antennas, out of multiple branches CR will opt the antenna's branch that has a large gain and release it to the detectors for further process.

- Probability of detection of multiple antennas based improved sensing detector can be defined as:

$$P_D^{MA\_ISD} = P_r \times P_d^{ED\_SAT} + \left(1 - P_r\right) \times P_d^{ED\_TAT} + \frac{P_r}{2} \tag{4}$$

$$P_D^{MA\_ISD} = P_r \left(\frac{1}{2} + P_d^{ED\_SAT} - P_d^{ED\_TAT} \right) + P_d^{ED\_TAT} \tag{5}$$

- Total Error Probability of multiple antennas based improved sensing detector can be defined as:

$$P_e^{MA\_ISD} = P_F^{MA\_ISD} + \left(1 - P_D^{MA\_ISD}\right) \tag{6}$$

$$P_e^{MA\_ISD} = P_r \left(P_f^{ED\_SAT} - P_f^{ED\_TAT} - P_d^{ED\_SAT} + P_d^{ED\_TAT} \right) + P_f^{ED\_TAT} - P_d^{ED_{TAT}} + 1 \tag{7}$$

Energy Detector With Single Adaptive Threshold (ED_SAT)

The most eminent and frequently exploited detector is ED, used by the researchers to find licensed user's signals. Figure 2 depicts conventional-ED, where incoming PU signal is received by band pass filter (BPF) and after filtration, it further surpasses it to analog to digital converter (ADC). Now, ADC generates bit patterns in binary format binary bit patterns in set to square law device (SLD), who enumerates the energy of the received input signals. Along with this SLD outcome is also received by integrator and integrate it up to T interval. At the end, decision-making device (DMD) gives overall decision with the help of single threshold value to verify whether PU is there or not.

Figure 1. Proposed Model: Multiple antennas based improved sensing detector (MA_ISD)

Where, $P_d^{ED\_SAT}$ is known as detection probability throughout of ED_SAT and $P_d^{ED\_TAT}$ acknowledged as ED_TAT detector. The false alarm probability for ED_SAT and ED_TAT detectors are $P_f^{ED\_SAT}$ and $P_f^{ED\_TAT}$ respectively. The probability factor P, varies $0 \leq P_r \leq 1$ and relies on SNR. The value of P_r less then 0.5 indicates extremely noisy channel and above from 0.5 shows noiseless channel.

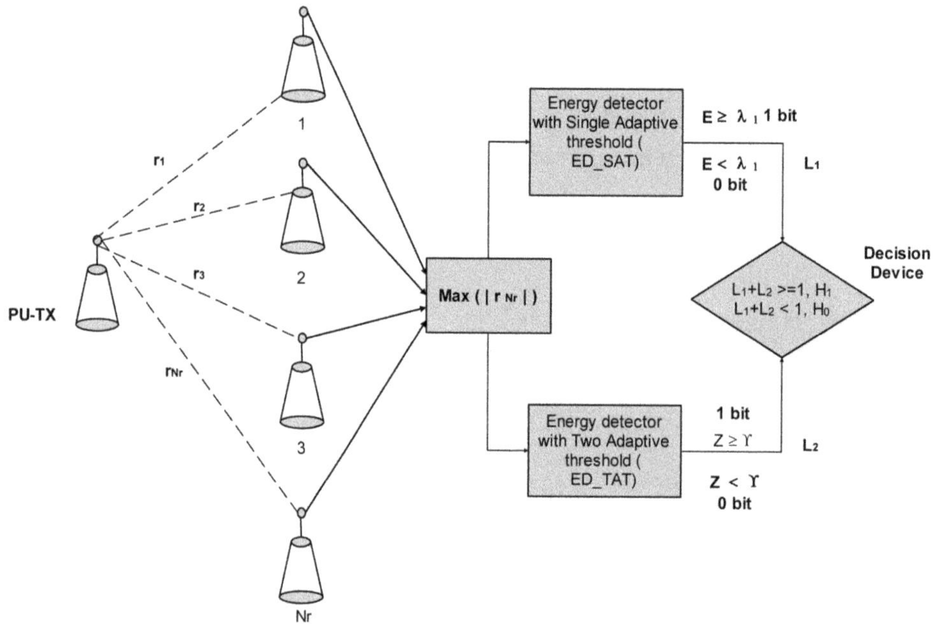

Figure 2. Energy detector with single adaptive threshold (ED_SAT)

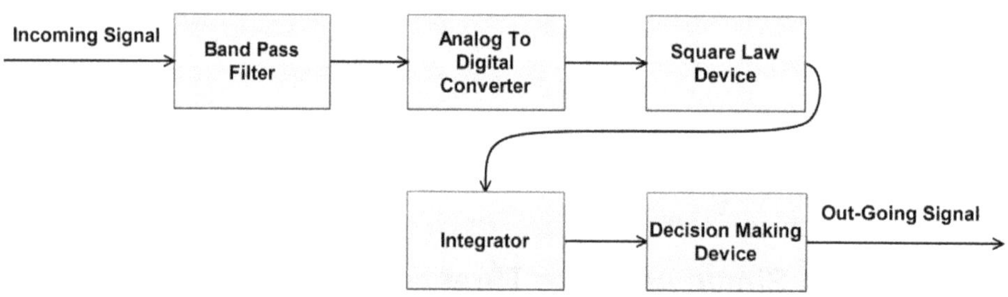

Expression of Single Adaptive Threshold

The mathematical expression of a single adaptive threshold ($\lambda 1$) can be defined as (Tandra, & Sahai, 2008):

$$\lambda_1 = \left[N \times \times \sigma_\omega^2 \left\{ Q^{-1}\left(P_f\right) \times \sqrt{\frac{2}{N}} + 1 \right\} \right] \tag{8}$$

where, N, $Q^{-1}()$, P_f and σ_ω^2 represents a number of samples, inverse-Q-function, false alarm probability, and noise variance respectively. Equation (8) defines the threshold ($\lambda 1_)$ which is proportional to noise variance (σ_ω^2), dependend on random nature noise signal which varies with respect to time, and because of this noise variance (σ_ω^2) also differs, and then threshold ($\lambda 1_)$ value also alter. Therefore, Adaptive threshold changes its value at every instance of time.

$$E = \frac{1}{N}\sum_{n=1}^{N}\left|r\left(n\right)\right|^2 \tag{9}$$

Above mentioned Equation (9) has $r(n)$ which is a received signals, total numbers of samples are N and observed energy of $r(n)$ *is* E. Finally, the local decision of ED_SAT detector can be written as:

$$ED\_SAT\,|_{o/p} = \begin{cases} E < \lambda_1, & bit0 = L_1 \\ E \geq \lambda_1, & bit1 = L_1 \end{cases} \tag{10}$$

Detection Probability for Conventional-ED

The final expression for probability of detection can be written as (Sobron, Diniz, Martins, & Velez, 2015):

$$P_d^{ED\_SAT} = Q\left[\sqrt{\frac{N}{2}} \times \left(\frac{\lambda'}{N} - 1\right)\right] \tag{11}$$

In Equation (11), numbers of samples, Gaussian tail probability Q-function is N and $Q()$ respectively. λ' is defined as $\lambda' = \frac{\lambda_1}{\left(\sigma_x^2 + \sigma_\omega^2\right)}$, where, λ_1 is a single adaptive threshold, σ_x^2 is PU signal variance, and σ_ω^2 is noise variance.

False Alarm Probability for Conventional-ED

The final mathematical expression of false alarm probability can be derived as (Sobron, Diniz, Martins, & Velez, 2015):

$$P_f^{ED\_SAT} = Q\left[\sqrt{\frac{N}{2}} \times \left(\frac{\lambda''}{N} - 1\right)\right] \tag{12}$$

In Equation (12), N is a number of samples, and λ'' is defined as $\lambda'' = \frac{\lambda_1}{\left(\sigma_\omega^2\right)}$.

Total Error Probability for Conventional-ED

The total error rate is the sum of the false alarm (P_f) and the missed-detection probability (P_m). Hence, the total error probability rate as follows (Tandra, & Sahai, 2008):

$$P_e^{ED\_SAT} = \left(P_m^{ED_{SAT}}\right) + P_f^{ED_{SAT}} \tag{13}$$

where, ($1\text{-}P_d$) shows the missed-detection probability (P_m), then:

$$P_e^{ED\_SAT} = Q\left[\sqrt{\frac{N}{2}} \times \left(\frac{\lambda''}{N} - 1\right)\right] + \left(1 - Q\left[\sqrt{\frac{N}{2}} \times \left(\frac{\lambda'}{N} - 1\right)\right]\right) \tag{14}$$

Energy Detector With Two Adaptive Thresholds (ED_TAT)

In CRN, it is a tough task for a detector to become aware about right signal although the PU signal and intersect or overlay mutually. This process is named as sensing failure problem and overlaid region is defined as the confused region, as discussed in (Liu, Hu, & Wang, 2012). Improved-ED-ADT detector is one of the advantageous solutions to control this issue.

Generally, in CRN, consists three segments defined as H_0 (only noise signal present), H_1 (only PU signal present) and *confused region* having both the noise and signal of PU.. Figure 3 (a) presents C-ED version where confused region is denoted either as zero or null. The remaining section splits into two parts (H_1 and H_0) using single threshold (γ) concept. H_1, if the observed energy is greater than or equal to γ, and $H_{0,}$ for the observed energy which is smaller than γ. Whereas, in Figure 2 (b) confused region itself devided into four parts, the area lower than γ_1 and higher than γ_2 called upper part (UP) of the detector, while the region comes between γ_1-γ and γ-γ_2 is known as lower part (LP).Therefore, the output of detector can be expressed as:

$$ED\_TAT\,|_{o/p} = UP + LP = Z \tag{15}$$

Let us assume, if observed energy is below than pre-defined threshold γ_1 it shows H_0, and if perceived energy is more than or equal to pre-defined threshold γ_2 it shows $H_{1.}$ But, in confused region, if observed energy lies in between (γ_1-γ) it gives 01 and later it transforms binary bits into decimal i.e. 1, similarly, if observed energy lies between (γ–γ_2), shows 10 and further its decimal value will be 2. Now, the pre-defined threshold (γ) can be written as (Tandra, & Sahai, 2008):

$$\gamma = \left[\left(\frac{N}{\sigma_\omega^2}\right) \times \left\{Q^{-1}\left(\frac{--}{P_f}\right) \times \sqrt{\frac{2}{N}} + 1\right\}\right] \tag{16}$$

Figure 3. (a) Single thresholddetection scheme, and (b) double threshold detection scheme

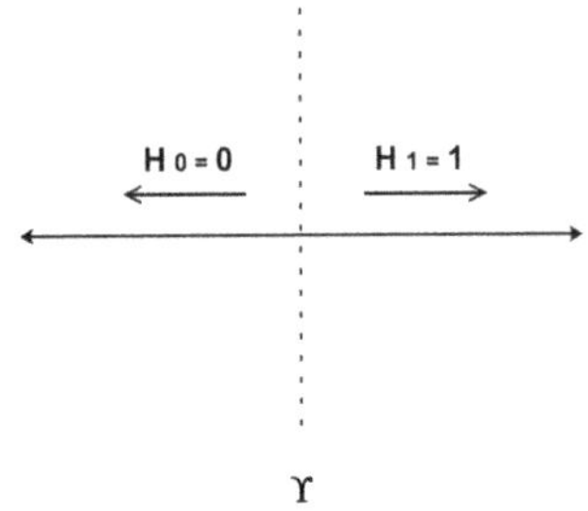

(a)

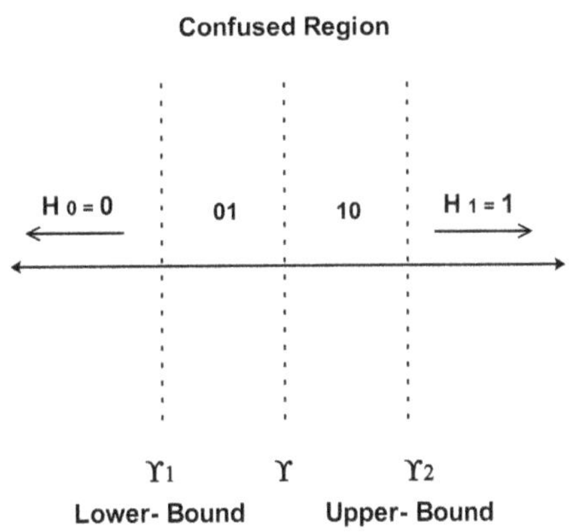

(b)

The value of lower threshold (γ_1) shows the minimum noise variance and the upper threshold (γ_2) defines the maximum noise variance. Now, the lower thresholds (γ_1) and upper threshold (γ_2) can be found as:

$$\gamma_1 = \left[\left(\frac{N}{\rho \times \sigma_\omega^2}\right) \times \left\{Q^{-1}\left(\frac{--}{P_f}\right) \times \sqrt{\frac{2}{N}} + 1\right\}\right] \tag{17}$$

$$\gamma_2 = \left[\left(N \times \rho \times \sigma_\omega^2\right) \times \left\{Q^{-1}\left(\frac{--}{P_f}\right) \times \sqrt{\frac{2}{N}} + 1\right\}\right] \tag{18}$$

In Equation (17) and (18) The mathematical expression of lower (γ_1) and upper(γ_2) threshold is represented. Now, considering the above equations i.e. (17) and (18) the thresholds (γ_1&γ_2) depend on the noise variance i.e. σ_ω^2, and the value of noise variance is mutable, so the values of thresholds also varies.

Therefore, the thresholds are also recognized as adaptive thresholds. The newly built sub-regions are $(\gamma_1 \gamma - \gamma\gamma_2)$ which comes under LP can be chosen as:

$$LP = \begin{cases} if\, \gamma_1 \leq E < \gamma, & represent\, bit\, 01 \\ if\, \gamma \leq E < \gamma_2, & represent\, bit\, 10 \end{cases} \tag{19}$$

$$UP = \begin{cases} if\, E < \gamma_1, & represent\, bit\, 0 \\ if\, \gamma_2 \leq E, & represent\, bit\, 1 \end{cases} \tag{20}$$

Hence, the combination of LP and UP represents the local decision of ED_TAT detector (Zhang, & Letaief, 2008):

$$ED\_TAT\,|_{o/p} = \begin{cases} \left(UP + LP\right) < \gamma, & bit\, 0 = L_2 \\ \left(UP + LP\right) \geq \gamma, & bit\, 1 = L_2 \end{cases} \tag{21}$$

Detection Probability for ED_TAT Detector

Suppose that $r(n)$ is a received sample whose normalized form is r_i. Now, the cumulative distribution function (CDF) of the ED-ADT detector can be written as:

$$f_{Z_i}\left(z\right) = P_r\left(\left|r_i\right| \leq \sqrt{\left(z^{\frac{2}{a}}\right)}\right) \tag{22}$$

In Equation (22), a variable contains value 2, is an arbitrary constant, $r(n)$ is zero-mean primary signal with average power σ_r^2 that does not depend on complex gaussian noise $w_i(n)$, and h_i is Rayleigh faded channel independent to gaussian noise. Hence, $|h_i|$is Rayleigh distributed with variance $\sigma_h^2 / 2$. Thus, the Probability distribution function (PDF) of the ED-ADT detector for H_j (where j = 0, 1) is given as:

$$f_{Z_i|H_j}\left(z\right) = \left[\frac{2 \times z^{\left(\frac{2}{a}\right)}}{\left(z \times a\right)}\right] \times f_{\left|r_i\right|^2|H_j}\left(z^{\left(\frac{2}{a}\right)}\right) \tag{23}$$

where, $f_{\left|r_i\right|^2|H_1}$ is exponentially distributed as follows:

$$f_{|r_i|^2|H_1}\left(z\right)=\left[\left(1+S\right)^{-1}\right]\times\exp\left[-z\times\left(1+S\right)^{-1}\right],z\geq 0 \tag{24}$$

Note that $S=\left(\sigma_h^2\times\sigma_x^2\right)/\sigma_w^2$ represents the average SNR of the sensing channel. Finally, by using Equation (23) and (24) we have:

$$f_{Zi|H_1}\left(z\right)=\left[\frac{2\times z^{\left(\frac{2}{a}\right)}\times\left(1+S\right)^{-1}}{\left(z\times a\right)}\right]\times\exp\left[-z^{\left(\frac{2}{a}\right)}\times\left(1+S\right)^{-1}\right],z\geq 0 \tag{25}$$

Now, the detection probability for ED_TAT can be obtained as:

$$P_d^{ED\_TAT}=\int_{\gamma}^{+\infty}f_{Z_i|H_1}\left(z\right)dz \tag{26}$$

Putting the value of PDF from Equation (25) to Equation (26), we get:

$$P_d^{ED\_TAT}=\int_{\gamma}^{+\infty}\left[\frac{2\times z^{\left(\frac{2}{a}\right)}}{\left(z\times a\right)\times\left(1+S\right)}\right]\times\exp\left[-\frac{z^{\left(\frac{2}{a}\right)}}{\left(1+S\right)}\right]dz \tag{27}$$

After solving Equation (27) the detection probability is:

$$P_d^{ED\_TAT}=\exp\left[-\frac{\left\{\left(\gamma\right)^{\frac{1}{a}}\right\}^2}{\left(1+S\right)}\right] \tag{28}$$

False Alarm Probability for ED_TAT Detector

Considering Equation (23), $f_{|r_i|^2|H_0}$ is exponentially distributed as follows:

$$f_{|r_i|^2|H_0}\left(z\right)=\exp\left[-z\right],z\geq 0 \tag{29}$$

Substituting $f_{|r_i|^2|H_0}\left(z\right)$ from Equation (23) to Equation (29), we have:

$$f_{Z_i|H_0}(z) = \left[\frac{2 \times z^{\left(\frac{2}{a}\right)}}{(z \times a)}\right] \times \exp\left(z^{\left(\frac{2}{a}\right)}\right), z \geq 0 \tag{30}$$

The false alarm probability will be calculated as:

$$P_f^{ED\_TAT} = \int_{\gamma}^{+\infty} f_{Z_i|H_o}(z)\,dz \tag{31}$$

Putting PDF value from Equation (30) to Equation (31), we get:

$$P_f^{ED\_TAT} = \int_{\gamma}^{+\infty} \left[\frac{2 \times z^{\left(\frac{2}{a}\right)}}{(z \times a)}\right] \times \exp\left(z^{\left(\frac{2}{a}\right)}\right) dz \tag{32}$$

After solving Equation (32) the false alarm probability is:

$$P_f^{ED\_TAT} = \exp\left[-\left\{(\gamma)^{\frac{1}{a}}\right\}^2\right] \tag{33}$$

Rate of Error Probability for ED_TAT Detector

As per IEEE 802.22, total error rate i.e $P_e^{ED\_TAT}$ based on false alarm (P_f) and missed-detection probability (P_m), defined as:

$$P_e^{ED\_TAT} = P_f^{ED\_TAT} + \left(1 - P_d^{ED\_TAT}\right) \tag{34}$$

Substituting Equations (28) and (33) in Equation (34), the final mathematical expression of total error probability is:

$$P_e^{ED\_TAT} = 1 + \exp\left[-\left\{(\gamma)^{\frac{1}{a}}\right\}^2\right] - \exp\left[-\frac{\left\{(\gamma)^{\frac{1}{a}}\right\}^2}{(1+S)}\right] \tag{35}$$

Now, the proposed sensing scheme can be described with the help of flow chart given in Figure 4.

Decision Device (DD)

Decision devise is responsible for identifying the presence and absence of PU signal. DD depends on the output of ED_SAT (L_1) and ED_TAT (L_2) as shown in Equations (10) and (21) respectively. Now, the L_1 and L_2 combination develops the decisive mathematical expression for the proposed model given as:

$$DD = \begin{cases} L_1 + L_2 \geq 1, & 1 \\ L_1 + L_2 < 1, & 0 \end{cases} \tag{36}$$

Figure 4 shows the flow of operation by using the flow chart. How MA_ISD technique works is illustrated in this figure where the received signals are sensed by the CR receiver and perform sensing operation according to it using ED_SAT and ED_TAT detectors, and at the end, DD will form the final decision regarding PU signal availability.

Simulation Model

The simulation environment is created using MATLAB. Following are the simulation steps described in Box 1.

Numerical Results and Analysis

During simulations, we make the use of OPSK modulation and Rayleigh channel for the assessment of system .Simultaneously, we compare the proposed MA_ISD scheme with conventional energy detection, ED and cyclostationary-2010 detection (Maleki, Pandharipande, & Leus, 2010), Adaptive spectrum sensing-2012 (Ejaz, Hasan, & Kim, 2012), and energy detection technique for adaptive spectrum sensing-2015 (EDT-ASS-2015) (Sobron, Diniz, Martins, & Velez, 2015). The parameters considered in simulation are given in Table 1.

In Figure 5, 1000 number of samples are considered to employ the MA_ISD technique, in the system the value of threshold should be set in order to achieve false alarm probability 0.1. In this simulation we have selected $\lambda 1 = 1.25$, $\gamma 1 = 0.9$, $\gamma 2 = 1.2$ and $\gamma = 1.014$ as trade-off value.

In simulation environment, there is detection performance comparison between proposed a multiple antennas based improved sensing detector (MA_ISD) scheme, EDT-ASS-2015 scheme, ED and cyclo-2010, adaptive SS-2012, and conventional-ED scheme. After examine Figure 5 we found that proposed MA_ISD scheme with number of antennas (N_r) = 3 surpasses N_r = 1, 2, EDT-ASS-2015 scheme, ED and cyclo-2010, adaptive SS-2012, and conventional-ED scheme 4.5%, 29.1%, 57.9%, 42.4%, & 54.1% at – 12 dB SNR in terms of detection probability respectively.

When the value of false alarm probability sets at 0.1 at that time acceptance value of detection probability should be 0.9 which show that PU signal presence at -12dB SNR (approximately) in a newly proposed MA_ISD scheme.

Figure 4. Flowchart of MA_ISD technique

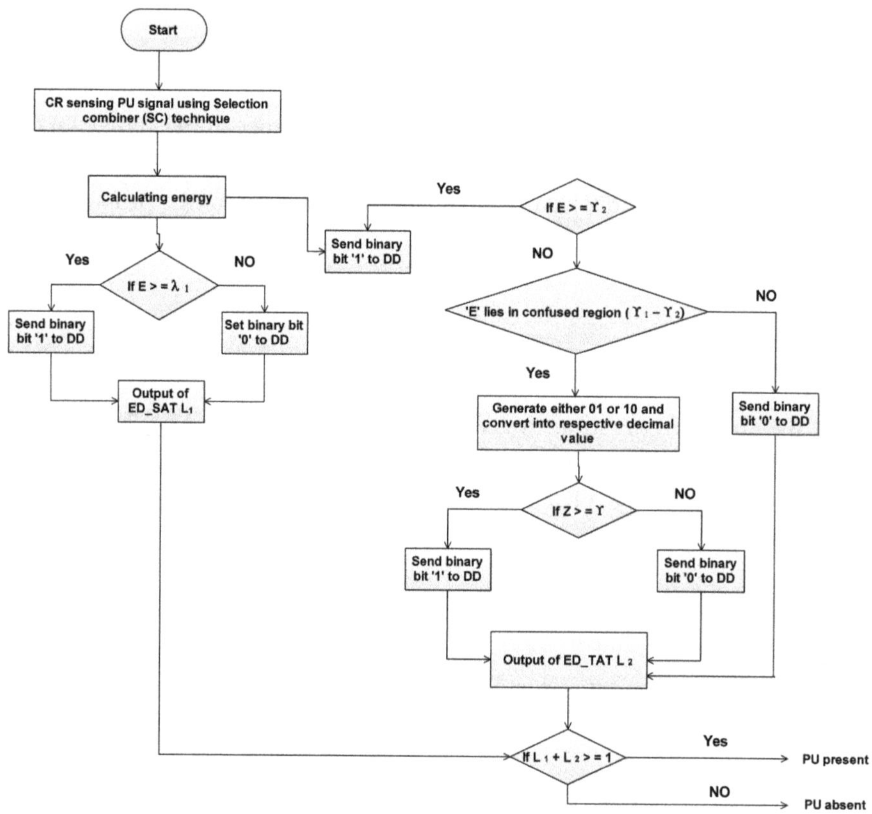

Figure 6 is showing the performance of proposed MA_ISD scheme having same total Error Probability parameter. This technique gives the lowest error rate i.e. 0.1 at – 10 dB SNR with $N_r = 3$ while other techniques have large SNR rate.

ROC is the Receiver Operating Characteristics and its curve shows the character of P_d with respect to P_f(Ling-ling, Jian-guo, & Cheng-kai, 2011). According to IEEE 802.22, it is mandatory that the value of detection probability should be high when P_f value becomes minimum. The next simulation step is shown in Figure 7, we employ QPSK modulation scheme to the system. and determine that how detection probability changes with respect to false alarm probability.

In this simulation work we are considering different false alarm probabilities while SNR is – 8, – 10, – 12, and – 14 dB with 1000 samples are inset to proposed MA_ISD scheme. This figure clearly shows that less false alarm probability gives more probability of detection. So, we come to a conclusion to carry our following simulations for $P_f = 0.1$, $N_r = 2$, and SNR – 12 dB, detection probability is . 0.9110, which is tolerable to licensed signal detection as per IEEE 802.22 rules (Cordeiro, Challapali, Birru, & Shankar, 2005).

Sensing time is the time taken by CR users to detect licensed frequency band. Sensing time can be calculated as:

$$T_{MA\_ISD} = T_{SC} + min. (T_{ED\_SAT}, T_{ED\_TAT}) + T_{DD} \tag{37}$$

Box 1.

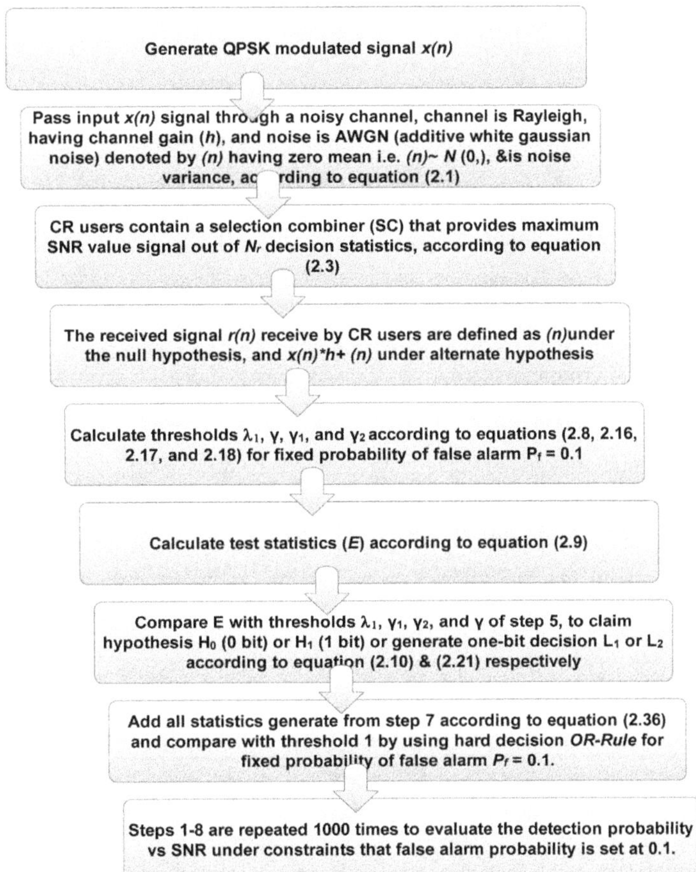

In Equation (37), T_{MA_ISD} is total time taken by proposed sensing technique for SS. T_{SC}, T_{ED_SAT} and T_{ED_TAT} are the selection combiner (SC), ED_SAT and ED_TAT detectors SS time respectively. In Figure 1, ED_SAT and ED_TAT are placed parallel, therefore, we take minimum time out of the two detectors as shown in Equation (37). Now, the sensing time for SC can be calculated as:

$$T_{SC} = C \times S_{SC} \tag{38}$$

In Equation (38), S_{SC} is the mean detection time for each channel and a number of sensed channels denoted by C.

$$S_{SC} = \frac{1}{2} \times \left(\frac{M_{SC}}{B} \right) \tag{39}$$

$M_{SC}$is the samples and B is the bandwidth of the channel for SC. Now, the ED_SAT sensing time can be calculated as:

Table 1. Data values for simulation

Parameters along with symbols	Value
Types of signal	QPSK
Channel (between PU and CR user)	Rayleigh
Total samples are (N)	1000
Total Antennas (N_r)	3 & 2
Threshold (λ_1)	1.25
Threshold (Υ)	1.014
Threshold (Υ_1)	0.9
Threshold (Υ_2)	1.2
Value of signal to noise ratio	-20 dB to 0 dB
Probability of false alarm for each detection scheme	0.1
Computer software	MATLAB R2012 a

Figure 5. Detection probability with respect to SNR values at $P_f = 0.1$

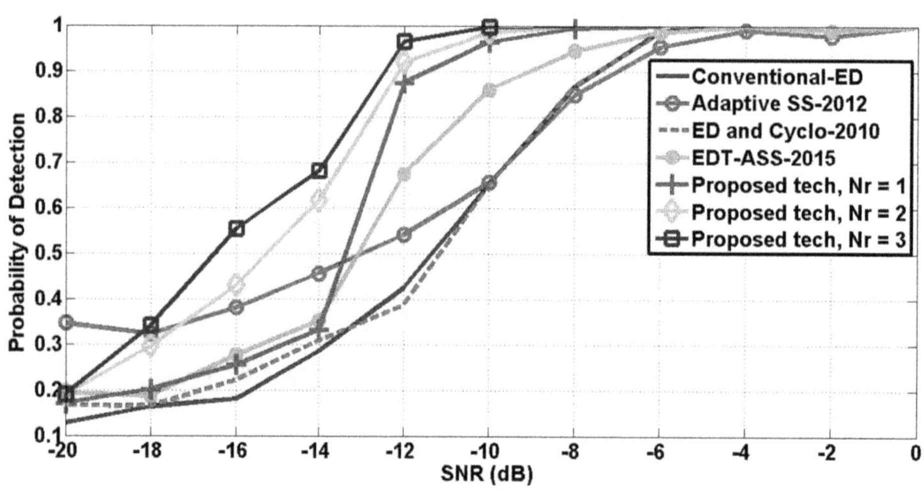

**For a more accurate representation see the electronic version.*

Figure 6. Probability of error with respect to SNR values

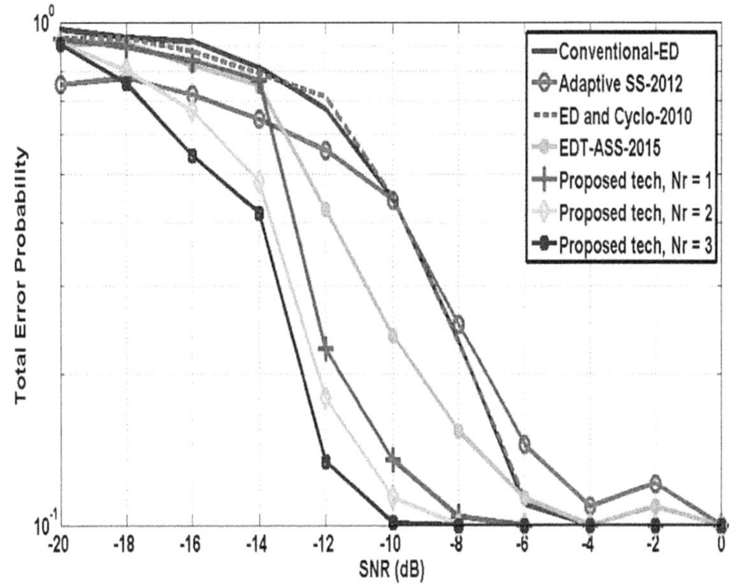

**For a more accurate representation see the electronic version.*

Figure 7. ROC Curve of MA_ISDspectrum sensing detector under different SNR values

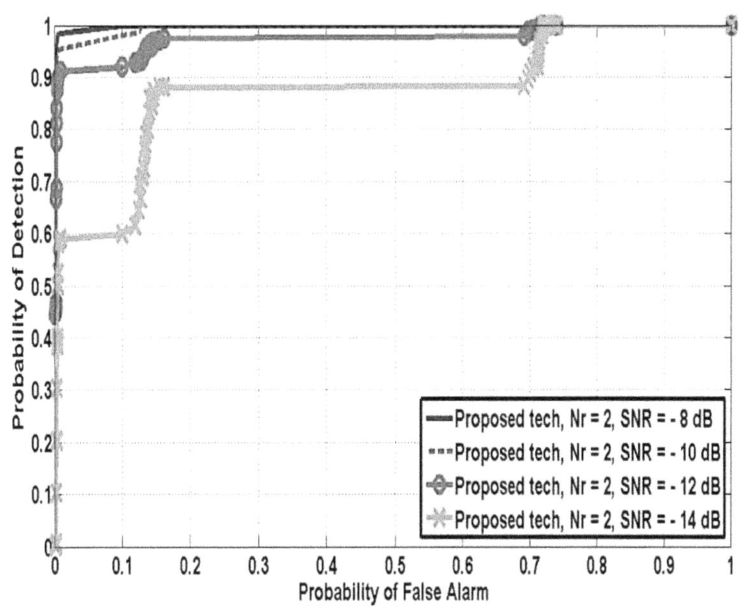

**For a more accurate representation see the electronic version.*

$$T_{ED\_SAT} = C \times P_r \times S_1 \tag{40}$$

In Equation (40), S_1 is the mean detection time for each channel, a number of sensed channels denoted by C and P_r is probability factor:

$$S_1 = \frac{1}{2} \times \left(\frac{M_{ED\_SAT}}{B} \right) \tag{41}$$

M_{ED_SAT} is the samples and B is the bandwidth of the channel for ED_SAT. Similarly, the ED_TAT detector sensing time can be calculated as:

$$T_{ED\_TAT} = C \times (1 - P_r) \times S_2 \tag{42}$$

In Equation (42), S_2 is the mean detection time for each channel, detected channel is C, and $(1 - P_r)$ is the probability factor for ED_TAT.

$$S_2 = \frac{1}{2} \times \left(\frac{M_{ED\_TAT}}{B} \right) \tag{43}$$

M_{ED_TAT} is the samples and the bandwidth of channel is B for the channel ED_TAT. Sensing time of Decision device can be equated as:

$$T_{DD} = C \times S_0 \tag{44}$$

In Equation (44), S_0 is the mean detection time for each channel and detected channels is denoted by C.

$$S_0 = \frac{1}{2} \times \left(\frac{M_0}{B} \right) \tag{45}$$

M_0 is the samples and B is the bandwidth of the channel for DD. Therefore, the overall spectrum sensing time can be computed using Equations (37), (38), (40), (42) and (44) as:

$$T_{MA\_ISD} = C \times S_{SC} + \min.\left\{ C \times P_r \times S_1, C \times \left(1 - P_r\right) \times S_2 \right\} + C \times S_0 \tag{46}$$

$$T_{MA\_ISD} = C \times \min.\left\{ P_r \times S_1, \left(1 - P_r\right) \times S_2 \right\} + C \times \left(S_{SC} + S_0 \right) \tag{47}$$

Figure 8 gives variation in sensing time (ms) with respect to SNR(dB) values. IEEE 802.22 stated that spectrum sensing time period should be small. Figure 8 concluded that the proposed detection scheme at $N_r = 2$ delivers least detection time as compared to present schemes. As the value of SNR goes up, it gives inverse effect on detection time. Equation (47) has been used for plotting the graph across the sensing time and SNR. Parameters along with their values are defined in table 1and used in Equation (47).

In proposed scheme, at -20 dB with $N_r = 2$ it takes about 46.7 ms time to spot PU signal. while existing schemes like EDT-ASS-2015, Adaptive SS-2012, ED and Cyclo-2010 needs sensing time approximately 47.0 ms, 49.0 ms, and 53.2 ms respectively.This graph shows that at -20dB sensing time (46.7 ms) taken by proposed schema is far better than previous schemes. The more availability of sensing and less time for data transmission responsible for poor throughput, and phenomenon is also sddressed as sensing efficiency problem (Lee, & Akyildiz, 2008; Liang, Zeng, Peh, & Hoang, 2008).

Using four different SNR values i.e. - 8 dB, - 10 dB, - 12 dB, and - 14 dB Figure 9 plots the curve between detection probability and Threshold value .

It concludes that the presented MA_ISD SS scheme can detect licensed signal at - 8 dB SNR at $N = 1000$, and $\lambda = 3.0$ with $Nr = 2$.

Figure 8. Sensing time with respect to SNR values

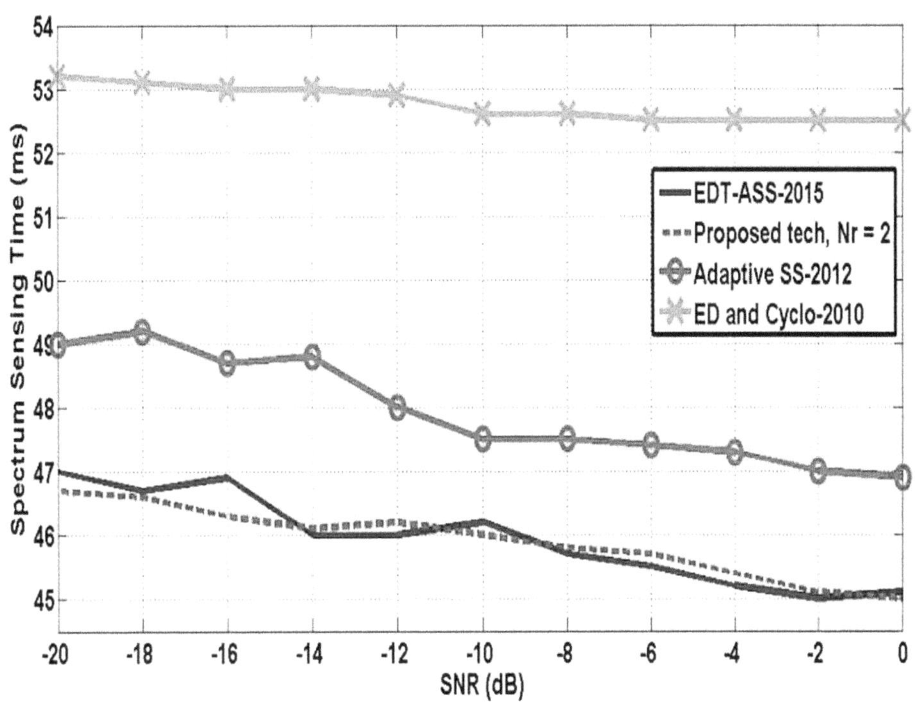

**For a more accurate representation see the electronic version.*

Figure 9. Detection probability with respect to threshold values for different SNRs

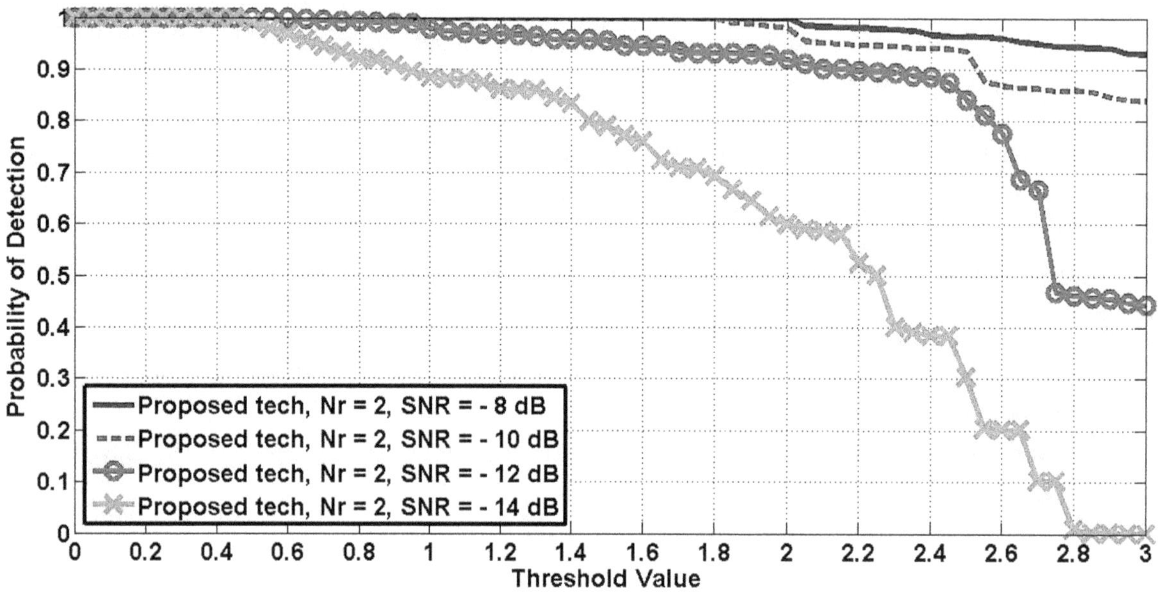

**For a more accurate representation see the electronic version.*

CONCLUSION AND FUTURE WORK

In this chapter, a new detector technique i.e multiple antennas based improved sensing has been introduced. This proposed scheme improves performance of signal detection and also decreases sensing time. Mathematical results proved that proposed MA_ISD scheme having $N_r = 2$ excelled other existing schemes (i.e. EDT-ASS-2015 scheme by 24.6%, ED and cyclo-2010 by 53.4%,, adaptive SS-2012 by 37.9%,, and conventional-ED scheme by 49.6%) at – 12 dB SNR . It is also verified that the proposed scheme bears less detection time than EDT-ASS-2015, Adaptive SS-2012, and ED and Cyclo-2010 scheme in the sequence of 47.0 ms, 49.0 ms, and 53.2 ms at - 20 dB SNR respectively.

ACKNOWLEDGMENT

The authors wish to thank their parents for supporting and motivating for this work because without their blessings and God's grace this was not possible.

REFERENCES

Bagwari, A., & Singh, B. (2012, November). Comparative performance evaluation of spectrum sensing techniques for cognitive radio networks. In *Computational Intelligence and Communication Networks (CICN), 2012 Fourth International Conference on* (pp. 98-105). IEEE. 10.1109/CICN.2012.66

Cordeiro, C., Challapali, K., Birru, D., & Shankar, S. (2005, November). IEEE 802.22: the first worldwide wireless standard based on cognitive radios. In *New Frontiers in Dynamic Spectrum Access Networks, 2005. DySPAN 2005. 2005 First IEEE International Symposium on* (pp. 328-337). IEEE.

Ejaz, W., Hasan, N. U., & Kim, H. S. (2012). SNR-based adaptive spectrum sensing for cognitive radio networks. *International Journal of Innovative Computing, Information, & Control*, *8*(9), 6095–6105.

Lee, W. Y., & Akyildiz, I. F. (2008). Optimal spectrum sensing framework for cognitive radio networks. *IEEE Transactions on Wireless Communications*, *7*(10).

Liang, Y. C., Zeng, Y., Peh, E. C., & Hoang, A. T. (2008). Sensing-throughput tradeoff for cognitive radio networks. *IEEE Transactions on Wireless Communications*, *7*(4), 1326–1337. doi:10.1109/TWC.2008.060869

Ling-ling, Z., Jian-guo, H., & Cheng-kai, T. (2011). Novel Energy Detection Scheme in Cognitive Radio. *IEEE conference on Signal Processing, Communications and Computing*, 1-4.

Liu, S. Q., Hu, B. J., & Wang, X. Y. (2012). Hierarchical cooperative spectrum sensing based on double thresholds energy detection. *IEEE Communications Letters*, *16*(7), 1096–1099. doi:10.1109/LCOMM.2012.050112.120765

Maleki, S., Pandharipande, A., & Leus, G. (2010, March). Two-stage spectrum sensing for cognitive radios. In *Acoustics Speech and Signal Processing (ICASSP), 2010 IEEE International Conference on* (pp. 2946-2949). IEEE. 10.1109/ICASSP.2010.5496149

Pandharipande, A., & Linnartz, J. P. (2007, June). Performance analysis of primary user detection in a multiple antenna cognitive radio. In *Communications, 2007. ICC'07. IEEE International Conference on* (pp. 6482-6486). IEEE. 10.1109/ICC.2007.1072

Sobron, I., Diniz, P. S., Martins, W. A., & Velez, M. (2015). Energy detection technique for adaptive spectrum sensing. *IEEE Transactions on Communications*, *63*(3), 617–627. doi:10.1109/TCOMM.2015.2394436

Taherpour, A., Nasiri-Kenari, M., & Gazor, S. (2010). Multiple antenna spectrum sensing in cognitive radios. *IEEE Transactions on Wireless Communications*, *9*(2), 814–823. doi:10.1109/TWC.2009.02.090385

Tandra, R., & Sahai, A. (2008). SNR walls for signal detection. *IEEE Journal of Selected Topics in Signal Processing*, *2*(1), 4–17. doi:10.1109/JSTSP.2007.914879

Zhang, W., & Letaief, K. B. (2008). Cooperative spectrum sensing with transmit and relay diversity in cognitive radio networks. *IEEE Transactions on Wireless Communications*, *7*(12), 4761–4766. doi:10.1109/T-WC.2008.060857

Chapter 10
Cooperative Spectrum Sensing Under Double Threshold With Censoring and Hybrid Spectrum Access Schemes in Cognitive Radio Network

Abhijit Bhowmick
VIT University, India

Sanjay Dhar Roy
NIT Durgapur, India

Sumit Kundu
NIT Durgapur, India

ABSTRACT

The spectrum sensing performance in cooperative cognitive radio (CR) network is studied under a double threshold (DTH)-based detection with censoring of CRs, and thereafter, the study is extended for a hybrid spectrum access scheme in presence of Rayleigh faded sensing (S) and reporting (R) channels. In spectrum sensing, a CR employs an energy detection to detect the presence of primary user (PU) and compares the received energy statistics with the DTH. The CRs with energy statistics lying in fuzzy zone are not allowed to send their sensing information to the fusion centre (FC). Further, the qualified CRs are censored (rank-based and threshold-based censoring) to report their decisions based on quality of R-channel. The incorporation of DTH-based sensing and censoring of CRs not only improves the detection performance but also reduces the transmission overhead. In spectrum access, two hybrid spectrum access schemes, namely conventional hybrid spectrum access scheme (CHSAS) and a modified hybrid spectrum access scheme (MHSAS) are studied and compared.

DOI: 10.4018/978-1-5225-5354-0.ch010

INTRODUCTION

Cognitive Radio (CR) has evolved as a smart technology to resolve the spectrum scarcity issue in the secondary network. In an overlay or underlay CR network, a CR node exploits the licensed spectrum band without creating harmful interference to the primary user (PU). In overlay network, a CR user accesses the PU band only if the PU is found to be absent, while in an underlay network, both the users (CR and the PU) share the spectrum. Thus, accuracy in sensing the presence of PU has a great importance in CR technology otherwise a PU has to face an undesirable interference (Haykin, 2005). In spectrum sensing (SS), a CR may employ an energy detector (ED) to detect PU in the given channel (Ghasemi, 2005). Each CR of a secondary network senses the status of PU and compares the received energy statistics with a predefined threshold, i.e., a single threshold based (STH) based detection (Ghasemi, 2005). Spectrum sensing can also be performed using double threshold (DTH) (Bhowmick, 2015a). It is observed that DTH based spectrum sensing reduces the transmission of sensing overhead at the cost of some performance loss (Bhowmick, 2015a; Chunhua, 2007). In DTH based sensing, the CRs with energy statistics lying in fuzzy zone (between two thresholds) are not allowed to send the sensing information to the fusion centre(FC) (Chunhua, 2007). If the R-channels are heavily faded, the overall decision, i.e., cooperative decision at FC may be affected (Bhowmick, 2015a). Thus, it is required to eliminate CRs having bad R-channels from the cooperation process. In (Nallagonda, 2013a), authors proposed censoring techniques, i.e., Threshold-based censoring (TBC) and Rank-based censoring (RBC) to combat with such type of adverse scenarios. Censoring of CRsis done in (Nallagonda, 2013a) on the basis of R-channel quality. In (Bhowmick, 2015a; Bhowmick, 2014), authors studied the censoring schemes under DTH. The incorporation of censoring of CRs with DTH based sensing improves the performance of SS and at the same time it reduces the overhead of the transmission (Bhowmick, 2015a; Bhowmick, 2014). It is shown in (Bhowmick, 2014), Reduction in overhead of the transmission improves the agility gain of the network. A cooperative SS (CSS) under DTH with censoring has been studied for perfect and imperfect estimation of channel and different fusion strategies such as Majority, and Maximal Ratio Combining (MRC) (Bhowmick, 2015a).

In spectrum access schemes (SAS), two HSAS: CHSAS and MHSAS are studied and compared (Bhowmick, 2016; Prasad, 2016). A CR uses an overlay mode in absence of PU while it transmits in an underlay mode of co-existence if the PU's presence is found in the spectrum. In overlay mode, a CR transmits with high power while in underlay mode, a CR has to control its transmission power to such that the interference remains within a tolerable limit at PU receiver. The channel state information (CSI) of the interfering linkhelps the CR system to control its transmission power (Prasad, 2016). In (Musavian, 2007; Stotas, 2011; Chen, 2012), the authors have studied the performance of CR considering the CSI of the link between CR and PU is perfect. In practice, the CSI is imperfect (Musavian, 2009; Suraweera, 2010). The manager of the band helps to know the link CSI (Peha, 2005) or via proper signaling between a CR and a PU (Kang, 2010). In CHSAS, a CR usesonly the transmission phase of a detection frame while in MHSAS, a CR uses the entire detection frame (sensing (S) phase plus transmission (T) phase) for its own transmission (Bhowmick, 2016). The outage performance of HSAS is studied in (Prasad, 2016).

A cooperative CR network is studied in two-fold. At first, the SA performance is studied under double threshold with censoring and thereafter the study is extended for a hybrid SAS in presence of Rayleigh faded sensing (S) and reporting (R) channels. Furthermore the performances of MHSAS and CHSAS schemes are compared and it is shown that MHSAS outperforms the CHSAS.

BRIEF INTRODUCTION TO THE DOUBLE THRESHOLD AND CENSORING SCHEMES

Figure 1 consists of aPU, *N* CRs and a FC (such as base station). The links between PU and CRs are known as S-channels and the links between CRs and FC are known as R-channels. During the sensing, the received signal at the input to the ED of a CR can be expressed as:

$$y_i(t) = \begin{cases} n_i(t) & H_0 \\ h\, s(t) + n_i(t) & H_1 \end{cases} \tag{1}$$

where i = 1, 2, 3…N; $s(k)$ is the PU signal; $n(k)$ represents the noise with zero mean and variance σ^2. Here, h is the S-channel coefficient and H_0 represents that the PU is absent, while H_1 is represents that the PU is present.

In spectrum sensing, a predefined threshold (λ) is used in an energy detector to find the PU status (present or absent) in the given channel. The energy statistics (W_i, i.e., $W_i = |y_i(t)|^2$) of the received signal after passing through an integrator is compared with the λ. The decision (denoted as D_i) corresponding to H_0 and H_1 depends on threshold λ can be expressed as:

$$D_i = \begin{cases} 0 & W_i < \lambda \\ 1 & W_i \geq \lambda \end{cases} \tag{2}$$

In the case of a STH based detection (Figure 2a), all the CRs send their local observations to FC, while in the case of a DTH based system, the local observations lying in the fuzzy zone ('no decision' region) are not reported to FC (Figure 2b). The observations are sent to FC may be hard decisions, i.e., 0 or 1 or soft decisions.

Figure 1. Cooperative CR network

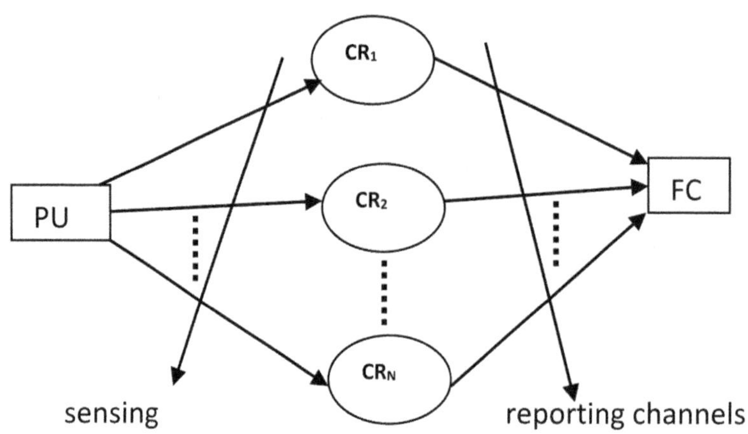

Figure 2. (a) Single threshold based detection. (b) Double threshold based detection.

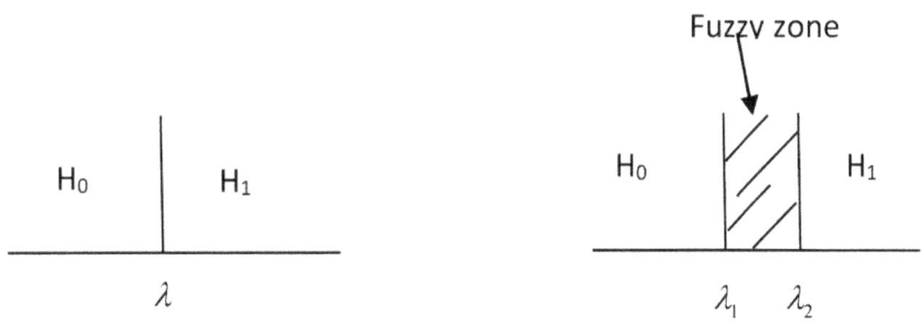

In double threshold (DTH) based detection, there are two thresholds, say λ_1 and λ_2 where $\lambda_2 > \lambda_1$ as shown in Figure 2b. A CR operating with DTH, decision bit D_i can be written as (Chunhua, 2007).

$$D_i = \begin{cases} 0 & W_i \le \lambda_1 \\ No\,decision & \lambda_1 < W_i < \lambda_2 \\ 1 & W_i \ge \lambda_2 \end{cases} \tag{3}$$

i.e., the sensing decision becomes '0' when the decision variable (W_i) is falls below the lower threshold (λ_1), the same becomes '1' when W_i exceeds the upper threshold (λ_2), and the CR takes no decision when the W_i lies in between λ_1 and λ_2.

Once the local observations are made, the CRs sent their decisions to FC for combining and generating the overall decision. Sometimes it may happen that a bad quality of R-channel affects severely a good quality local decision which in turns affects the overall decision of the CR system. Thus, it is required to eliminate the bad quality R-channels from the decision process to improve the sensing performance. In such situation, censoring of CRs can expedite the sensing performance.

CRs are censored on the basis of theCSI of the estimated R-channel coefficient. A CR sends a training symbol to FC for estimation of the channel coefficient before sending its local observation. We consider an MMSE based estimation andit can be imperfect or perfect. RBC and TBC can be used for selecting the CRs with best quality R-channel (Nallagonda, 2013a).

In RBC, FC first estimates the channel coefficient values of the R-channels; thereafter FC selects a subset of CRs which has highest channel coefficient values. On the other hand, in TBC, FC selects the CRs which have higher R-channel coefficient values as compared to a threshold which is predefined.

The censoring of CRs in CSSfor RBC and TBC schemes have been studied underSTH based ED in (Nallagonda, 2013b; Nallagonda, 2013c). These censoring schemes are investigated in (Nallagonda, 2013b; Nallagonda, 2013c) considering STH and conventional energy detector. In this chapter, we consider censoring schemes with DTH based energy detector. Algorithm 1 shows the sensing and censoring process of considered CSS.

Algorithm 1

```
Step 1: Initialize the number of CRs = N;
Step 2: for i = 1 to N
Step 3: Received energy statistics of a CR is compared with both the thresh-
olds.
Step 4: If the CR is allowed after DTH, the R-channel of that CR is estimated.
Step 5: Censoring of R-channel is done on the basis of quality.
Step 6: end
Step 7: Allowed CRs report their local observations to FC after DTH based
sensing and Censoring.
Step 8: All the local observations are fused at FC and it takes the overall
decision.
```

ANALYTICAL DISCUSSION ON DTH BASED SENSING WITH CENSORING SCHEMES

In DTH based cooperative spectrum sensing (CSS), the CRs whose observation statistics about the spectrum are either fall below λ_1 or greater than λ_2 are eligible to convey their hard decisions to base station, i.e.,FC via their corresponding R-channel. Thus, the number CRs who are eligible for transmission of local decision may be equal or less than the number of cooperating CRs. Let us consider that *N*CRs are cooperating, in which N_i CRs are allowed for transmitting their local decision to FC after DTH based sensing where $N_i \le N$. The censoring technique is applied on the N_i number of eligible CRs. When we apply R-channel based censoring on CRs with DTH based sensing, we are essentially applying two stages, namely (A) *censoring phase* and (B) *decision transmission phase*; in the first phase, a set of CRs whose local decision statistics are outside the fuzzy zone are considered, and during the second phase, a subset of the above CRs whose R-channels are good are allowed to convey their local observations to FC. Thus, on the average, *K* CRs out of N_i CRs are selected to convey their local observations to FC once the censoring is done since their R-channels quality are good where $K \le N_i$. Hence, it is quite clear that if the censoring technique is incorporated after DTH, sensing overheads can be reduced significantly.

In double threshold based sensing there is chance that the test statistic of a CR may lie in the fuzzy zone, the probability of such type of scenario is represented as no decision probability. In this chapter, the no decision probability is considered as $\Delta_{0,1}$ under hypotheses $H_{0,1}$ where $\Delta_0 = P\left(\lambda_1 \le W_i \le \lambda_2 \mid H_0\right)$ and $\Delta_1 = P\left(\lambda_1 \le W_i \le \lambda_2 \mid H_1\right)$.In DTH based SS, if all the observed statistics (sensing information) of CRs lie in the fuzzy zone, i.e., $N_i = 0$, means none of the CRs are eligible for reporting their observation to the FC. Such an event is referred to as failed sensing event and the probability related to the sensing is called fail sensing probability. If failed sensing event occurs, the FC put request for sensing to the CRs once again. Let $\beta_{0,1}$ be the failed sensing probability under hypotheses $H_{0,1}$ where

$\beta_{0,1} = P\left(\lambda_1 \leq W_i \leq \lambda_2, N_i = 0 \mid H_{0,1}\right)$. The fail sensing probability and no decision probability are related as $\beta_{0,1} = \left(\Delta_{0,1}\right)^N$ (Bhowmick, 2015a). On the other hand, if $N_i \neq 0$, a number of CRs report to FC.

Sensing of Spectrum at CR

In the considered system model N_i CRs are eligible for transmitting their local observations in terms of hard decisions to FC after DTH based sensing. Let us consider that there is no fading in S-channels as well as in R-channels. The sensing performance matrices, such as detection probability (DP), missed detection probability (MDP)and the false alarm probability (FAP)of i-*th* CR can be represented as follows (Bhowmick, 2014):

$$P_{d,i} = P(W_i > \lambda_2 \mid H_1) = Q_u\left[\sqrt{2\gamma_i}, \sqrt{\lambda_2}\right] \tag{4}$$

$$P_{m,i} = P(W_i < \lambda_1 \mid H_1) = 1 - Q_u\left[\sqrt{2\gamma_i}, \sqrt{\lambda_1}\right] \tag{5}$$

$$P_{f,i} = P(W_i > \lambda_2 \mid H_0) = \Gamma(u, \lambda_2 / 2) / \Gamma(u) \tag{6}$$

where $\Gamma(.,.)$ and $Q_u\left[.,.\right]$ are is incomplete gamma functionand*Q*-functionrespectively. To avoid the mathematical complexity, an identical average SNR $(\gamma_i = \bar{\gamma}; \forall i)$ is considered for all CRs. Here, the probability of deciding '1' successfully is detection probability ($P_{d,i}$), similarly, the probability of deciding '0' successfully is no false alarm ($P_{nf,i}$) which can be expressed as:

$$P_{nf,i} = P(W_i < \lambda_1 \mid H_0) = 1 - \Gamma(u, \lambda_1 / 2) / \Gamma(u) \tag{7}$$

In the present chapter, Rayleigh fading is considered for both S-channel and R-channel. The average DP under fading environment can be expressed as:

$$\overline{P}_{dR} = \int_0^\infty P_{d,i} f_\gamma\left(\gamma\right) d\gamma \tag{8}$$

where $f_\gamma(\gamma)$ is the probability distribution function (PDF) of SNR under fading. For Rayleigh fading, the SNR γ follows an exponential PDF, i.e., $f_\gamma(\gamma) = \left(1 / \bar{\gamma}\right)\exp\left(-\gamma / \bar{\gamma}\right)$ for $\gamma \geq 0$ (Bhowmick, 2014).

Therefore, the average DP under Rayleigh faded S-channel, i.e., $\overline{P}_{dR,i}$, can be as (Bhowmick, 2015b; Bhowmick, 2014),

$$\overline{P}_{dR,i} = e^{-\lambda_2/2}\sum_{n=0}^{u-2}\frac{1}{n!}\left(\frac{\lambda_2}{2}\right)^n + \left(\frac{1+\overline{\gamma}}{\overline{\gamma}}\right)^{u-1}\left[e^{-\frac{\lambda_2}{2(1+\overline{\gamma})}} - e^{-\frac{\lambda_2}{2}}\sum_{n=0}^{u-2}\frac{1}{n!}\left(\frac{\lambda_2\overline{\gamma}}{2(1+\overline{\gamma})}\right)^n\right] \tag{9}$$

The eligible N_i CRs whose test statistics are not lying in between λ_1 and λ_2 are censored. In this chapter, two censoring schemes (RBC and TBC) are considered which are discussed below.

- **Rank Based Censoring (RBC):** In RBC, R-channel coefficients are organized in decreasing order according to their estimated values and a set of CRs associated with the good quality R-channels (according to the order statistics of the R-channels) are selected to improve the overall spectrum sensing performance. In the considered system model, the amplitudes of N_i estimated R-channel coefficients are sorted. The FC selects K_R CRs out of N_i CRs which have best estimated R-channel coefficients are sorted in order, i.e., CRs associated with highest channel coefficients sorted in order to achieve the targeted sensing performance.
- **Threshold Based Censoring (TBC):** In TBC, a CR is selected for transmission if the estimated R-channel fading coefficient amplitude exceeds a predefined threshold C_{th}. A CRis selected with a probability $\overline{p} = P\left(\left|\hat{h}_{r,i}\right| > C_{th}\right)$ (Nallagonda, 2013b; Nallagonda, 2013c). Let K_T number of CRs are selected after Threshold based censoring for transmitting their decision. Here, K_T CRs are selected with the probability $P\left(K_T\right)$follows the binomial distribution as discussed in (Nallagonda, 2013b; Nallagonda, 2013c), $P\left(K_T\right) = \binom{N_i}{K_T}\overline{p}^{K_T}\left(1-\overline{p}\right)^{\left(N_i-K_T\right)}$.

Two phases of the censoring schemes are discussed below:

1. **Channel Estimation Phase:** After DTH based detection, each of N_i number of available CRs send their training symbol to enable the FC to estimate the channel coefficients of R-channels existing between FC and N_i CRs. The FC uses MMSEto estimate the channel coefficient. Hence, the estimated channel coefficient can be defined as:

$$\hat{h}_{r,i} = E\left[h_{r,i} \mid y_{rt,i}\right] = \frac{\sqrt{E_b}}{E_b + \sigma_n^2}y_{rt,i} \tag{10}$$

where $h_{r,i}$ is the actual channel coefficient of *i*-th CR and it is i.i.d, $\hat{h}_{r,i}$ is its estimated value. Here, $y_{rt,i}$ is the training symbol received at FC from *i*-th CR user, $y_{rt,i} = h_{r,i}S_{t,i} + n_{r,i}$. For the sake of simplicity, it isconsidered that the training symbol is $S_{t,i} = \sqrt{E_b}$, where E_b is the energy of a bit. Here, the variance of R-channel coefficient is σ^2, i.e., $h_{r,i} \sim CN\left(0,\sigma^2\right)$ as in (Bhowmick, 2015b; Bhowmick, 2014) the channel noise as $n_{r,i} \sim CN\left(0,\sigma_n^2\right)$ and both are independent. The distribution of estimation error follows the distribution of actual co-efficient with zero mean and has the variance following (Bhowmick, 2015b; Bhowmick, 2014):

$$\sigma_{he}^2 = \left(\frac{E_b}{\sigma_n^2} + 1\right)$$

where h_e is the estimation error, $h_e = h_{r,i} - \hat{h}_{r,i}$.

2. **Decision Transmission Phase:** After the channel estimation phase, let K CRs are selected for decision transmission (T). Here, Kbe an average value. For RBC, $K = K_R$ and for TBC, $K = K_T$. After BPSK modulation, all the selected K CRs send their local observations to FC through R-channel. The received signal (decision signal) at the FC from i-th selected CR can be expressed as (Bhowmick, 2015b; Bhowmick, 2014):

$$y_{i,d} = m_i h_{r,i} + n_{i,d} \tag{11}$$

where i=1, 2,.. K_{avg}, noise in channel, $n_{i,d} \sim CN(0,\sigma_n^2)$ and $m_i \in \{+\sqrt{E_b}, -\sqrt{E_b}\}$ is the binary decision obtained after BPSK modulation. FC uses a fusion rule to combine the local decisions reported to it and obtain the overall decision. In this chapter two fusion rules, one hard fusion rule such as Majority and one soft fusion rule such as MRC are investigated. For both type of fusion schemes, a predefined threshold λ_T is used at FC for overall decision.

- **Majority Fusion Rule:** At FC, the decision from i-th CR is compared with λ_T and d_i is generated as:

$$d_i = \begin{cases} 0 & if \quad R(y_{i,d}) < \lambda_T \\ 1 & if \quad R(y_{i,d}) \geq \lambda_T \end{cases} \tag{12}$$

where $R(.)$ represents the real part. Let the overall decision is D_{ov}. For Majority fusion rule, the overall decision can be described as:

$$D_{ov} = \begin{cases} 0 & if \quad \sum_{i=1}^{K} d_i < \frac{K}{2} \\ random\ choice & if \quad \sum_{i=1}^{K} d_i = \frac{K}{2} \\ 1 & if \quad \sum_{i=1}^{K} d_i \geq \frac{K}{2} \end{cases} \tag{13}$$

In RBC, the overall DP and FAP for Majority fusion rule can be written as,

$$Q^{R}_{d,Maj}\left(K\right)=\sum_{t=K/2}^{K}\binom{K}{t}\left(P_{d,FC}\right)^{t}\left(1-P_{d,FC}\right)^{K-t} \tag{14}$$

$$Q^{R}_{f,Maj}\left(K\right)=\sum_{t=K/2}^{K}\binom{K}{t}P^{t}_{f,FC}\left(1-P_{f,FC}\right)^{K-t} \tag{15}$$

where R is used for RBC and MDP, $Q^{R}_{m,Maj}=1-Q^{R}_{d,Maj}$. Here, $P^{i}_{d,FC}$ and $P^{i}_{f,FC}$ are the DP and FAP of i-th node at FC for faded R-channel and can be expressed as:

$$P^{i}_{d,FC}=P^{1}_{d,R}P_{d,i}+P^{0}_{d,R}\left(1-P_{d,i}\right) \tag{16}$$

$$P^{i}_{f,FC}=P^{1}_{d,R}P_{f,i}+P^{0}_{d,R}\left(1-P_{f,i}\right) \tag{17}$$

where $P^{1}_{d,R}$ and $P^{0}_{d,R}$ can be expressed as (Bhowmick, 2016b):

$$P^{1}_{d,R}=\left[erfc\left(-\sqrt{\frac{1}{2\sigma^{2}_{n}}}\right)-\exp\left(\frac{1}{\gamma_{r}}-2\sqrt{\frac{1}{2\sigma^{2}_{n}\gamma_{r}}}\right)+\sqrt{\frac{1}{2\sigma^{2}_{n}}}\exp\left(\frac{1}{\gamma_{r}}\right)\int_{0}^{1}t^{3/2}\exp\left(-\frac{1}{2\sigma^{2}_{n}t}\right)\exp\left(\frac{t}{\gamma_{r}}\right)dt\right] \tag{18}$$

$$P^{0}_{d,R}=\left[erfc\left(-\sqrt{\frac{1}{2\sigma^{2}_{n}}}\right)+\exp\left(\frac{1}{\gamma_{r}}-2\sqrt{\frac{1}{2\sigma^{2}_{n}\gamma_{r}}}\right)-\sqrt{\frac{1}{2\sigma^{2}_{n}}}\exp\left(\frac{1}{\gamma_{r}}\right)\int_{0}^{1}t^{3/2}\exp\left(-\frac{1}{2\sigma^{2}_{n}t}\right)\exp\left(\frac{t}{\gamma_{r}}\right)dt\right] \tag{19}$$

In TBC, the overall DPand FAPcan be written as,

$$Q^{T}_{d,Maj}\left(K\right)=\sum_{j=1}^{K}Q_{d}\left(j\right)P\left(j\right) \tag{20}$$

$$Q^{T}_{f,Maj}\left(K\right)=\sum_{j=1}^{K}Q_{f}\left(j\right)P\left(j\right) \tag{21}$$

where $P\left(j\right)$ is the probability of selecting j- number of user, $Q_{d}\left(j\right)$ and $Q_{f}\left(j\right)$ can be described as:

$$Q_d(j) = \sum_{t=j/2}^{j} \binom{j}{t} \left(P_{d,FC}\right)^t \left(1 - P_{d,FC}\right)^{j-t}$$

and

$$Q_f(j) = \sum_{t=j/2}^{j} \binom{j}{t} \left(P_{f,FC}\right)^t \left(1 - P_{f,FC}\right)^{j-t}$$

respectively. Hence, the total error rate of system can be written as (Nallagonda, 2017),

$$T_e = P(H_0)Q_m + P(H_1)Q_f \tag{22}$$

Here, $Q_m = Q^R_{m,Maj}\left(K\right) = 1 - Q^R_{d,Maj}\left(K\right)$ and $Q_f = Q^R_{f,Maj}\left(K\right)$ for Rank based censoring and $Q_m = Q^T_{m,Maj}\left(K\right) = 1 - Q^T_{d,Maj}\left(K\right)$ and $Q_f = Q^T_{f,Maj}\left(K\right)$ for Threshold based censoring.

- **MRC Fusion Rule:** In MRC fusion scheme, a MRC combiner combines all the BPSK modulated decision signals which are to be reported at the FC. The combined signal can be expressed as:

$$\Lambda_{MRC} = \sum_{t=1}^{K} \mathrm{Re}\left(y_{i,d} \widehat{h^*_{r,i}}\right) \tag{23}$$

where $\widehat{h^*_{r,i}}$ is the complex conjugate of estimated channel coefficient.

In the case of RBC, the overall DP ($Q^R_{d,MRC}$), the overall MDP ($Q^R_{m,MRC}$), and the FAP ($Q^R_{f,MRC}$) can be expressed as:

$$Q^R_{d,MRC} = P\left(\Lambda_{MRC} > \lambda_T \mid H_1\right) \tag{24}$$

$$Q^R_{m,MRC} = P\left(\Lambda_{MRC} < \lambda_T \mid H_1\right) \tag{25}$$

$$Q^R_{f,MRC} = P\left(\Lambda_{MRC} > \lambda_T \mid H_0\right) \tag{26}$$

Similarly, in the case of TBC, the same sensing metrics are as:

$$Q_{d,MRC}^{T} = \sum_{k}^{K} Q_{d,MRC}(k)P\left(k\right) \tag{27}$$

$$Q_{m,MRC}^{T} = \sum_{k}^{K} Q_{m,MRC}(k)P\left(k\right) \tag{28}$$

$$Q_{f,MRC}^{T} = \sum_{k}^{K} Q_{f,MRC}(k)P\left(k\right) \tag{29}$$

where $P(k)$ follows the same distribution as $P\left(K_T\right)$, $Q_{d,MRC}(k) = P\left(\sum_{t=1}^{k} \mathrm{Re}\left(y_{i,d}\widehat{h_{r,i}^{*}}\right) > \lambda_T \mid H_1\right)$, $Q_{m,MRC}(k) = 1 - Q_{d,MRC}(k)$, and $Q_{f,MRC}(k) = P\left(\sum_{t=1}^{k} \mathrm{Re}\left(y_{i,d}\widehat{h_{r,i}^{*}}\right) > \lambda_T \mid H_0\right)$.

ANALYSIS OF TRANSMISSION OVERHEAD AND AGILITY GAIN

In this chapter, transmission overhead and agility gain (AG) are analyzed for hard (Majority logic) fusion scheme. In the case of soft fusion scheme, a CR sends its local observation to FC as a soft data. Both the transmission overhead and agility gain can easily be analyzed for hard decision fusion where both are considered as a function of average number of transmitted bits. Thus, only hard decision fusion is considered for these analyses. In hard decision fusion, each CR sends its sensing observation as a one bit (either 1 or 0) to FC. In DTH based sensing, N_i CRs out of N CRs send decisions to FC. Let k_D be the number of normalized transmission overhead to FC which can be expressed as (Chunhua, 2007):

$$k_D = \frac{N_i}{N} = 1 - P\left(H_0\right)\Delta_0 - P\left(H_1\right)\Delta_1 \tag{30}$$

While censoring is applied after DTH, K CRs out of N_i CRs report to FC. Thus, the normalized transmission overhead for DTH with censoring can be represented as (Bhowmick, 2015b):

$$k_{D,C} = \frac{K}{N} \tag{31}$$

From Equations (30) and (31), one can easily find that $k_{D,C} \le k_D$ means if censoring is incorporated with DTH, the transmission overhead reduces since $K \le N_i$. The agility gain for DTH based sensing is defined as (Bhowmick, 2015b),

$$\mu_D = \frac{T_{STH}}{T_{DTH}} \tag{32}$$

where $T_{STH} = \tau + NT_{PC}$ and $T_{DTH} = \tau + N_iT_{PC}$. Here, τ is the local sensing time and T_{PC} is the time for polling each CR. Similarly, the agility gain for DTH based sensing with censoring can be defined as (Bhowmick, 2015b),

$$\mu_{D,C} = \frac{T_{STH}}{T_{DTH,Cen}} \tag{33}$$

where $T_{DTH,Cen} = \tau + KT_{PC}$.

Results and Discussions

Simulation results for DTH based censoring as obtained are discussed below. For all the subsequent plots default values of parameters are used as $P_f = 0.01$, $P(H_0) = P(H_1) = 0.5$, $T_{PC} / \tau = 0.05$and u = 5 if they are not mentioned. The performance of SS has been discussed in terms of missed detection probability, and network agility gain with respect to cooperative CR numbers and threshold of censoring, i.e., C_{th}. In the results,*S- SNR* and *R- SNR* indicateS-channel SNR and R-channel SNR. Analysis of results obtained for perfect channel estimation (PE) and imperfect channel estimation (IE) are also included.

The MDP for several failed sensing probabilities, *S-SNR* and several values of *R-SNR* has been investigated in Figure 3. The MDP decreases as the number of selected CR increases. The MDP can be reduced if secondary nodes employ DTH (b_0= 0.0001) based EDs instead of STH (b_0=0). The MDP decreases as *R-SNR* increases for a particular value of number of selected CR. Out of two different fusion schemes, soft fusion scheme is better as compared to the hard fusion scheme. In Figure 4, the agility gain has been investigated. The network agility gain improves as the probability of failed sensing increases. The MDP for Threshold based sensing has been studied in Figure 5. It is observed that there is an optimal value of threshold for which the probability of MDP is minimized. Agility gain for Threshold based censoring is studied in Figure 6 and observed that as the number of cooperating CRs increases network agility gain improves.

Once the sensing is done, if PU is found to be absent, CR transmits its own data. Thus, after sensing, a CR needs to access the spectrum. In this chapter, two SASs namely CHSAS and MHSAS are studied.

SPECTRUM ACCESS SCHEMES: A BRIEF INTRODUCTION TO UNDERLAY OVERLAY MODE

A CR node transmits in underlay mode else in overlay mode. In underlay mode, a CR shares the PU spectrum band. Thus, it has to control the power of transmission based on estimated CSI of the link of interference. On the other hand, in absence of PU means mode of operation is overlay, a CR exploits the spectrum band exclusively and transmits with high power.

Figure 3. MDPis as function of number of selected CRs on for RBC
Bhowmick, 2015b.

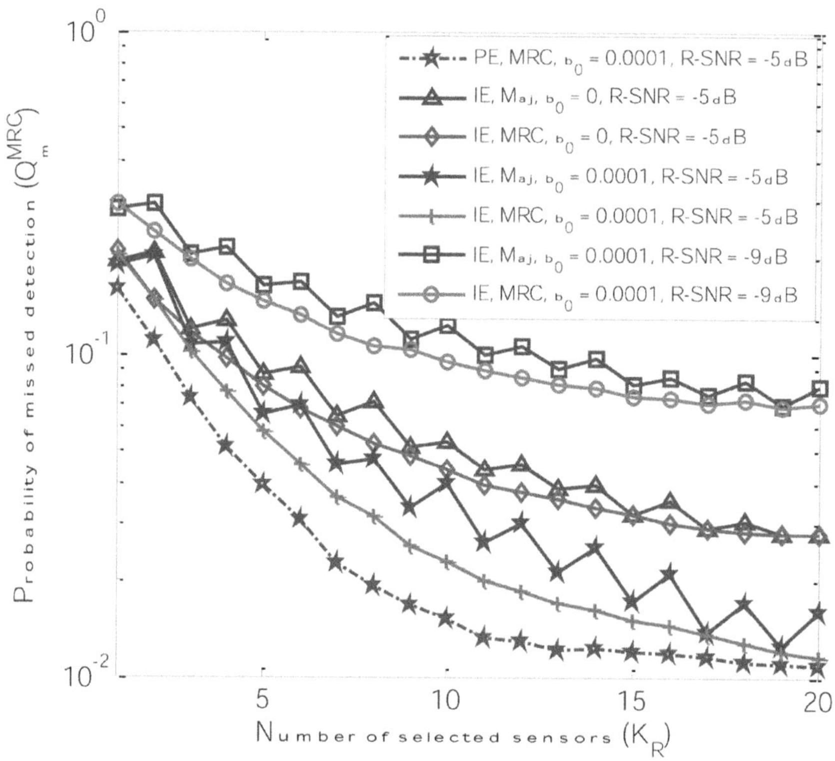

**For a more accurate representation see the electronic version.*

CHSAS and MHSAS Schemes

There are two phases in a detection cycle: S-phase and T-phase. In CHSAS, a CR uses the T- phase only for its information transmission, and the mode of transmission is either underlay or overlay. In MHSAS, a CR uses the spectrum for entire detection cycle. During the S-phase in MHSAS, sensing of PU and transmission are done simultaneously in underlay mode. It remains in same mode if PU is found to be present otherwise it switches to overlay transmission mode. Let us consider, the length of the detection cycle is T_f. A CR uses τ time for sensing and $\left(T_f - \tau\right)$ time for transmission purpose. Table 1 shows the mode of transmission of a CR for the considered schemes. In table, sensing probabilities are denoted by $P(x|y)$ where x is the observed data and y is the actual data. Here, in the case of $P(H_0|H_1)$, a CR transmits in overlay mode due to wrong observation and collide with PU data. Thus, the throughput for this case is not useful and it is avoided in the present chapter.

Analysis of Throughput and Outage Performance

Underlay Scheme

In underlay mode, CSI of the interfering link is used for allocation of transmission power to a CR such that the interference to PU remains within an acceptable limit. In a practice, due to error in estimation,

Figure 4. Impact of failed sensing probability on agility gain for RBC
Bhowmick, 2015b.

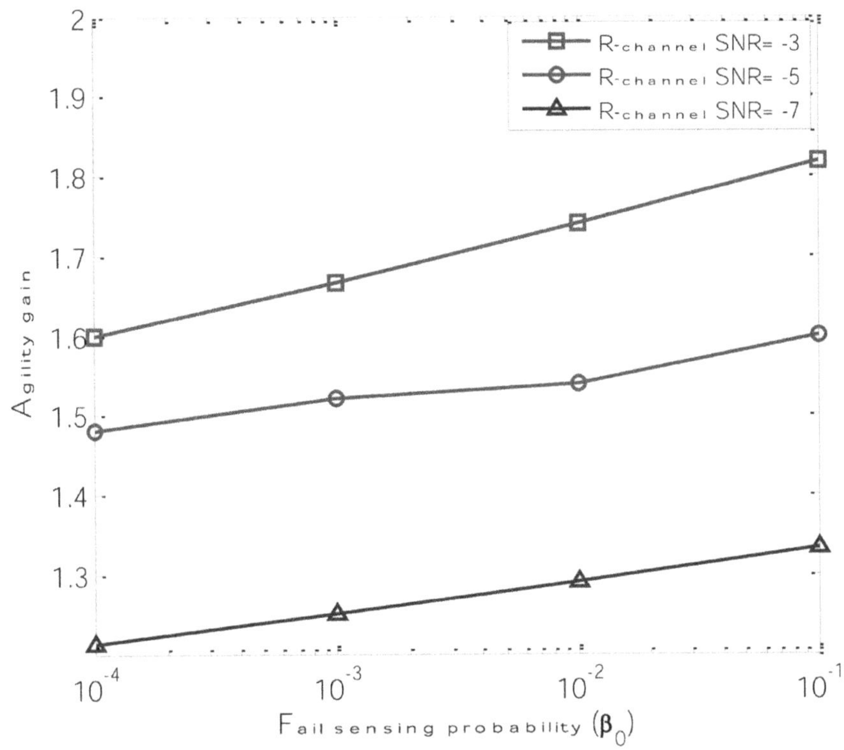

**For a more accurate representation see the electronic version.*

delay in feedback and mobility make it almost impossible for the system to get the perfect knowledge of CSI (h_{sp}) of the interfering link (Prasad, 2014). In this chapter, an imperfect channel estimation (Prasad, 2014) is considered where the channel coefficient is estimated imperfectly ($\hat{h}_{sp}$) and can be expressed as:

$$\hat{h}_{sp} = \rho h_{sp} + \left(\sqrt{1-\rho^2}\right)\varepsilon \tag{34}$$

Here, $\hat{h}_{sp}$ is the estimated CSI and h_{sp} is the perfect CSI of the interfering link, ε is with $CN \sim \left(0, \sigma_\varepsilon^{\ 2}\right)$ and is not correlated with h_{sp}. In (11), ρ is considered as correlation coefficient ($0 \leq \rho \leq 1$). For example, when $\rho = 0$, CSI at CR transmitter is imperfect completely while link CSI is perfect for $\rho = 1$. It may be noted that the gain of the estimated channel, i.e., $\hat{g}_{sp} = \left|\hat{h}_{sp}\right|^2$ is distributed exponentially with mean λ_{sp}.

As the information of imperfect CSI is used to decide CR transmission power, the maximum allowable instantaneous transmission power of a CR transmitter under a given threshold limit of peak interference (I_{th}) imposed by PU is given by,

Figure 5. Impact of threshold of censoring on probability of missed detection (TBC)
Bhowmick, 2015b.

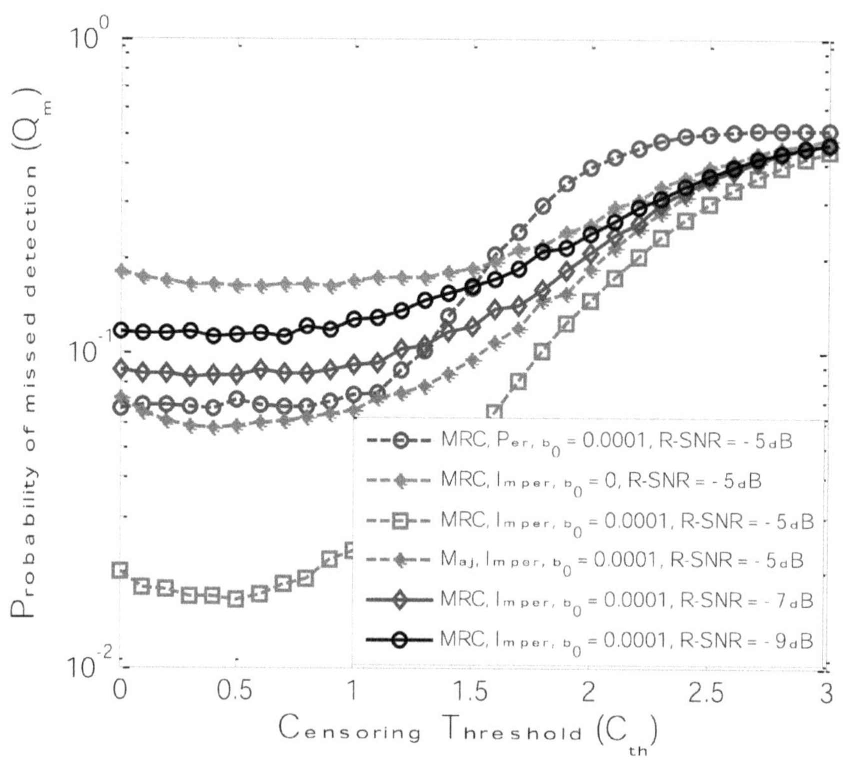

**For a more accurate representation see the electronic version.*

$$p_s = I_{th} / \hat{g}_{sp} \tag{35}$$

Including a maximum transmission power (P_{max}) constraint of CR, the CR transmit power can be modeled as (Prasad, 2016):

$$P_s = \min\left(\eta I_{th} / \hat{g}_{sp}, P_{\max}\right) \tag{36}$$

where η ($0 \leq \eta \leq 1$) is the scaling factor and it is used to control the CR transmit power and maintain the QoS of PU in terms of outage constraint. The transmit power of a CR may become high if the estimation of CSI is imperfect. In such case, the actual interference power may high and creates interference to PU which is unacceptable. In the case of imperfect CSI, the actual interference falling on PU receiver is $I_s = p_s g_{sp} = I_{th} g_{sp} / \hat{g}_{sp}$ where g_{sp} is the error free CSI of the interfering link. Thus, considering imperfect link CSI, the maximum allowable instantaneous transmit power of CR indicated in (36) can be modified as $p_s = \eta I_{th} / \hat{g}_{sp}$.

If CR transmission is governed by (35),the outage probability of PU is given by (Bhowmick, 2016a):

Figure 6. Variation of network agility gain for a number of CRs in TBC
Bhowmick, 2015b.

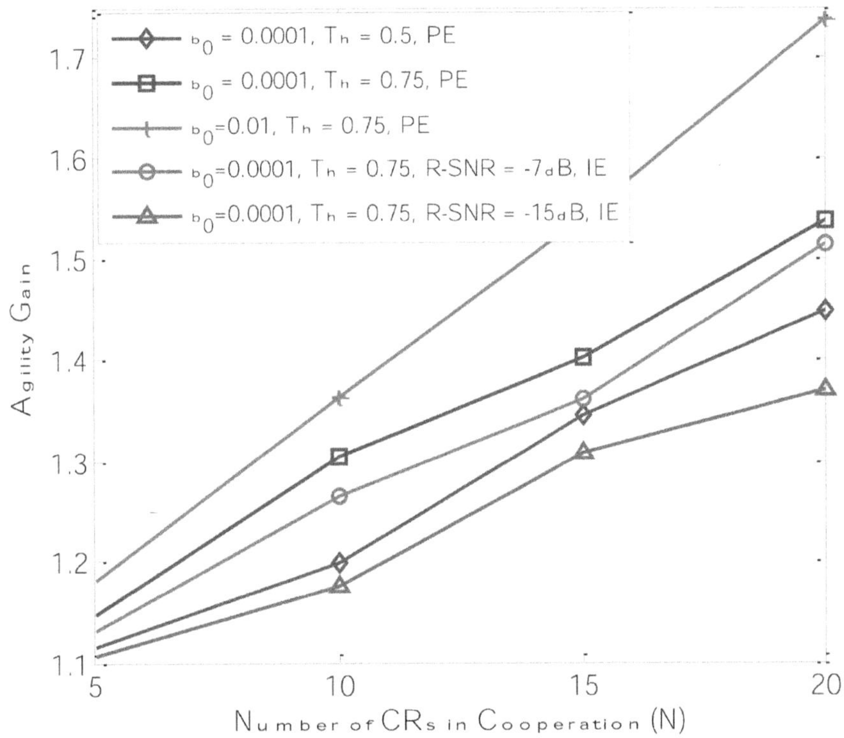

**For a more accurate representation see the electronic version.*

Table 1. Spectrum access mode for two schemes (CHSAS and MHSAS)

Spectrum Access Scheme	Sensing Time (τ)	Transmission Time (T_f-τ)			
		$P(H_0\|H_0)$	$P(H_1\|H_0)$	$P(H_0\|H_1)$	$P(H_1\|H_1)$
CHSAS	CR performs sensing only	Overlay mode	Underlay mode	X	Underlay mode
MHSAS	CR performs sensing as well as underlay transmission simultaneously	Overlay mode	Underlay mode	X	Underlay mode

$$\begin{aligned} P_{out}^{pu} &= P\{p_s g_{sp} > I_{th}\} = P\left\{\frac{\eta I_{th}}{\hat{g}_{sp}} g_{sp} > I_{th}\right\} \\ &= 1 - P\left\{\frac{g_{sp}}{\hat{g}_{sp}} \le \frac{1}{\eta}\right\} = 1 - F_{\frac{g_{sp}}{\hat{g}_{sp}}}\left(\frac{1}{\eta}\right) = \frac{1}{2}\left(1 - \frac{1-\eta}{\sqrt{\left(1+\eta\right)^2 - 4\rho^2\eta}}\right) \end{aligned} \tag{37}$$

where $F_x(.)$ denotes CDF of *x*. To keep the PU outage within a desirable limit, the probability of PU outage should maintain the condition as follows,

$$P_{out}^{pu} \leq \delta \tag{38}$$

For a given ρ, δ, and solving (37), one can obtain the maximum value of η .Thus, the average throughput of CR in underlay mode is given by,

$$R_{un} = \int_0^\infty \log\left(1 + p_s g_{ss} / \sigma_0^2\right) f_{g_{ss}}\left(g_{ss}\right) dg_{ss} = \int_0^\infty \log\left(1 + \min\left(\frac{\eta I_{th}}{\hat{g}_{sp}}, P_{\max}\right) g_{ss}\right) f_{g_{ss}}\left(g_{ss}\right) dg_{ss}$$

$$= \left[D_2 - D_3\right] / \log 2 \tag{39}$$

where

$$D_2 = \left(A + \frac{1}{(\alpha - 1)} \exp\left(-\frac{\eta I_{th}}{P_{\max}}\right)\right) \exp\left(\frac{\sigma_0^2}{P_{\max}}\right) Ei\left(-\frac{\sigma_0^2}{P_{\max}}\right),$$

$$D_3 = \frac{\exp\left(-\dfrac{(\eta I_{th} - \sigma_0^2 / \alpha)}{P_{\max}}\right)}{\alpha - 1} Ei\left(-\frac{\sigma_0^2}{\alpha P_{\max}}\right)$$

$$A = \left(1 - \exp\left(-\left(\eta I_{th}\right) / P_m\right)\right),$$

$$\alpha = \sigma_0^2 / \eta I_{th},$$

and $Ei(x)$ is an exponential integral function (Gradshteyn, 2007).

In underlay mode of transmission, the CR transmits for $(T_f - \tau)$ duration with a controlled transmission power. Hence, the average throughput can be expressed as,

$$R_1 = \left(1 - \tau / T_f\right) R_{un} \left\{P(H_1)\overline{Q}_d + P(H_0) Q_f\right\} \tag{40}$$

where the overall detection probability is given as, $Q_d = Q_{d,Maj}^R$ or $Q_{d,MRC}^R$ for Rank based censoring and $Q_d = Q_{d,Maj}^T$ or $Q_{d,MRC}^T$ for Threshold based censoring which are described in Equations (14), (24), (20) and (27). In (40), the targeted detection probability is considered, i.e., $\overline{Q_d}$. Here, Q_f is the false alarm probability where $Q_f = Q^R_{f,Maj}$ or $Q^R_{f,MRC}$ for Rank based censoring and $Q_f = Q^T_{f,Maj}$ or $Q^T_{f,MRC}$ for Threshold based censoring.

The outage probability of a CR considering the constraint in outage of PU is given by,

$$P_{out,un} = P\left\{\frac{T_f - \tau}{T_f}\log\left(1 + p_s g_{ss} / \sigma_0^2\right) < R_{th}\right\} \tag{41}$$

where R_{th} is the targeted rate of transmission. The CR-Rx suffers from outage while the capacity of channel exists between CR-Tx to CR-Rx falls below the targeted rate due to severe fading, it is considered that. Thus, the outage mentioned in (41), reduces to:

$$P_{out,un} = P\left\{p_s g_{ss} < R_{th}\right\} = P\left\{\min(\eta I_{th} / \hat{g}_{sp}) g_{ss} < R_{th}\right\} \tag{42}$$

where $R_{th} = \left(2^{RT_f/(T_f-\tau)} - 1\right)\sigma_0^2$, after some algebra and reducing (42), the probability of outage can be written as:

$$P_{out,un} = 1 - \exp\left(-\frac{R_{th}}{P_{\max}}\right)\left(1 - \exp\left(-\frac{\eta I_{th}}{P_{\max}}\right)\right) - \frac{1}{\left(D_1 + 1\right)}\exp\left(-\frac{\left(D_1 + 1\right)\eta I_{th}}{P_{\max}}\right) \tag{43}$$

where $D_1 = R_{th} / \eta I_{th}$. Hence,the outage in averagecan be written as,

$$P_{out,1} = P_{out,un}\left\{P(H_1)\overline{Q}_d + P(H_0)Q_f\right\} \tag{44}$$

Overlay Scheme

In overlay mode, CR transmission power is maximum, i.e., $P_{\max}$ over the entire transmission phase, i.e., $(T_f - \tau)$. The useful throughput can be obtained only in the case of $P(H_0 \mid H_0)$, i.e., $\left(1 - Q_f\right)$ while the throughput for the case of missed detection probability (Q_m) is useless since there is a collision between CR data and PU data. Thus, the throughput over the duration $(T_f - \tau)$ is given as (Liang, 2008):

$$R_2 = \left(1 - \tau / T_f\right) R_0 P(H_0)(1 - Q_f) \tag{45}$$

where

$R_0 = \left[\left(T_f - \tau\right) / T_f\right]\log_2\left(1 + \gamma\left|h_{ss}\right|^2\right)$ and $\gamma = P_s / \sigma_0^2$ and $P_s = \sigma_s^2$.

The outage probability at CR receiver is given by,

$$P_{out}^{CR_o} = P\left\{\frac{T_f - \tau}{T_f}\log_2\left(1 + P_s g_{ss} / \sigma_0^2\right) \le R_s\right\} = 1 - \exp\left[\left(2^{R_s T_f/(T_f - \tau)} - 1\right)\sigma_0^2 / P_{\max}\right] \quad (46)$$

Thus, the outage in average for overlay transmission mode is given by,

$$P_{out,2} = P(H_0)(1 - P_f(\tau))P_{out}^{CR_o} \quad (47)$$

Analysis of Overall Outage and Throughput for CHSAS

In CHSAS, the overall average throughput (combination of average throughput for both the schemes) is given by,

$$R_{CHSAS} = \left(1 - \tau / T_f\right)\left[R_{un}\left\{P(H_1)\overline{Q}_d + P(H_0)Q_f\right\} + R_0 P(H_0)(1 - Q_f)\right] \quad (48)$$

and the overall average outage can be expressed as,

$$P_{out,CHSAS} = P_{out,un}\left\{P(H_1)\overline{Q}_d + P(H_0)Q_f\right\} + P(H_0)(1 - P_f(\tau))P_{out}^{CR_o} \quad (49)$$

Analysis of Overall Outage and Throughput for MHSAS

In MHSAS, the CR uses both the phases of a detection cycle, i.e., S-phase and T-phase for its transmission. In the previous section, transmission of information (data) during T-phase is discussed already. During S-phase, since the CR is not aware about the status of PU, it controls its transmit power (i.e., the CR works in underlay mode of transmission). Thus, the throughput for the S-phase can be expressed as,

$$R_3 = R_{un} \quad (50)$$

Thus, the throughput achievable in sensing phase and transmission phase can be combined to get the overall average throughput for MHSAS. Hence, in MHSAS, the overall average throughput is given by,

$$R_{MHSAS} = R_{CHSAS} + R_3 \quad (51)$$

Similarly, the outage at CR-Rx during S-phase is given by,

$$P_{out,3} = P_{out}^{CR_u} \tag{52}$$

Thus, at CR-Rx the overall outage can be written as,

$$P_{out,MHSAS} = P(\tau)P_{out,3} + P(T_f - \tau)P_{out,CHSAS} \tag{53}$$

where it is assumed that $P\left(\tau\right)$ and $P\left(T_f - \tau\right)$ follow the linear distribution and $P(\tau) = \tau / T_f$, is the probability that a CR remain in the S-phase and $P(T_f - \tau) = (T_f - \tau) / T_f$, is the probability that a CR remain in the T-phase. Thus, overall outage indicated in Equation (53) reduces to:

$$P_{out,MHSAS} = \frac{\tau}{T_f} P_{out,3} + \frac{(T_f - \tau)}{T_f} P_{out,CHSAS} \tag{54}$$

Results and Discussions

A simulation is carried out to study the outage and throughput performance of the considered network scenario. The parameters used are: $f_s = 6MHz$, $\sigma_0^{\ 2} = 1$, $\sigma_s^{\ 2} = 1$, $\alpha_0 = 0.64$, $\alpha_1 = 0.36$, $T_f = 100ms$, $R_{th} = R_s = 1$, $\bar{P}_d = 0.9$, outage at PU receiver is 1%, i.e., $P_{out}^{pu} = 0.01$, SNR = -16 dB and $R_0 = 6.6582$.

Figure 7 shows that the outage at CR receiver decreases as the sensing time increases. For a particular sensing time, the outage probability decreases as the number of CRs participating in double threshold based detection and the value of scaling factor increases. In Figure 8, outage probability is shown as a function of interfering threshold. It is found that as the value of interfering threshold increases, the outage probability decreases. If interfering threshold increases, the CR transmit power increases which in turn reduces the outage.

Figure 9 shows the outage performance for several values of ρ. As ρ increases, the outage decreases. In Figure 10, throughput performance for MHSAS scheme is studied for several values of scaling factor. For particular value of scaling factor, the network throughput increases as the number of CRs after censoring increases. Similarly, for a fixed number of CRs, increase scaling factor improves throughput.

Figure 11 shows the throughput as a function of sensing time. The network throughput initially increases as the sensing time increases; it reaches to a maximum value; thereafter it decreases as the sensing time increases further. It is also found that the network throughput increases as the number of CRs taking part in censoring increases. As the number of CRs increases, sensing error decreases which results in throughput increment. It is observed for both the performance metrics (outage and throughput), MHSAS outperforms the CHSAS.

Figure 7. Outage performance with respect to sensing time

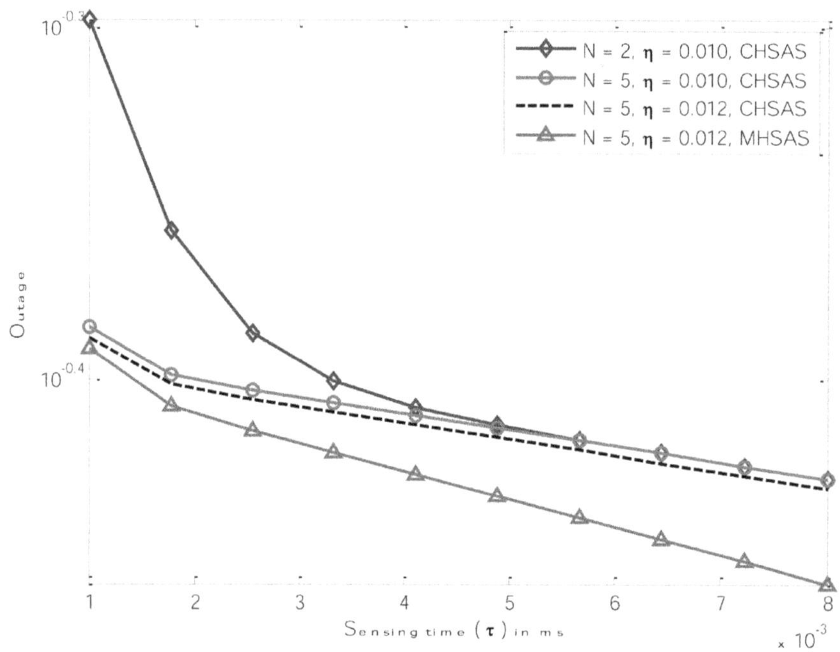

**For a more accurate representation see the electronic version.*

CONCLUSION

DTH based sensing with censoring of CRs for a CSS has been studied where the S-channels and R-channels are Rayleigh faded. Each CRs perform SS with DTH based EDs and are censored on the basis of R-channel quality. Two censoring schemes namely *RBC* and *TBC* have been studied. In spectrum access, two hybrid spectrum access schemes (CHSAS and MHSAS) are studied. It is observed that in fusion schemes, the performance of MRC is better as compared to the majority fusion rule and in spectrum access, MHSAS outperforms the CHSAS. If the failed sensing probability, number of CRs, and SNR of R-channel increases, the sensing performance improves. Imperfection in CSI has significant effect on detection performance. The use of DTH also improves the network agility. It is found, outage probability decreases as the sensing time, number of CRs in cooperation, scaling factor, correlation coefficient, and interfering threshold increases. It is also observed that the network throughput improves if the number CRs taking part in censoring and scaling factor increases. The study is useful in designing a HSAS based CR network employing an efficient spectrum sensing scheme.

Figure 8. Outage performances as a function of interfering threshold

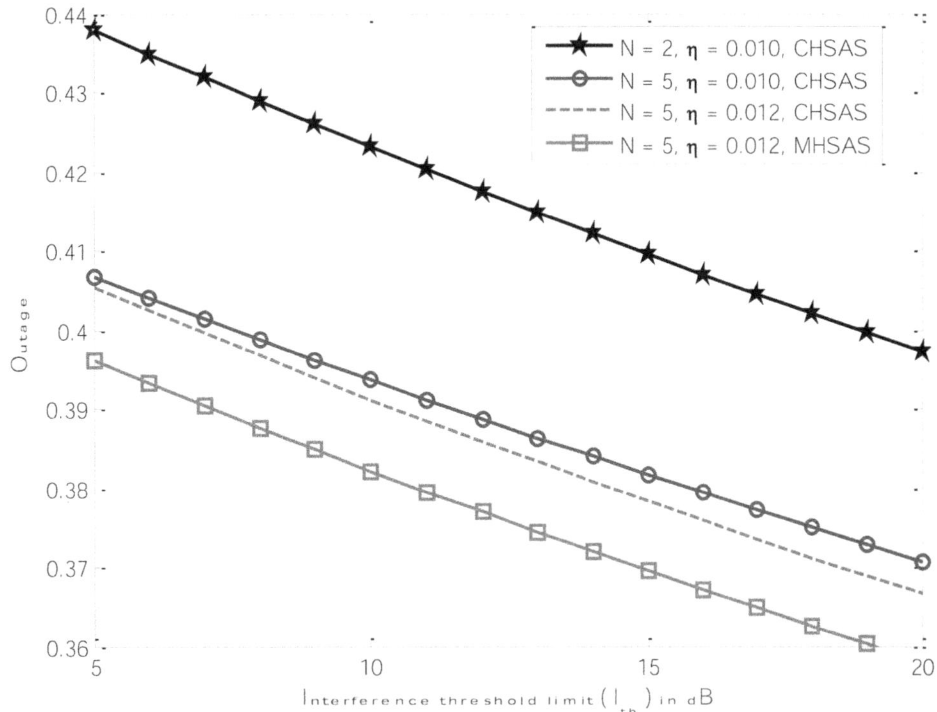

**For a more accurate representation see the electronic version.*

Figure 9. Outage performances for several correlation coefficients

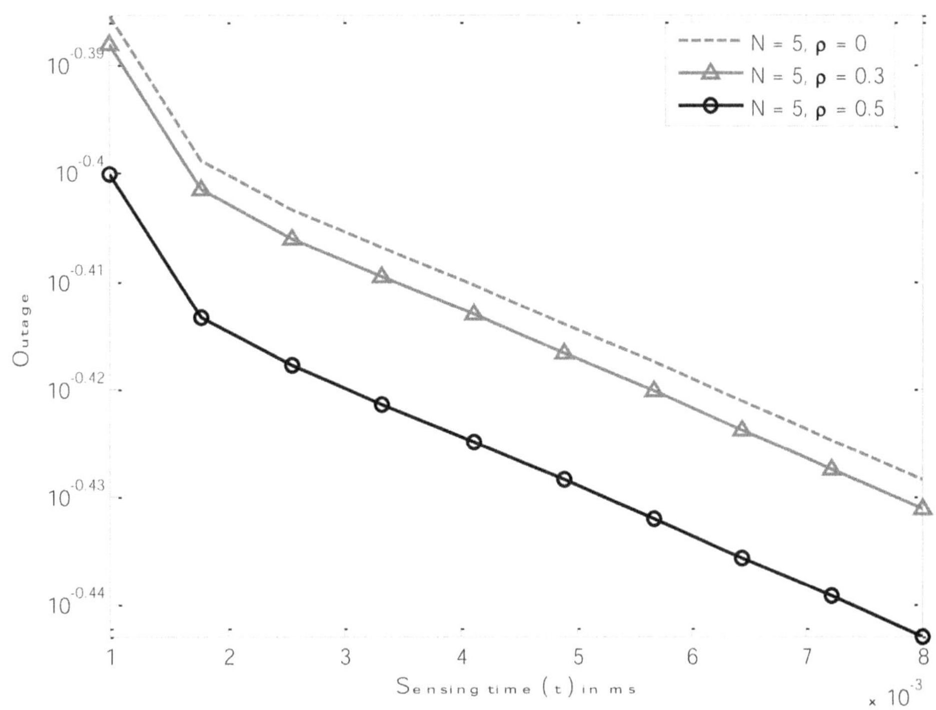

**For a more accurate representation see the electronic version.*

Figure 10. Variation in throughput for a number of CRs after censoring

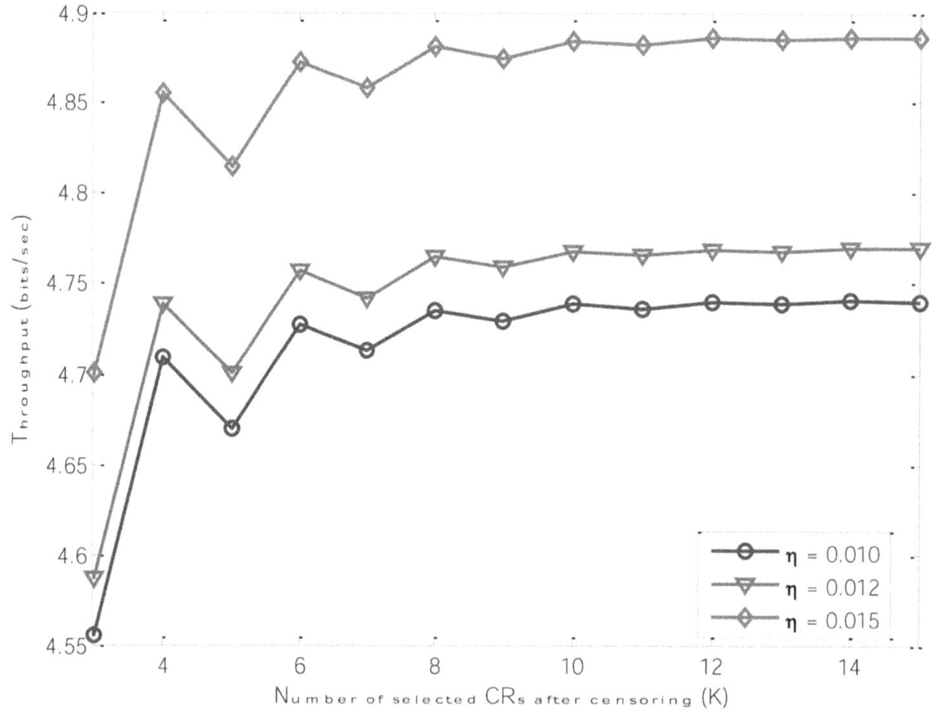

**For a more accurate representation see the electronic version.*

Figure 11. Variation of throughput with sensing time

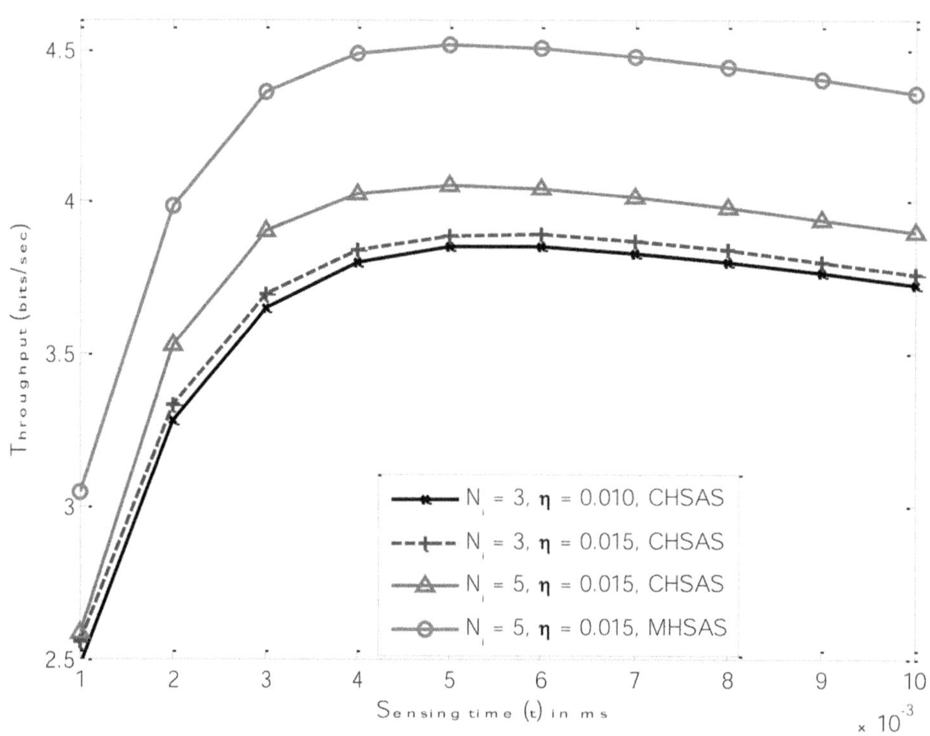

**For a more accurate representation see the electronic version.*

REFERENCES

Bhowmick, Abhijit, Nallagonda, Srinivas, Sanjay, & Sumit. (2014). Spectrum sensing with censoring of double threshold based cognitive radios in rayleigh fading. In *Communications (NCC), 2014 Twentieth National Conference on* (pp. 1-6). IEEE.

Bhowmick, A., & Sanjay, D. R. (2016). Sensing throughput trade-off for an energy efficient cognitive radio network under faded sensing and reporting channel. *International Journal of Communication Systems*, *29*(7), 1208–1218. doi:10.1002/dac.3087

Bhowmick, Abhijit, Binod, Sanjay, & Sumit. (2016). Performance of Cognitive Radio Network with Novel Hybrid Spectrum Access Schemes. *Wireless Personal Communications*, *91*(2), 541–560. doi:10.100711277-016-3476-5

Bhowmick, Abhijit, Nallagonda, Srinivas, Sanjay, & Sumit. (2015). Cooperative Spectrum Sensing with Double Threshold and Censoring in Rayleigh Faded Cognitive Radio Network. *Wireless Personal Communications*, *84*(1), 251–271. doi:10.100711277-015-2606-9

Chen, X., Jing, T., Huo, Y., Li, W., & Cheng, T. (2012). Achievable Transmission Capacity of Cognitive radio Network with cooperative relaying. *Proc. IEEE CROWNCOM*, 1-6.

Chunhua, S., & Wei, Z., & Ben.(2007). Cooperative spectrum sensing for cognitive radios under bandwidth constraints. In *Wireless Communications and Networking Conference, 2007.WCNC 2007. IEEE* (pp. 1-5). IEEE.

Dhar Roy, S., Bhowmick, A., Kundu, S., & Chandra, A. (2015). Double threshold-based cooperative spectrum sensing for a cognitive radio network with improved energy detectors. *IET Communications*, *9*(18), 2216–2226. doi:10.1049/iet-com.2014.1098

Ghasemi, A., & Sousa, E. S. (2005). Collaborative spectrum sensing for opportunistic access in fading environments. *Proc. of 1st IEEE Symp. New Frontiers in Dynamic Spectrum Access Networks*, 131-136.

Gradshteyn, I. S., & Ryzhik, I. M. (2007). *Table of integrals, series and products*. Elsevier.

Haykin, S. (2005). Cognitive radio: Brain empowered wireless communication. *IEEE Journal on Selected Areas in Communications*, *23*(2), 201–220. doi:10.1109/JSAC.2004.839380

Kang, X., Garg, H., Liang, Y.-C., & Zhang, R. (2010). Optimal power allocation for OFDM-based cognitive radio with new primary transmission protection criteria. *IEEE Transactions on Wireless Communications*, *9*(6), 2066–2075. doi:10.1109/TWC.2010.06.090912

Liang, Y.-C. (2008). Sensing-throughput tradeoff for cognitive radio networks. *IEEE Transactions on Wireless Communications*, *7*(4), 1326–1337. doi:10.1109/TWC.2008.060869

Musavian, L., & Aïssa, S. (2007). Ergodic and outage capacities of spectrum sharing systems in fading channels. *Proc. IEEE GLOBECOM*, 3327–3331.

Musavian, L., & Aissa, S. (2009). Fundamental capacity limits of cognitive radio in fading environments with imperfect channel information. *IEEE Transactions on Communications*, *57*(11), 3472–3480. doi:10.1109/TCOMM.2009.11.070410

Nallagonda, Sanjay, Sumit, Ferrari, & Raheli. (2013a). Cooperative spectrum sensing with censoring of cognitive radios in Rayleigh fading under majority logic fusion. *Proc. IEEE NCC*, 1-5. 10.1109/NCC.2013.6487926

Nallagonda, Sanjay, Sumit, Ferrari, & Raheli. (2013b). Performance of MRC fusion based cooperative spectrum sensing with censoring of cognitive radios in Rayleigh fading channel. *Proc. IEEE IWCMC*, 30- 35.

Nallagonda, S., Roy, S. D., & Kundu, S. (2013). Performance evaluation of cooperative spectrum sensing scheme with censoring of cognitive radios in Rayleigh fading channel. *Wireless Personal Communications*, *70*(4), 1409–1424. doi:10.100711277-012-0756-6

Nallagonda, Srinivas, & Sanjay, Sumit, Ferrari, & Raheli. (2017). Censoring-based Cooperative Spectrum Sensing with Improved Energy Detectors and Multiple Antennas in Fading Channels. *IEEE Transactions on Aerospace and Electronic Systems*.

Peha, J. A. (2005). Approaches to spectrum sharing. *IEEE Communications Magazine*, *43*(2), 10–12. doi:10.1109/MCOM.2005.1391490

Prasad, B. Sanjay, & Sumit. (2014). Outage and SEP of secondary user with imperfect channel estimation and primary user interference. *Electronics, Computing and Communication Technologies (IEEE CONECCT), 2014 IEEE International Conference on*, 1-6.

Prasad, B., Bhowmick, A., Dhar Roy, S., & Kundu, S. (2016). Performance of cognitive relay network with novel hybrid spectrum access schemes with imperfect CSI. *International Journal of Communication Systems*, *29*(11), 1761–1776. doi:10.1002/dac.3142

Stotas, S., & Nallanathan, A. (2011, October). Enhancing the capacity of spectrum sharing cognitive radio networks. *IEEE Transactions on Vehicular Technology*, *60*(8), 3768–3779. doi:10.1109/TVT.2011.2165306

Suraweera, H. A., Smith, P. J., & Shafi, M. (2010). Capacity limits and performance analysis of cognitive radio with imperfect channel knowledge. *IEEE Transactions on Vehicular Technology*, *59*(4), 1811–1822. doi:10.1109/TVT.2010.2043454

Chapter 11
Spectrum Sensing and Identification Techniques of Cognitive Radio Networks

Rashmi M. Kittali
Basaveshwar Engineering College, India

Ashok V. Sutagundar
Basaveshwar Engineering College, India

ABSTRACT

Cognitive radio emerged as one of the key enabling technologies of dynamic spectrum access, addressing the problem of inefficient usage of the available radio spectrum. It aims at providing more effective and convenient communication capabilities. Cognitive radio which is considered as the key technology for future mobile computing and wireless communications can form cognitive radio networks (CRN) by extending the features of radio link to network layer functions and above. The cognitive radio systems have the ability to exploit the spectrum holes by detecting and using them in an opportunistic manner through spectrum sensing. This chapter will brief the various architectures of cognitive radio networks, terminal capabilities of cognitive radio as nodes of CRN highlighting the spectrum sensing ability and different methods of it, and some applications of CRN.

INTRODUCTION

Today, the production of new radio to get advancements is restricted by the deficiency of the accessible radio spectrum. These new innovations are ending up noticeably demanding because of their higher rate prerequisites. Cognitive radio systems and spectrum detecting strategies are a characteristic approach to enable these new advancements to be employed.

The need for higher information rates is expanding therefore of the move from voice-communication to sight and sound sort applications. Given the constraints on the characteristic frequency range, it winds up plainly indicating that the current static frequency allotment plans cannot oblige the prerequisites of an expanding number of higher information rate gadgets. Thus, creative procedures that can offer bet-

DOI: 10.4018/978-1-5225-5354-0.ch011

ter approaches for effectively utilizing the accessible spectrum are required. Cognitive radio emerges to be a fascinating answer for the ghostly blockage issue by presenting entrepreneurial utilization of the spectrum groups that are not vigorously involved by authorized clients (Mitola, & Maguire, 1999; FCC, 2003). While there is no concession to the formal meaning of cognitive radio starting at now, the idea has advanced as of late to incorporate different implications in a few situations. In this chapter, the author utilizes the definition embraced by Federal Correspondences Commission (FCC):

Cognitive Radio: A radio or system that senses its operational electromagnetic environment and can dynamically and autonomously adjust its radio operating parameters to modify system operation, such as maximize throughput, mitigate interference, facilitate interoperability, access secondary markets. (FCC, 2003)

Henceforth, one primary part of cognitive radio is identified with independently making use of locally unused spectrum i.e., spectrum holes as in Figure 1 to give new ways for utilizing it in a better way.

A standout amongst the most significant component of the cognitive radio idea is the capacity to quantify, sense, learn, and know about the parameters identified with the radio channel qualities, accessibility of range and power, radio's working condition, client prerequisites and applications, accessible systems (foundations) and hubs, neighborhood arrangements and other working confinements. In cognitive radio diction, primary clients can be characterized as the clients who have higher need or inheritance rights on the use of a particular piece of the spectrum whereas secondary clients, who have down need, employ this range of frequencies without disturbing primary clients. In this way, secondary (optional) clients need cognitive radio abilities, for example, detecting the spectrum dependably to check whether it is being utilized by a primary client and to change the radio parameters to exploit the unused piece of the spectrum.

Being the main objective of this chapter, spectrum detection by a wide margin is the most vital segment for the foundation of cognitive radio. Spectrum sensing is nothing but the process of acquiring knowledge about the spectrum utilization and presence of primary clients in a terrestrial range. This information can be acquired by utilizing geolocation and database, by utilizing reference signals, or by neighborhood spectrum detecting at cognitive radios (Marcus, 2005; Zhao, Morales, Gaeddert, Bae, Um, & Reed, 2007). Whenever reference signals are utilized, the transmitted data can be mixture of spectrum and also other propelled components, for example, channel quality.

Figure 1. Illustration of spectrum holes

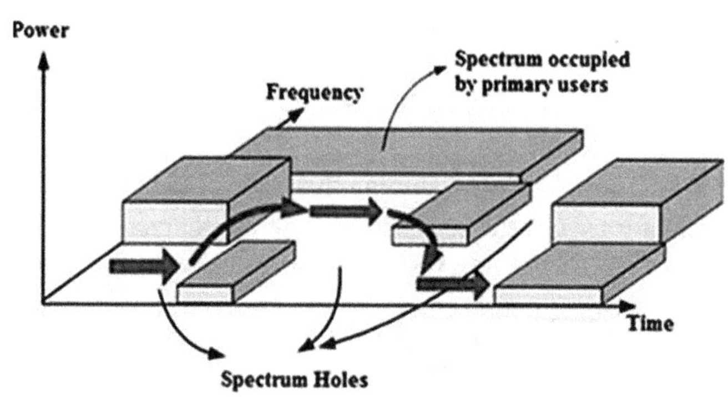

In spite of the fact that spectrum detecting is generally comprehended as measuring the range of spectrum available or on the other hand measuring the radio frequency energy over the range; at the point when cognitive radio is thought of it as, is a more broad term that includes acquiring the range use qualities over numerous measurements, for example, time, space, frequency, and code. It likewise includes figuring out what sorts of signals the spectrum is composed of along with the type of modulation, waveform, data transmission, frequency, and so on. Despite, this requires more capable signal examining systems with extra computational intricacy.

To avoid the problem of interference, cognitive radio networks have to follow some set of policies brought by regulatory agencies (Haykin, 2005; Akyildiz, Lee, Vuran, & Mohanty, 2006). These mainly deal with the problem of interference management by regulating the amount of disturbance the secondary system can incur on primary ones. To treat this, solution can be given in two ways: receiver centric and transmitter centric interference management.

Receiver Centric Interference Management

In this, a threshold value which is nothing but the worst interference value that will no longer harm the operation of receiver is set for the interference at the receiver side to restrict on the power transmissions around it. Though this method seems useful but the job of selecting appropriate threshold is a tricky job and depends on different locations, scheme of coding, modulation etc.

Transmitter Centric Interference Management

Here, the focus is on interference source. Although it is not possible for the transmitter to locate the exact position of receiver, some effort is done by spectrum sensing to identify how much of distortion these receivers can tolerate. Thus, sensing acts as a novel solution which allows secondary systems to communicate only when primary channel is empty.

In this chapter we mainly focus on spectrum sensing for cognitive radio networks. Firstly, we will begin by giving brief background knowledge regarding the problem identified. Then we move on to various architectures of cognitive radio networks. Later terminal capabilities are discussed highlighting the spectrum sensing ability. Lastly, different methods of it are discussed in detail.

BACKGROUND

Spectrum detecting is the key usefulness of cognitive radio systems. Numerous narrowband range detecting calculations, including energy recognition, matched filtering and cyclostationay highlight discovery calculations have been highlighted in (Yucek & Arslan, 2009). Energy recognition is the most widely recognized and surely understood spectrum detecting technique (Sun, Laurenson, & Wang, 2010). All these narrowband spectrum sensing methods plan to distinguish between narrow and wide frequency ranges to take correct decision of complete spectrum Cognitive radio systems at last requires to exploit large frequency values extended from many megahertz to a few gigahertz for accomplishing higher throughput. This request prompts wideband range detecting strategies. Ordinary wideband spectrum detecting systems in view of standard simple to-advanced converter (ADC) prompt high examining rate or complexity of implementation (Sun, Nallanathan, Wang, & Chen, 2013); accordingly, progressive wideband range detecting procedures, for example, sub-Nyquist wideband range detecting or compressive detecting turn out to be progressively significant for future cellular network generation.

Cognitive Radio Networks Architecture

The architecture of the CRNs comprises of mobile station (MS), base station/access point (BS/ AP) and backbone/core networks. These three basic components together will compose three kinds of network architectures in CRNs:

1. Network architectures
2. Links in CRN
3. IP mobility management in CRN

In this chapter, emphasis will be mainly on three different network architectures:

1. **Infrastructure-Based:** A typical infrastructure architecture of cognitive radio networks is shown in Figure 2 where the access to BS/AP by a MS is in a single hop manner. Good communication is achieved for MSs in the transmission range of BS/AP whereas for other cells backbone networks come into picture. A terminal in such network is capable of managing various types of communication systems through BS capable of handling single or multiple protocols at a time.
2. **Ad-Hoc Architecture:** This infrastructure less architecture where network setup is done on the fly as depicted in Figure 3. It can be formed by many MSs which will communicate through standard protocols like Wi-Fi, Bluetooth or by dynamically utilizing the spectrum holes. The communication link between them is provided by different technologies.
3. **Mesh Architecture:** It is the combination of both the architectures discussed before as shown in Figure 4. It resembles the hybrid wireless mesh network where BSs/APs work as routers and backbones. MSs can communicate either by a single hop communication through BS or by multi hop through relay nodes.

Figure 2. Infrastructure architecture of CRN

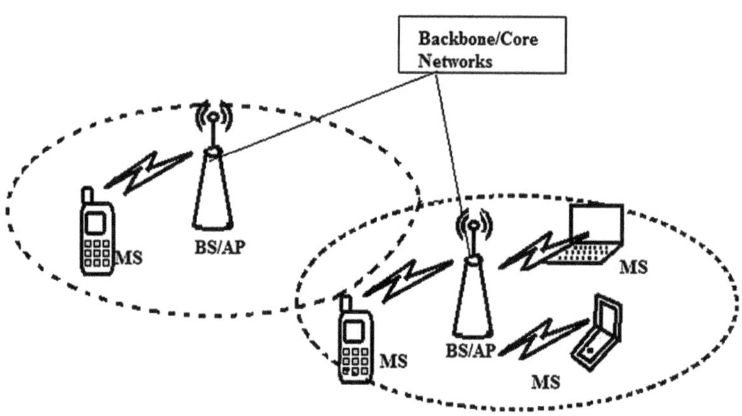

Figure 3. Ad-hoc architecture of CRN

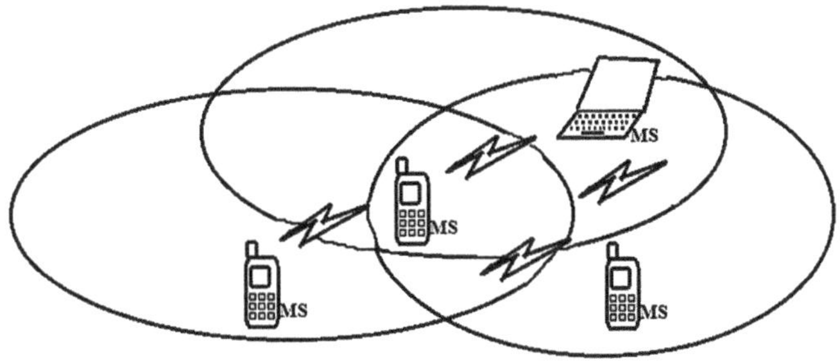

Figure 4. Mesh architecture of CRN

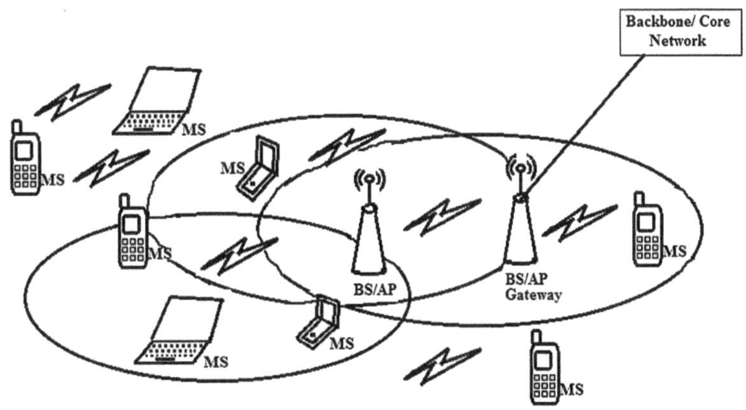

TERMINAL CAPABILITIES OF CRN

In a cognitive radio cycle as described in Figure 5, a cognitive radio screens spectrum groups and catches their data and afterward recognizes spectrum spaces. The qualities of these spaces are evaluated and proper band is picked by its attributes and client necessities. Once working spectrum band is resolved, the communication can be performed over this band of frequencies.

From the definition of cognitive radio networks, three main characteristics of cognitive radio can be defined:

1. **Cognitive Capability:** Interacting with radio environment in real time helps in identifying the unused spectrum portions or holes. Consequently, best usage can be made by properly selecting the spectrum, sharing with others without causing any interference to primary user. The four main points in this characteristic are as follows:
    a. **Spectrum Sensing:** It refers to the process of detecting the unused spectrum and sharing it without causing interference to already existing users. The main motto of any cognitive network is to identify the spectrum holes and primary users in a more efficient way.
    b. **Spectrum Management:** Selecting the perfect available spectrum to deal with the communication requirements of consumers.

Figure 5. Cognitive radio cycle

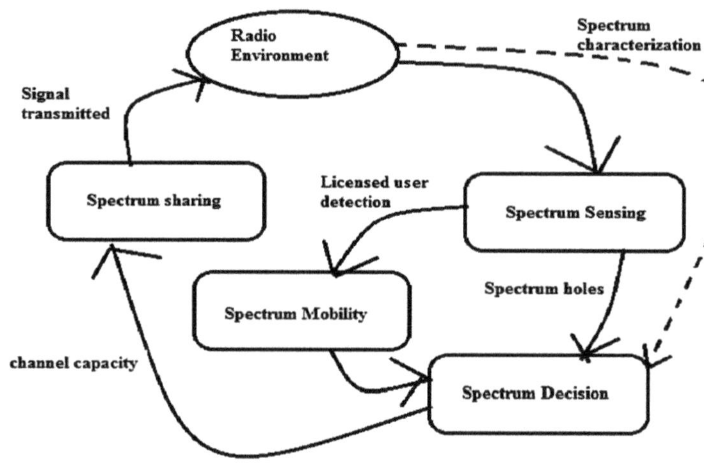

   c. **Spectrum Mobility:** The process where the user in network is capable of exchanging the frequency of operation accordingly with changing network environment.
   d. **Spectrum Sharing:** Distributing the spectrum among the registered users in a scheduled manner. In fact, it the open challenge in the field of spectrum usage.
2. **Reconfigurability:** It is possible to program the cognitive radio to operate on a wide range of frequencies adopting different technologies for accessing in accordance to its hardware design. Through this feature, it is possible to select the finest band of frequencies and suitable operating parameters which can be reconfigured.
3. **Self – Organized Capability:** As intelligence is imparted into all communication devices, cognitive radios must be in a position to organize them based on varying sensing and reconfigurable functions.

SPECTRUM SENSING

Spectrum sensing is the ability to sense, measure and get to know the parameters related to characteristics of radio channel, its operating environment, spectrum availability, transmits power, interference and noise, user requirements for different applications, availability of nodes and networks (infrastructures), local policies and other operating restrictions. This process can be carried out across frequency, time, phase, code and geographical space.

Spectrum sensing methods can be broadly classified as: transmitter/non-cooperative detection, cooperative detection and interference based detection (Cabric, Mishra, & Brodersen, 2004). Each of these methods in turn has various techniques being used.

Before getting into thick of things, let's know some of the challenges posed in the process of spectrum sensing for cognitive radio.

1. Hardware Requirements

There is a need for high speed signal conditioners and processors, analog to digital converters with high resolution, more dynamic range and high sampling rates. Normally noise and interference estimation problem is not an issue for receivers that are tuned for a particular bandwidth. However, in cognitive radio, it is not the case as end terminals need to process very large band of frequencies imposing additional need of huge antennas and power amplifiers. Further, to handle large computational complexity DSPs and FPGAs with high speed are necessary.

2. Hidden User Problem

This is similar to node problem in case of Carrier Sense Multiple Access (CSMA) which can be caused by multipath fading to a greater extent or by shadowing caused by secondary users while detecting the primary user region of operation. Figure 6 shows hidden node problem where the operating ranges are shown by dashed lines. CR causes huge unwanted interference due to receivers inability to detect transmitter signal because of problem of location of devices. This problem is proposed in literature (Ganesan & Li, 2005; Cabric, Tkachenko, & Brodersen, 2006).

3. Detecting Primary Users of Spread Spectrum

It is very difficult to identify the primary users who make use of spread spectrum technology as their power is distributed over a wide range of frequencies. There is solution to this problem only if there is known hopping pattern and accurate synchronization among signals.

4. Sensing Duration and Frequency

The time taken by cognitive radio to identify the presence of primary users and vacant the occupied frequency values should be as small as possible to avoid interference with license owners. This time

Figure 6. Illustration of hidden user problem

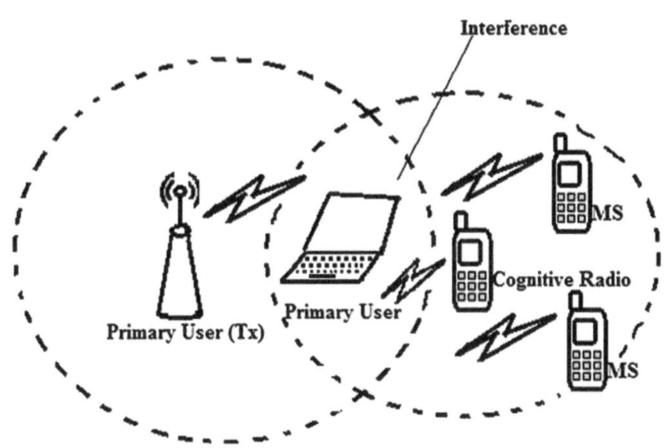

limit poses a major restriction on full fledge utilization of applied sensing technique and leads to a new challenge in this area. Tradeoff between time for sensing and reliability depends on sensing parameters.

Choosing of temporal characteristics of licensed user in changing environment depends on optimum capabilities of cognitive radio.

5. Security

In cognitive radio, there are high chances where an unintended user can capture the primary user properties misleading the spectrum sensing process and also creating missed opportunities for secondary users. This leads to various attacks on layers of cognitive radio as depicted in Figure 7.

We can classify the spectrum sensing techniques into two broad categories: One is direct method also termed as frequency domain where for frequency estimation signal is directly approached. The other is indirect method (time domain) in which estimation is done through autocorrelation of signal. Figure 8 gives the overall view of types of techniques present.

6. Narrow Band Sensing Method

Here opportunities for secondary users to sense are over a narrow frequency range which means the demanding bandwidth is less than the channel bandwidth. It consists of three types as explained in following sections.

a. Energy Detection

It is the blind detection technique which does not require any knowledge about primary user. The principle involved here is detecting the existence of any user signal by comparing received energy with threshold value set. Assumption is made that detecting signal energy will be always higher than the noise energy. But, the process of fixing the threshold value is a tricky job.

Figure 7. Cognitive radio attacks by layer

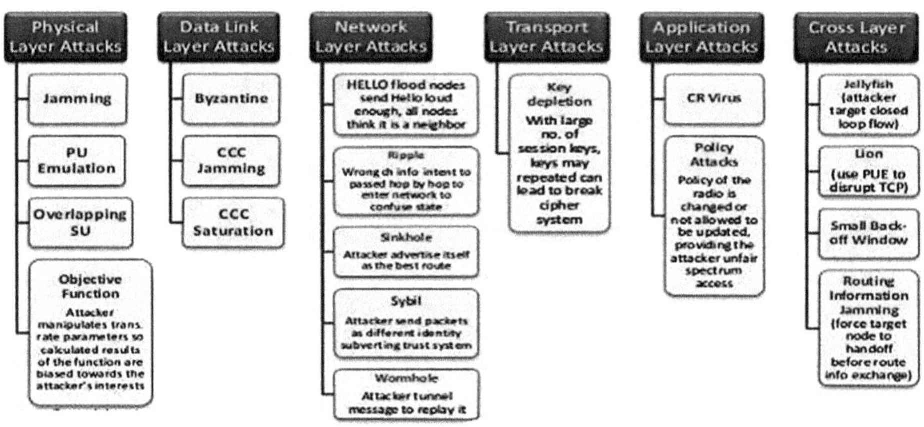

Figure 8. Varieties of spectrum sensing methods

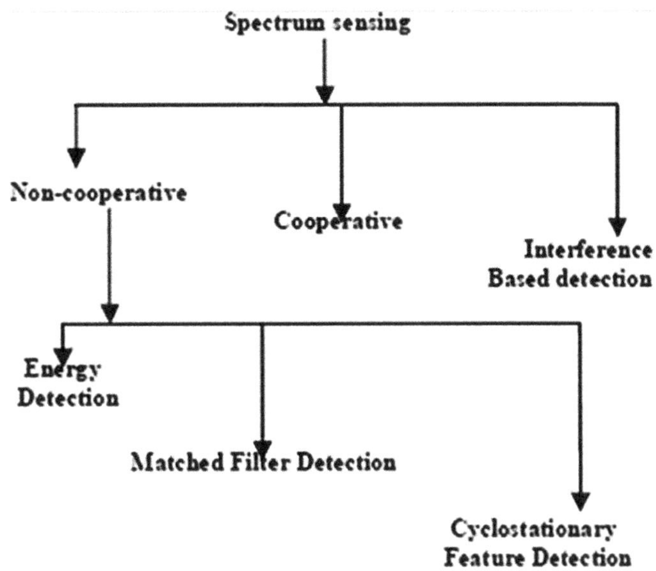

The block diagram shown in Figure 9 has four blocks. Band pass filter which passes only specific band of frequencies intended for use. Squaring device gives out the energy which is moved to integrator block to identify interval T. Output from integrator is compared with threshold value to detect the existence of licensed user.

Advantages

- No need of prior information of primary user
- Less cost
- Ease of implementation

Disadvantages

- More time consuming to reach better results
- Failed to differentiate among users
- Not suitable for low SNR values

Figure 9. Energy Detection Block Diagram

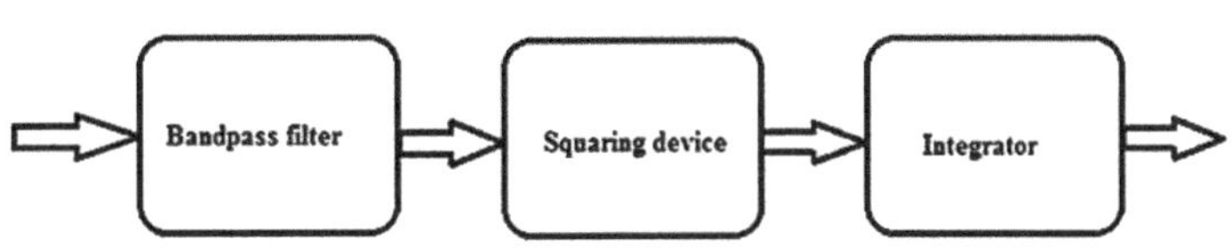

b. Matched Filter Detection

It is non- blind decision-making circuit which will separate out signal from noise components by suppressing them. This develops a contrasting output of noise and signal leading to correct segregation with minimized errors. If a signal is present then there appears a peak in the output and such a filter which do highlight signal over noise is to be designed. Matched filter is one such solution which maximizes signal amplitude by squaring over noise amplitude. This requires the prior knowledge of primary user.

From Figure 10 we observe that first and last blocks are similar to that of energy detection method which does the same job of passing out only required frequency signals and comparing the filter outputs with threshold values respectively. The functioning of matched filter is in resemblance with the correlation process where unknown signal is correlated with filter impulse response and time shifted version of original signal as expressed in Equation 1.

$$Y[n] = \Sigma^{\infty}_{k=-\infty} h[n-k] * x[k] \quad (1)$$

where 'x' is unknown signal and h is the impulse response of matched filter.

Advantages

- Good method when noise is present
- Less detection time
- Ideal performance

Disadvantages

- Needs information of primary signal well in advance
- Requires dedicated radio for each user

c. Cyclostationary Feature Detection

It relays on the principle that since all signals possess some of the characteristics that are periodic in nature as in case of modulations, carriers, prefixes and other features. But, noise fails to show such properties and hence the signal can be easily identified. This also requires information of primary signal well in advance. It also has the ability to differentiate among variety of primary signals by virtue of their cyclic properties.

The block diagram in Figure 11 consists of five blocks: BPF, N point FFT, correlate, average over T and feature detection. This technique is very robust compared to energy detection and works well in

Figure 10. Matched filter Block Diagram

Figure 11. Cyclostationary feature detection block diagram

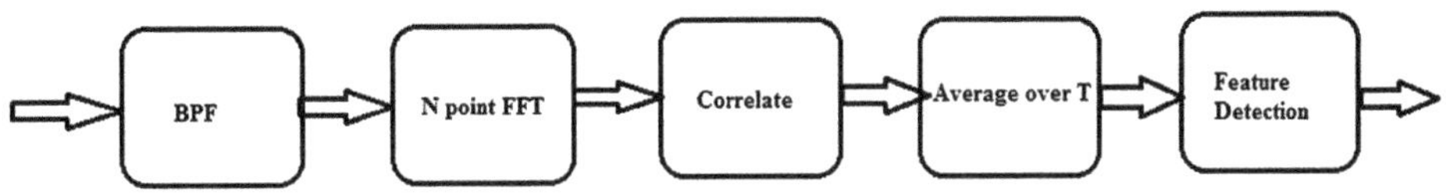

noise and low SNR conditions. No need of synchronization. As complexity involved in correlation with time shifting is more, it is less often used.

Advantages

- Works fine with low SNR regions
- Robust performance

Disadvantages

- High cost requirements
- Requires information of primary signal well in advance
- Applicable only for certain signals with periodic characteristics

7. Cooperative Sensing Method

This method arises as the solution to comply with the drawbacks of earlier discussed non- Cooperative sensing by allowing cooperative operation for secondary users in spectrum sensing efforts. Thus, actual availability of spectrum will be known. Also, the hidden node problem can be resolved by cooperative sensing. There are three approaches in this method:

a. Decentralized Uncoordinated Technique

In this, cognitive radio alone identifies the channel for transmissions and will leave it whenever licensed user is recognized without giving any information to neighborhood users. So, the chances of interference are too high due to some unwanted assumptions made while selecting channel. This method is therefore not widely used compared to others.

b. Centralized Coordinated Technique

A controller to control cognitive radio is present in here. Whenever a primary user is being detected by any cognitive radio it informs to controller about it. Later, it is job of this controlling device to broadcast the same message to neighbor radio devices. The method can be further treated as partially cooperative where in cooperation from nodes is only for sensing the channel and totally cooperative where information exchange takes place in addition to sensing.

c. Decentralized Coordinated Technique

The need for a central controller for coordination is removed by forming the network of cognitive radios. Different algorithms like clustering or gossiping can be applied resulting in self coordination among the network nodes. The problem of need for control channel can be resolved by use of dedicated frequency or by ultra-wide band channel.

8. Interference Based Detection Technique

Interference based detection relays on primary receiver detection and interference temperature management.

a. Primary Receiver Detection

It is behavior of any receiver to emit leakage power due to presence of local oscillator while receiving any data from the transmitter. This is made use to determine the presence of any primary transmitter with the help of a sensor mounted near to primary receiver. The signal from this sensor helps in identifying the vacant spectrum places.

b. Interference Temperature Management

The interference generally occurring at transmitter side can be controlled by either locating the individual transmitters or through power radiation. But, interference usually is at receiver side. So, the FCC recently introduced a way to restrict interference to some extent by interference temperature (FCC, 2003). This model restricts the interference at the receiver through an interference temperature bound, which is the measure of new impedance the collector could endure. For whatever length of time that CR clients try not to surpass this cutoff, they can utilize the spectrum band. The problem here is to properly determine the interference temperature limit.

CONCLUSION

Spectrum is an extremely profitable asset in wireless communication frameworks, and it has been a point of research and improvement endeavors in the course of time. Cognitive radio, which is one of the solutions to use the accessible spectrum more strategically through entrepreneurial spectrum utilization, has turned into an energizing and promising idea. One of the main components of cognitive radio is detecting the accessible spectrum availability. In this chapter, the spectrum sensing concepts are covered in detail along with various methods in use. Estimation of spectrum utilization in various measurements including time, frequency, space, angle, and what's more code; enables openings in these measurements; and forming new algorithms for forecast into the future utilizing past data can be considered as a portion of the open research area.

REFERENCES

Akyildiz, I. F., Lee, W. Y., Vuran, M. C., & Mohanty, S. (2006). NeXt generation/dynamic spectrum access/cognitive radio wireless networks: A survey. *Computer Networks*, *50*(13), 2127–2159. doi:10.1016/j.comnet.2006.05.001

Cabric, D., Mishra, S. M., & Brodersen, R. W. (2004, November). Implementation issues in spectrum sensing for cognitive radios. In *Signals, systems and computers, 2004. Conference record of the thirty-eighth Asilomar conference on* (Vol. 1, pp. 772-776). IEEE. 10.1109/ACSSC.2004.1399240

Cabric, D., Tkachenko, A., & Brodersen, R. W. (2006, October). Spectrum sensing measurements of pilot, energy, and collaborative detection. In Military communications conference, 2006. MILCOM 2006. IEEE (pp. 1-7). IEEE. doi:10.1109/MILCOM.2006.301994

FCC. (2003). *Docket No 03-222 Notice of proposed rule making and order*. FCC.

Ganesan, G., & Li, Y. (2005, December). Agility improvement through cooperative diversity in cognitive radio. In *Global Telecommunications Conference, 2005. GLOBECOM'05. IEEE* (Vol. 5, pp. 5-pp). IEEE. 10.1109/GLOCOM.2005.1578213

Haykin, S. (2005). Cognitive radio: Brain-empowered wireless communications. *IEEE Journal on Selected Areas in Communications*, *23*(2), 201–220. doi:10.1109/JSAC.2004.839380

Marcus, M. J. (2005). Unlicensed cognitive sharing of TV spectrum: The controversy at the federal communications commission. *IEEE Communications Magazine*, *43*(5), 24–25. doi:10.1109/MCOM.2005.1453413

Mitola, J., & Maguire, G. Q. (1999). Cognitive radio: making software radios more personal. *IEEE Personal Communications, 6*(4), 13-18.

Neel, J. O. D. (2006). *Analysis and design of cognitive radio networks and distributed radio resource management algorithms* (Doctoral dissertation).

Sun, H., Laurenson, D. I., & Wang, C. X. (2010). Computationally tractable model of energy detection performance over slow fading channels. *IEEE Communications Letters*, *14*(10), 924–926. doi:10.1109/LCOMM.2010.090710.100934

Sun, H., Nallanathan, A., Wang, C. X., & Chen, Y. (2013). Wideband spectrum sensing for cognitive radio networks: A survey. *IEEE Wireless Communications*, *20*(2), 74–81. doi:10.1109/MWC.2013.6507397

Yucek, T., & Arslan, H. (2009). A survey of spectrum sensing algorithms for cognitive radio applications. *IEEE Communications Surveys and Tutorials*, *11*(1), 116–130. doi:10.1109/SURV.2009.090109

Zhao, Y., Morales, L., Gaeddert, J., Bae, K. K., Um, J. S., & Reed, J. H. (2007, April). Applying radio environment maps to cognitive wireless regional area networks. In *New Frontiers in Dynamic Spectrum Access Networks, 2007. DySPAN 2007. 2nd IEEE International Symposium on* (pp. 115-118). IEEE. 10.1109/DYSPAN.2007.22

KEY TERMS AND DEFINITIONS

Channel Capacity: The rate (in bps) at which information is transmitted over communication link.

Cognitive Radio: It is an innovative intelligent technology that can automatically detect the unused licensed spectrum for allocation to non-licensed users.

Interference: Process which changes or disrupts a signal while transmitting between transmitter and receiver.

Primary User: The user who is allocated with licensed spectrum for communication over network.

Secondary User: Unlicensed user who always keep track of activities of primary user.

Spectrum Holes: Band of frequencies where secondary user can transmit his data without affecting primary user.

Spectrum Sensing: The process of continuous monitoring of a particular frequency band to occupy it if unused.

Wireless Networks: It is a computer network which uses wireless connections for data transfer between nodes.

Chapter 12
Compressive Spectrum Sensing:
Wavelet-Based Compressive Spectrum Sensing in Cognitive Radio

Said E. El-Khamy
Alexandria University, Egypt

Mina B. Abd el-Malek
Alexandria University, Egypt

Sara H. Kamel
Alexandria University, Egypt

ABSTRACT

In this chapter, the authors discuss how compressive sensing can be used in wideband spectrum sensing in cognitive radio systems. Compressive sensing helps decrease the complexity and processing time and allows for higher data rates to be used, since it makes it possible for the signal to be sampled at rates lower than the Nyquist rate and still be reconstructed with high accuracy. Different sparsifying bases for compressive sensing are presented in this chapter and their performance is compared. The design of these matrices is based on different types of wavelet transforms, including the discrete wavelet transform and the stationary wavelet transform; the latter having shown a clear improvement in performance over the former. The authors present different ways of implementing these transforms in a compressive sensing framework. Additionally, different types of reconstruction methods including the genetic algorithm and the auxiliary function method are also presented and their impact on the overall performance is discussed.

INTRODUCTION

The problem of spectrum scarcity has become of critical importance with the increase in the number of wideband communication systems. The main goal of cognitive radio is to efficiently utilize the spectrum by allowing unlicensed (secondary) users to opportunistically access licensed parts of the spectrum that are momentarily unoccupied by licensed (primary) users. In order to avoid interference with primary users, secondary users must sense the spectrum in order to determine which parts of the spectrum are

DOI: 10.4018/978-1-5225-5354-0.ch012

unoccupied by primary users, these unoccupied frequency bands are called spectrum holes (Haykin, 2005; Ma & Juang, 2009). The goal of spectrum sensing is to identify these spectrum holes. Spectrum sensing methods have evolved from using tunable band pass filters to sense one narrowband frequency at a time to more efficient methods like wavelet edge detection and compressive sensing.

For narrowband systems, a tunable band pass filter (BPF) operating at a narrowband range can be used to sense the spectrum, one band at a time. For wideband systems, an early approach was to use a series of narrowband BPFs. In this case the filter range of each BPF is preset. However, this architecture requires a large number of components and therefore increases the cost and complexity of the system (Divakaran, Manikandan, & Hari, 2011; Polo, 2011). Another method is the windowed Fourier transform, otherwise known as the Short Time Fourier Transform (STFT). The Fourier transform gives time averaged values for the frequency components of the signal but does not provide information on the time of occurrence of these frequencies, but using the STFT, the signal is first divided into smaller time segments using a suitable window function, then the Fourier transform of each of these segments is obtained. Breaking the signal into smaller segments and identifying the frequency components of each time segment provides time resolution, however, it suffers low frequency resolution, high variance of estimated power spectrum as well as side lobes of large magnitude (Divakaran, Manikandan, & Hari, 2011; Merry & Steinbuch, 2005).

The periodogram is another method quite similar to the STFT, where the spectrum signal is truncated using a rectangular window function, and the spectral density plot is obtained by squaring the Fast Fourier Transform (FFT) of the truncated signal. This method suffers from spectral leakage and lack of time-frequency localization due to the truncation of the signals which results in a Dirichlet Kernel in the frequency domain (Haggstrom, 2016). To alleviate the shortcomings of the periodogram; multi-taper spectral estimation can be used as an alternative, since it reduces spectral leakage by using multiple orthogonal filters. Multi-taper refers to the method used for estimating power spectra using an orthogonal set of data tapers, such as Slepian sequences, which allow us to trade spectral resolution for reduced variance of the estimate without compromising its bias (Divakaran, Manikandan, & Hari, 2011; Wang & Zhang, 2009). However, this method involves excessive computations and cannot completely eliminate the spectral leakage problem while better methods can.

When the information of the primary user signal is known to the cognitive user, the optimal detector in stationary Gaussian noise is the matched filter, which maximizes the received signal-to-noise ratio (SNR); this is known as coherent detection. Such information includes the preamble, signaling information for synchronization, pilot patterns for channel estimation, and modulation orders of the transmitted signal. The matched filter method is fast and achieves optimum performance but its main disadvantage is its dependence on primary user information as well as hardware complexity. Moreover, in case of channel fading, time dispersion and Doppler shifts may affect synchronization and ultimately the performance of the matched filter (Bagwari & Singh, 2012; Divakaran, Manikandan, & Hari, 2011).

The cyclostationary features (or periodicity) of the signal can be exploited to determine the presence of primary users in a frequency band, this type of sensing is called cyclostationary feature detection. Almost any transmitted signal has periodic features as a result of being coupled with sinusoidal carriers for modulation, cyclic prefixes or repeated spreading code sequences, which have built-in periodicity (Bagwari & Singh, 2012). This periodic behavior can be observed in the mean, autocorrelation and other statistical features of the signal. Cyclostationary feature detection provides accurate information about the spectral occupancy even in high noise levels, but the location of the band edges cannot be precisely determined (Adoum & Jeoti, 2010; Divakaran, Manikandan, & Hari, 2011).

Another method is Quadrature Mirror Filter Banks (QMFB) which splits up the entire wideband spectrum into M pre-defined bands where user traffic is monitored using n stages of quadrature mirror filters tuned for a particular band, such that $M=2^n$. As is the case with the matched filter method, prior knowledge about the primary user is required. One advantage of the QMFB is the reusability of the filters; since the filters used for spectrum sensing can also be used to extract the channels once they are detected (Divakaran, Manikandan, & Hari, 2011; Narendar, Vinod, Madhukumar, & Krishna, 2010).

Wavelet edge detection is considered one of the most efficient methods of spectrum sensing, since the wavelet transform is known for its ability to detect local singularities in a signal, and has therefore become a powerful tool for edge detection. This method applies the wavelet transform to the wideband spectrum signal in order to locate the occupied channel edges, which are characterized by abrupt changes in the power spectral density (PSD) of the signal as the amplitude changes from a vacant band to an occupied one and vice versa. Once the locations of the edges are determined, the signal power between each two consecutive edges is measured in order to identify the vacant bands available for secondary users (Tian & Giannakis, 2006). In (Nason & Silverman, 2005), the stationary wavelet transform was introduced as a modified version of the wavelet transform and its statistical applications were discussed. The stationary wavelet transform proved to be a reliable and efficient method that provides more accurate estimates in edge detection than the discrete wavelet transform (Kamel, El-Khamy & Abdel-Malek, 2014). In this chapter the authors discuss how the wavelet transform, with its different forms and types, plays an important role in compressive spectrum sensing.

Compressive Sensing is an emerging technique that makes it possible to sample signals at rates lower than the Nyquist rate, provided the signals are sparse or have a sparse representation in some domain; which is the case with most signals in practice, including PSD signals (Baraniuk, 2007). It was for spectrum estimation in cognitive radio in (Tian & Giannakis, 2007). In this chapter the authors discuss the mathematical concepts behind this technique and its application to wideband spectrum sensing in cognitive radio, with emphasis on how such a technique can be practically implemented in said application. Two main design aspects of compressive sensing are presented: the basis matrix, whose choice depends on the sparsifying domain, and the algorithms used to reconstruct the signal from the samples obtained at a sub-Nyquist sampling rate. The most commonly used method for reconstruction in compressive sensing is the Basis Pursuit algorithm, which is a linear program that uses ℓ_1-norm minimization. Another alternative to the BP algorithm is the Orthogonal Matching Pursuit (OMP) algorithm which is a greedy algorithm whose results are comparable to that of the BP and in some cases faster (Baraniuk, 2007; Tropp & Gilbert, 2007). Other reconstruction algorithms include the iterative reweighted ℓ_1-minimization (Needell, 2009) and the Multiplicative Update Selector and Estimator (MUSE) algorithm based on game theory (Jafarpour, Schapire, & Cevher, 2011). The two algorithms discussed in this chapter for reconstruction are the genetic algorithm (El-Khamy, Abdel-Malek, & Kamel, 2014) and an optimization algorithm known as the auxiliary function method based on the Lagrange multiplier (Tsutsu & Morikawa, 2012). The motivation for both algorithms is the fact that they are suitable for ℓ_p-minimization for any value of p; and ℓ_p-norms for values of p that are less than one provides better measures of sparsity and hence better objective functions for the minimization algorithms.

The rest of the chapter is organized as follows: first, the mathematical foundation of compressive sensing is presented, followed by the reconstruction algorithms and the different measures of sparsity that are used as objective functions for the optimization algorithms. The different types and modifications of the wavelet transform are discussed, focusing mainly on how these different forms can be implemented in compressive sensing as a sparsifying basis, why they are considered for wideband spectrum sensing and how they affect the overall performance of the system and the accuracy of edge detection. Simulation results are presented at the end of the chapter to verify the performance of the methods discussed throughout the chapter.

COMPRESSIVE SENSING

The Nyquist sampling theorem states that in order to perfectly reconstruct a signal from a set of discrete samples, the sampling rate must be equal to twice the highest frequency component in the signal (Shannon, 1949). For wideband signals, operating at the Nyquist rate means increased complexity, processing time and the need to store a huge number of samples. Cognitive radio users need to quickly and efficiently sense the presence of primary users across the entire bandwidth, which is quite a challenge when using rates equal to or greater than the Nyquist rate.

Compressive sensing allows us to reconstruct signals accurately while using rates well below the Nyquist rate, provided that the signal is sparse or has a sparse representation in some domain, and such is the case for most signals in practice, and more importantly, wideband spectrum PSD signals (Baraniuk, 2007; Donoho, 2006).

Signal Representation

Consider the discrete signal represented by the $N\times1$ vector $\mathbf{x} \in \mathbb{R}^N$, any such signal can be written as a linear combination of N basis vectors ψ_i of size $N\times1$ each. This representation can also be viewed as a linear transformation to the $\mathbf{\Psi}$-domain using the *basis matrix* $\mathbf{\Psi}$ whose columns are the N basis vectors. The signal $\mathbf{x}$ can therefore be expressed as:

$$x = \mathbf{\Psi} z \tag{1}$$

where $\mathbf{z}$ is the $N\times 1$ column vector of weighting coefficients. The vectors $\mathbf{x}$ and $\mathbf{z}$ are equivalent representations of the same signal; each in a different domain (Baraniuk, 2007).

The domains of interest here are those in which the equivalent signal $\mathbf{z}$ is sparse i.e. the weighting coefficients are mostly zeroes; if $\mathbf{z}$ is a K-sparse signal, then it has only K non-zero elements and the rest of the $(N - K)$ elements are equal to zero (or of negligible value), where $K << N$.

Regular Compression and Compressive Sensing

Transform coding, or regular compression, is very different from compressive sensing. In regular compression, the signal $\mathbf{x}$ is transformed into its sparse equivalent $\mathbf{z}$, then the $(N - K)$ elements of negligible value are discarded and only the K non-zero elements are stored or sent. Upon reconstruction, the values of the K non-zero elements are needed as well as their original locations in the vector $\mathbf{z}$, using this information $\mathbf{z}$ can be formed, and by using the inverse transformation, the original signal $\mathbf{x}$ can be obtained (Baraniuk, 2007). This is illustrated in Figure 1(a) and it is often referred to as the sample-then-compress framework. In compressive sensing, on the other hand, the signal is not transformed into its sparse equivalent, a transformation that involves are random matrix of specific properties (which shall be discussed later in detail) is simply applied to obtain a vector **y** of length K called the measurement vector, as shown in Figure 1(b). Using these measurements and a suitable algorithm the original signal can be reconstructed with great accuracy.

The advantage of compressive sensing is that it eliminates the need to evaluate and store all the N elements of the sparse vector **z** before reducing them to K elements by discarding the negligible elements as well as the need to store the locations of the discarded elements in order to be able to reconstruct the signal. This translates into reduced processing time, reduced cost and complexity.

Compressive Sensing Framework

In compressive sensing, our goal is to obtain a set of measurements in the form of a vector $\mathbf{y} \in \mathbb{R}^M$ from the original signal **x** where $M < N$ such that $M \approx K$, and use these measurements to reconstruct the signal or its equivalent sparse representation **z**. The relation between **y** and **x** is defined as follows:

$$y = \Phi x = \Phi \Psi z = Gz \tag{2}$$

where $\mathbf{\Phi}$ is an $M \times N$ matrix called the measurement matrix and $\mathbf{G} = \mathbf{\Phi\Psi}$ (also an $M \times N$ matrix). The choice of the measurement matrix $\mathbf{\Phi}$ and the basis matrix $\mathbf{\Psi}$ should ensure that **G** has the restricted isometry property (RIP), which means that:

$$1-\varepsilon \leq \frac{\left\|Gv\right\|_2}{\left\|v\right\|_2} \leq 1+\varepsilon \tag{3}$$

where $\varepsilon > 0$ and **v** is any K-sparse vector. This means that the linear transformation $\mathbf{v} \rightarrow \mathbf{Gv}$ must preserve the length of the K-sparse vector **v** (Baraniuk, 2007). The verification of the RIP is combinatorially complex for any given matrix **G**, but it has been found that certain choices for $\mathbf{\Phi}$ ensure that for any basis matrix $\mathbf{\Psi}$, **G** will have the RIP. A suitable choice for the measurement matrix, which has been proven to satisfy the RIP for any basis matrix, is a random matrix. Gaussian matrices are one example of random matrices, whose elements are drawn as independent and identically distributed (i.i.d) zero mean normally distributed random variables. Another example is the random permutation (or selection) matrix. A permutation matrix $\mathbf{S}_{N\times N}$ is defined as an identity matrix whose rows have been permuted ac-

Figure 1. The difference between (a) regular compression and (b) compressive sensing

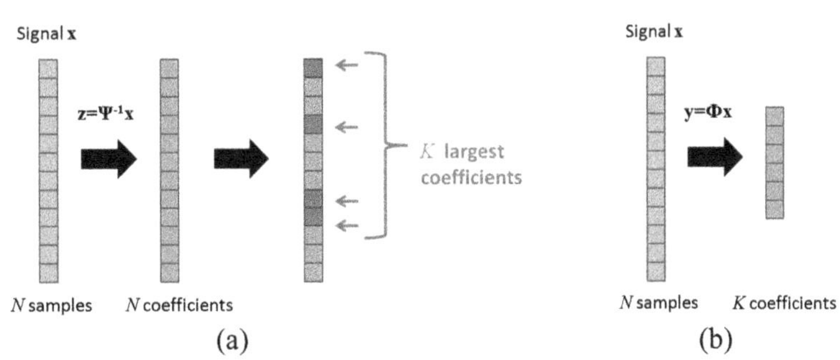

cording to a random permutation of the numbers from 1 to N. It contains exactly one 1 in each row and column such that, when multiplied to a length-N vector, the elements of the vector are permuted in the same order as the location of the 1's in each row (Blair-Stahn, 2000). For example, consider a length-6 vector **x** and a permutation matrix designed to rearrange the elements in the following order: {1, 3, 6, 2, 4, 5}. Such a permutation matrix would be constructed as follows:

$$\mathrm{S} = \begin{bmatrix} \mathbf{1} & 0 & 0 & 0 & 0 & 0 \\ 0 & 0 & \mathbf{1} & 0 & 0 & 0 \\ 0 & 0 & 0 & 0 & 0 & \mathbf{1} \\ 0 & \mathbf{1} & 0 & 0 & 0 & 0 \\ 0 & 0 & 0 & \mathbf{1} & 0 & 0 \\ 0 & 0 & 0 & 0 & \mathbf{1} & 0 \end{bmatrix} \tag{4}$$

If this matrix is multiplied by the vector $\mathbf{x} = [x_1\ x_2\ x_3\ x_4\ x_5\ x_6]^{\mathrm{T}}$, the resulting vector $\mathbf{x}_p = \mathbf{Sx}$ will be $[x_1\ x_3\ x_6\ x_2\ x_4\ x_5]^{\mathrm{T}}$. As a measurement matrix, the permutation matrix is truncated to the size $K \times N$. The application of the permutation matrix directly to the PSD signal (the non-sparse vector) can be practically implemented by sub-sampling the spectrum (i.e. sampling it at a sub-Nyquist rate) followed by interleaving the resulting samples (El-Khamy, Abdel-Malek, & Kamel, 2017). This is what makes the permutation matrix a practical choice for compressive spectrum sensing.

Equation (2) represents an underdetermined system of linear equations where **x** (or equivalently **z**) is the vector of unknowns, so there are infinitely many vectors that satisfy the equation, but our aim is to find a sparse solution, since **z** is known to be K-sparse. The best measure of sparsity is the ℓ_0-norm of the vector, which is equal to the number of non-zero elements in it. The problem, however, is that it is combinatorially complex to find the solution with the least ℓ_0-norm. On the other hand, the ℓ_1-norm - which is equal to the sum of the absolute values of the elements of the vector - is a suitable and efficient choice, since finding the solution with the least ℓ_1-norm is a convex optimization problem can be reduced to a simple linear program known as basis pursuit (Candes & Romberg, 2005), and therefore the optimization problem becomes:

$$\hat{\boldsymbol{z}} = \arg\min \left\| \boldsymbol{z} \right\|_1 \text{ such that } \boldsymbol{y} = \boldsymbol{G}\boldsymbol{z} \tag{5}$$

where $\|.\|_1$ denotes the ℓ_1-norm. The RIP of the matrix **G** guarantees that the K-sparse solution, which is the N-length vector **z**, can be found with high probability (Baraniuk, 2007).

RECONSTRUCTION METHODS

The ℓ_1-norm minimization, or the basis pursuit algorithm, is the most widely used reconstruction method in compressive sensing. However, there are other norms that, theoretically, provide better solutions. The ℓ_p-norm of a length-N vector **x** is defined as:

$$\|\boldsymbol{x}\|_p = \left(\sum_{i=1}^{N} |x_i|^p \right)^{1/p} \tag{6}$$

where $p > 0$. Using the ℓ_p-norm for $0 < p < 1$ as a measure of sparsity should produce better results than using the ℓ_1-norm since these norms promote the sparsification of the solution (Carothers, 2000; Migliore, 2014). The geometrical representation of the ℓ_p-norm minimization for different values of p helps explain why. The solution of the underdetermined system of equations $\mathbf{Gz} = \mathbf{y}$ is represented geometrically by a hyperplane of dimension $(N - K)$; considering that the vector $\mathbf{z}$ of length N is known to be K-sparse. So our only focus is on solutions that lie in an $(N - K)$-dimensional hyperplane. To visualize this, consider the case of a two-dimensional vector as shown in Figure 2. The straight line represents all the possible solutions to the equation $\mathbf{y} = \mathbf{Gz}$. The circle depicted in Figure 2(a) represents the ℓ_2-ball while the rhombus in Figure 2(b) represents the ℓ_1-ball. The points that lie on the surface of these balls have the same norm; for example, the points on the ℓ_2-ball – which is a circle – are all at the same Euclidean distance from the origin (which is equal to the radius of the circle in this case), and the Euclidean distance is the same as the ℓ_2-norm so it can be said that all the points on the circumference of the circle have the same ℓ_2-norm, the same goes for any ℓ_p-ball but the norm used to measure the distance is calculated using Equation (6). The minimization process (or finding the solution with the least norm) can be visualized by enlarging the balls from the origin (at which the norm is equal to zero) until they touch the hyperplane; for example, in case of the ℓ_2-ball, the radius of the circle is gradually increased until the circle touches the line. The point of intersection between the straight and the circle represents the solution with the minimum ℓ_2-norm. It is clear that the solution $\hat{\mathbf{z}}$ in case of the ℓ_1-ball is indeed the sparsest solution while the solution reached by the ℓ_2-ball is far from sparse compared to the ℓ_1-norm case.

Figure 3 shows the ℓ_p-balls for four different values of p in $\mathbb{R}^3$. It is clear that, as the value of p decreases, the surface of the ball grows closer to the axes. At $p = 0$ the ball coincides with the axes, while as $p \rightarrow \infty$ the ball tends to a cube. This implies that the minimization procedure will yield sparser solutions as the value of p decreases since the point of intersection between the ball and the hyperplane, wherever that hyperplane may be, is more likely to be closer to the axes, and the closer the point is to the axes, the sparser it is. So using ℓ_p-norms as an objective function with $0 < p < 1$ should yield better results. Two different algorithms for ℓ_p-norm minimization are presented in the following section.

Figure 2. Geometrical Interpretation of ℓ_p-norm Minimization in $\mathbb{R}^2$ for (a) p=2 and (b) p=1

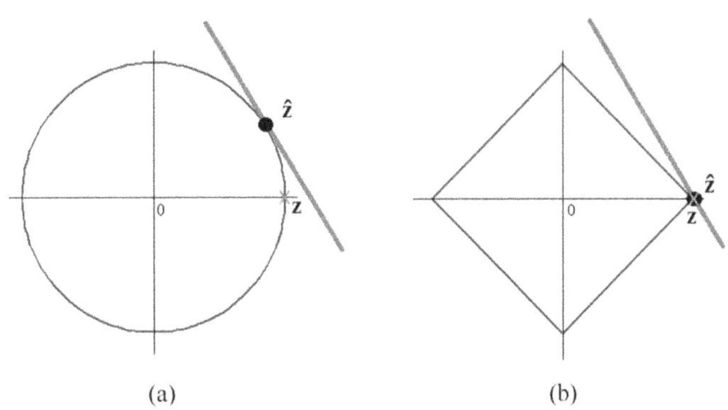

Figure 3. ℓ_p-ball shapes in $\mathbb{R}^3$ for (a) p=0.5, (b) p=1, (c) p=2 and (d) p=5
El-Khamy, Abdel-Malek, & Kamel, 2014.

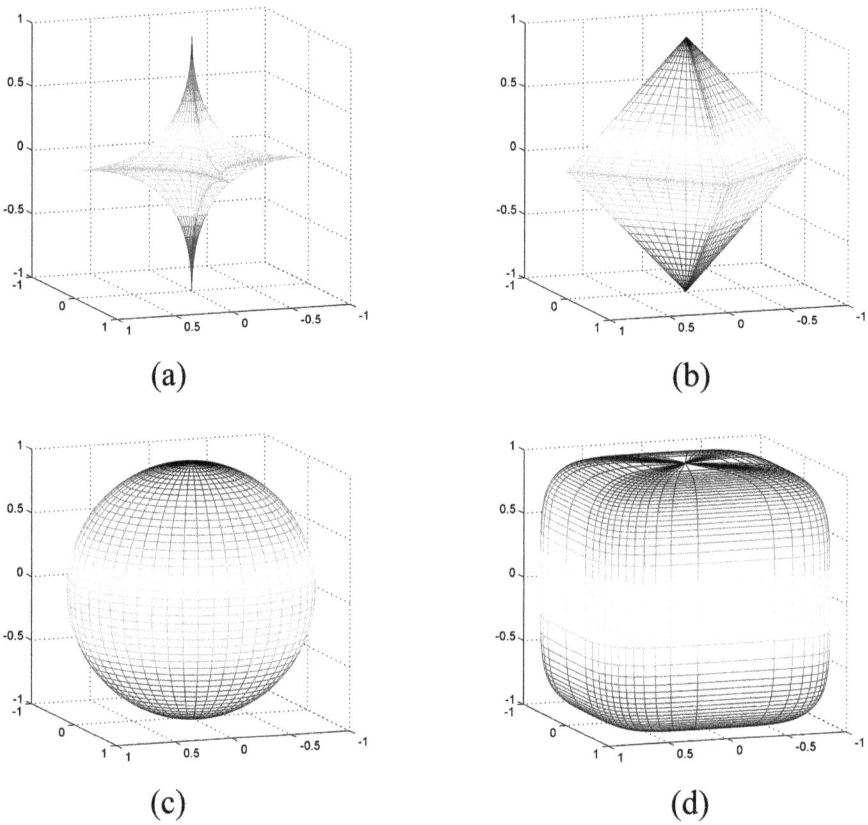

Reconstruction Using the Auxiliary Function Method for ℓ_p-Norm Minimization

In (Tsutsu & Morikawa, 2012), a fast algorithm was proposed for ℓ_p-norm minimization that is suitable for values of p that are less than one. This algorithm, which is based on the auxiliary function method, was proven to be faster than the ℓ_1-minimization algorithm. In this chapter, the authors test the performance of this algorithm in wideband compressive sensing. The auxiliary function method iteratively decreases a given objective function by using a generalization of the iterative reweighted least squares method proposed in (Gorodnitsky & Rao, 1997) for ℓ_1 minimization to ℓ_p minimization. The relation between an objective function $L(\mathbf{x})$ and its auxiliary function $L^+(\mathbf{x},\mathbf{a})$ is:

$$L(\boldsymbol{x}) = \min_{a} L^{+}(\boldsymbol{x}, \boldsymbol{a}) \tag{7}$$

where the vector $\mathbf{a}$ represents the auxiliary variable. In case of ℓ_p minimization the objective function $L(\mathbf{x}) = (\|\mathbf{x}\|_p)^p$ and the corresponding auxiliary function that satisfies Equation (7) is:

$$L^{+}(\boldsymbol{x},\boldsymbol{a})=\frac{1}{2}\sum_{n}\frac{x_{n}^{2}}{a_{n}}+\sum_{n}\left(1-\frac{p}{2}\right)\left(pa_{n}\right)^{p/(2-p)}a_{n}>0,0<p<2 \tag{8}$$

where a_n and x_n are the nth elements of the vectors **a** and **x** respectively. In vector form Equation (8) becomes:

$$L^{+}(\boldsymbol{x},\boldsymbol{a})=\frac{1}{2}\boldsymbol{x}^{T}diag^{-1}(\boldsymbol{a})\boldsymbol{x}+1^{T}\gamma(\boldsymbol{a}) \tag{9}$$

where diag(**a**) denotes a square matrix whose diagonal elements are the elements of the vector **a**, the vector **1** is a column vector whose elements are all ones and:

$$\gamma(\boldsymbol{a})=\left(1-\frac{p}{2}\right)\left[(pa_{1})^{p/(2-p)},\ldots,(pa_{n})^{p/(2-p)}\right]^{T} \tag{10}$$

In order to find the vector **x** that minimizes the objective function L(**x**) (and hence finds the minimum ℓ_p-norm), the Lagrange multiplier method is used to minimize the auxiliary function subject to the constraint **y** = **Gz**, so the Lagrange function becomes:

$$J=\frac{1}{2}\boldsymbol{x}^{T}diag^{-1}(\boldsymbol{a})\boldsymbol{x}+1^{T}\gamma(\boldsymbol{a})+\boldsymbol{\lambda}^{T}(\boldsymbol{y}-\boldsymbol{G}\boldsymbol{z}) \tag{11}$$

where **λ** is the Lagrange multiplier vector. By getting the partial derivatives of J with respect to **x**, **a** and **λ** and equating each of them to zero, the following equations are obtained:

$$[\boldsymbol{G}\,diag(\boldsymbol{a})\boldsymbol{G}^{T}]\boldsymbol{\lambda}=\boldsymbol{y} \tag{12}$$

$$\boldsymbol{x}=diag(\boldsymbol{a})\boldsymbol{G}^{T}\boldsymbol{\lambda} \tag{13}$$

$$\boldsymbol{a}=\frac{1}{p}\left|\boldsymbol{x}\right|^{2-p} \tag{14}$$

The minimization procedure is carried out by first initializing **a** to **1** (which corresponds to putting p = 2), and then solving Equation (12) to get **λ** and substituting in Equation (13) to get **x**. The vector **x** is then used to update **a** using Equation (14) and the process is repeated until the relative mean square error between the successive iterations becomes less than a predetermined threshold (Tsutsu & Morikawa, 2012).

Reconstruction Using the Genetic Algorithm

Another method for minimization that works with almost any ℓ_p-norm is the genetic algorithm, as proposed in (El-Khamy, Abdel-Malek, & Kamel, 2014). In this case the objective function can either be the ℓ_p-norm of the sparse vector **z** for any value of p under the constraint $\mathbf{y} = \mathbf{Gz}$, or it can simply be the ℓ_p-norm of $(\mathbf{y} - \mathbf{Gz})$, which is ideally zero; so it can also be a suitable objective function. The flow chart of the algorithm is shown in Figure 4. The algorithm starts with an initial guess, which is either the least squares solution of the equation $\mathbf{y} = \mathbf{Gz}$ (using the pseudo-inverse of G) or simply $\mathbf{z}_0 = \mathbf{G}^T\mathbf{y}$ as shown in Figure 5. This initial vector is used to create a population by randomly adding noise to the entries of the vector to create more vectors. The objective function is then calculated for each of the members of the population to create an intermediate population consisting of the vectors for which the objective function gives the lowest values. Crossover and mutation are carried out to the vectors in the intermediate population and then the stopping criteria are checked; that is, if the maximum number of generations (or iterations) is reached, or if the objective function reaches a predetermined minimum value.

WAVELET-BASED SPARSIFYING MATRICES

The first step in setting up the compressive sensing problem is to choose a basis (or sparsifying) matrix $\boldsymbol{\Psi}$, which – in spectrum sensing – depends on the domain in which the spectrum PSD signal is sparse. One of the most efficient choices in this case is the wavelet transform domain. The wavelet transform of a PSD signal takes the form of peaks that appear at the edges of every occupied channel as shown in Figure 6. The discrete wavelet transform (DWT) is applied to the signal shown in Figure 6(a) and the result is shown in Figure 6(b) for four dyadic scales; $s = 2^j$ for $j = 0, 1, 2$ and 3. The transform can clearly be considered a sparse signal which is why the wavelet domain is considered for compressive spectrum sensing.

Another reason that makes the wavelet transform a good choice is the fact that the goal of spectrum sensing is to identify spectrum holes which can be easily done by identifying the edges of the occupied bands then measuring the signal power between each two consecutive edges to determine whether the band is occupied or not, so the required signal is actually the sparse signal **z**, which can be obtained directly through the reconstruction algorithm, rather than the original signal **x**, which eliminates the need to transform the sparse signal back to the original signal, since all the information about the location of the edges is in the sparse signal.

THE DISCRETE WAVELET TRANSFORM

The DWT can be demonstrated through its filter bank representation. As can be seen in Figure 7, the DWT decomposes the input signal into two sets of coefficients; approximation coefficients and detail coefficients obtained by passing the signal through a low pass filter $g(n)$ and a high pass filter $h(n)$ respectively, followed by down sampling the output of the filters by a factor of 2. By repeatedly filtering the approximation coefficients at every stage and down sampling the outputs, the wavelet coefficients at higher levels can be obtained.

Figure 4. Genetic algorithm flowchart
El-Khamy, Abdel-Malek, & Kamel, 2014.

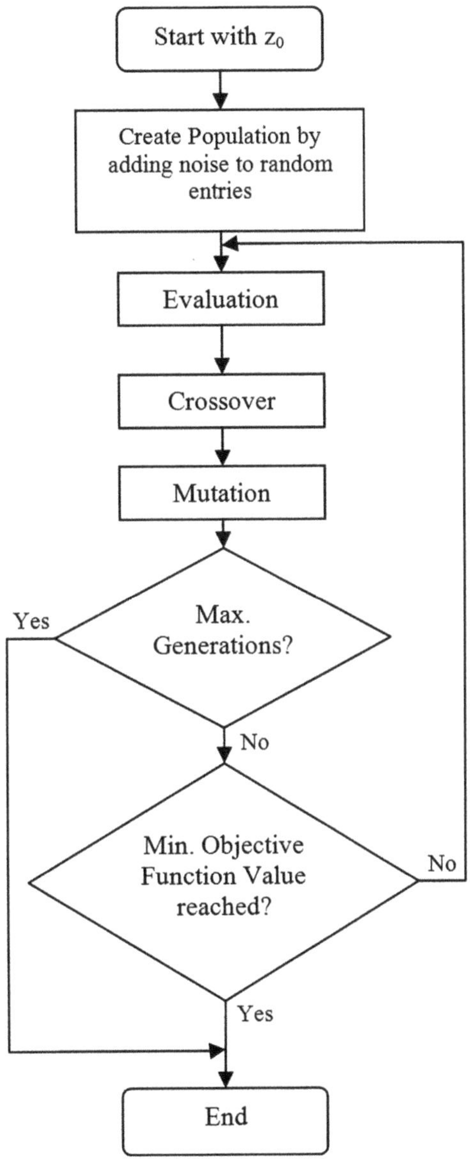

Figure 5. Creating the initial population

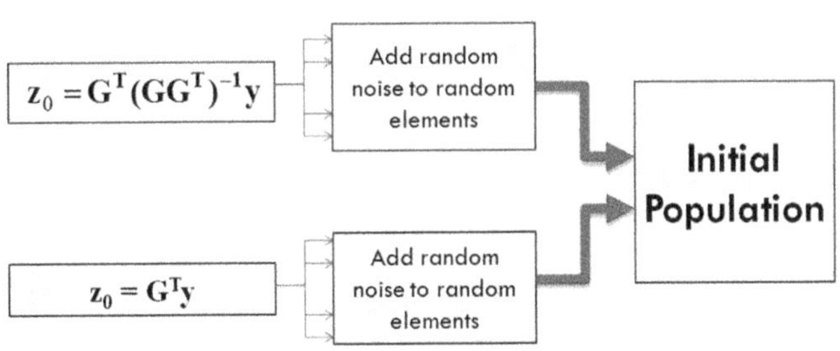

Figure 6. The discrete wavelet transform: (a) A PSD signal and (b) its DWT at different scales

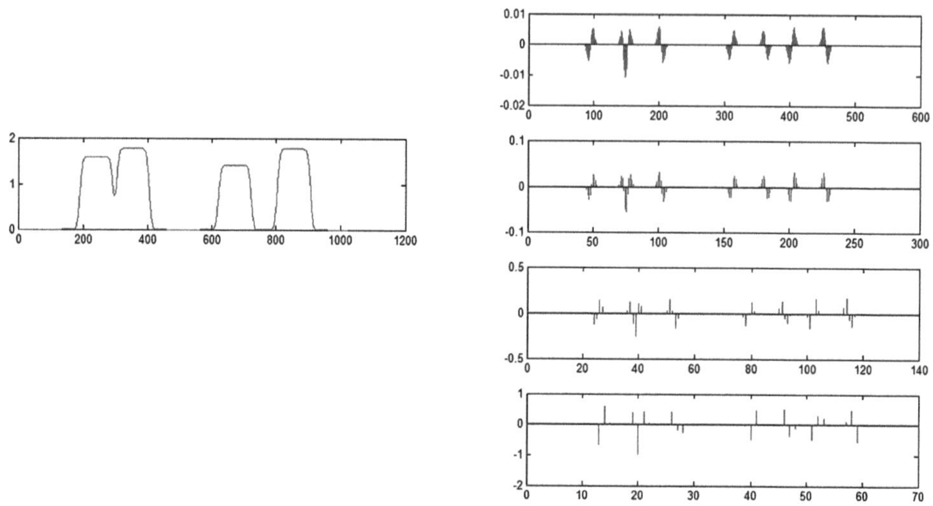

(a) (b)

Figure 7. The filter bank representation of the DWT
El-Khamy, Abdel-Malek, & Kamel, 2017.

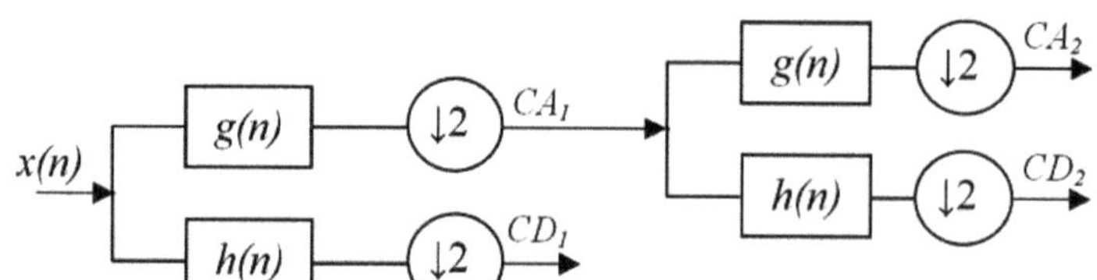

The detail and approximation coefficients are denoted by CD_i and CA_i respectively for the ith level. In edge detection, only the information provided by the detail coefficients is of interest, since the detail coefficients represent the channel edges if the input signal is the PSD signal, as explained earlier.

In order to use the DWT in compressive sensing, a basis matrix that transforms the signal into its detail coefficients is needed. Since passing a signal through a filter is equivalent to the convolution of said signal with the coefficients of the filter then this convolution process can be done using matrix multiplication by employing a Toeplitz matrix (Gray, 2006). For example, if the coefficients of the high pass filter $h(n)$ are h_1, h_2, h_3 and h_4, and the signal vector $\mathbf{x}$ is composed of 8 samples, then a convolution matrix $\mathbf{H}$ is formed as follows:

$$\boldsymbol{H} = \begin{bmatrix} h_1 & h_2 & h_3 & h_4 & 0 & 0 & 0 & 0 \\ 0 & h_1 & h_2 & h_3 & h_4 & 0 & 0 & 0 \\ 0 & 0 & h_1 & h_2 & h_3 & h_4 & 0 & 0 \\ 0 & 0 & 0 & h_1 & h_2 & h_3 & h_4 & 0 \\ 0 & 0 & 0 & 0 & h_1 & h_2 & h_3 & h_4 \end{bmatrix} \quad (15)$$

such that:

$$d = h * x = Hx \tag{16}$$

where **d** is the vector of detail coefficients, (*) is the convolution operator and **Hx** is a matrix multiplication operation equivalent to the convolution operation. If the high pass filter coefficients are those of the wavelet transform, the matrix **H** is the wavelet transform matrix (El-Khamy, Abdel-Malek, & Kamel, 2017). Note that, from Equation (1), the sparse vector $\mathbf{z} = \mathbf{\Psi}^{-1}\mathbf{x}$. In spectrum sensing, **z** represents the detail coefficients of the wavelet transform and **x** represents the PSD signal, which means that $\mathbf{z} = \mathbf{Wx}$ where **W** denotes the wavelet transform matrix, so $\mathbf{\Psi} = \mathbf{W}^{-1}$. This means that henceforth all the wavelet matrices are equivalent to $\mathbf{\Psi}^{-1}$.

Multiscale Sum Implementation of the Wavelet Transform

In order to enhance edge detection, the detail coefficients at different scales are often combined either by addition or multiplication. If the coefficients are multiplied, the edges (peaks) become more prominent and any spurious edges due to noise will be eliminated; since it is unlikely that a spurious edge or peak will be repeated in more than one scale, so when multiplying the coefficients, these erroneous peaks will be multiplied by zero (or values very close to zero) and will therefore disappear. The problem with the multiscale product is that if at any scale one of the edges is not detected due to noise, there is a good chance that it will be lost through multiplication, and this is why addition is preferred. When adding the coefficients, the actual edges are strengthened and at the same time spurious edges will become relatively small through addition since it is quite unlikely that the same spurious edge will be repeated at every scale, also if an actual edge is missed in one of the scales it will appear in other scales but it won't be lost as in case of multiplication (Divakaran, Manikandan, & Hari, 2011; El-Khamy, El-Mahallawy, & Youssef, 2013).

To implement multiscale wavelet sum, a matrix that mimics the process of adding the detail coefficients at different scales is needed. Notice that in order to add coefficients from different scales in the DWT, they need to have the same number of samples (and as vectors, the same number of elements), so down sampling is followed by zero insertion (El-Khamy, Abdel-Malek, & Kamel, 2013). The basis matrix is therefore defined as:

$$\boldsymbol{DW} = \left(\sum_k \boldsymbol{D}_k\right)\boldsymbol{W}, k = 1,2,4,8\ldots \tag{17}$$

where **W** is the wavelet matrix formed using the high pass filter as shown in Equations (15) and (16) and D_k is the down sampling matrix defined as:

$$\begin{aligned} \boldsymbol{D}_k &= diag\left(d_{i,j}\right) \\ d_{i,j} &= \begin{cases} 1 & i = j = nk + 1, n \in Z \\ 0 & otherwise \end{cases} \end{aligned} \tag{18}$$

where $i, j \in [1, N]$ and N is the length of the vector to which $\mathbf{D}_k$ is multiplied (El-Khamy, Abdel-Malek, & Kamel, 2017).

The Stationary Wavelet Transform

The stationary wavelet transform (SWT) can be considered as a modified version of the DWT. The SWT enhances the performance of the system and improves the accuracy of the estimated edges. Figure 8 shows the filter bank representation of the SWT. As the block diagram shows, the main difference between the SWT and the DWT is that in the SWT, there is no down sampling; instead, the filter coefficients of both the high pass and the low pass filter are up sampled by a factor of two (Kamel, El-Khamy, & Abdel-Malek, 2014; Nason & Silverman, 1995).

To clarify why the SWT performs better than DWT, both transforms are applied to the simple PSD signal is depicted in Figure 9(a). The signal shows two occupied bands, which means that four edges should be detected by applying the transform; one at each falling or rising end. The 2-level DWT decomposition (detail coefficients) of the signal is shown in Figure 9(b) while the 2-level SWT decomposition is shown in Figure 9(c). In case of the DWT, down sampling eliminates every other element of the output of the high pass filter, so an edge might be eliminated, which is what happens in this particular example. On the other hand, the SWT does not down sample the outputs of the filters so no edges are lost. This means that the probability of misdetection when using the SWT should be lower than the DWT, which is what the simulation results confirm, as will be seen later in this chapter (Kamel, El-Khamy, & Abdel-Malek, 2014).

Constructing the SWT Basis Matrix

The SWT basis matrix construction is very similar to that of the DWT. To obtain the SWT at any given scale, simply use the coefficients of the high pass filter at that level to form the convolution (Toeplitz) matrix as shown in Equation (15). So for example, to get the detail coefficients at the third decomposition level, the coefficients of the filter at the first level are up sampled twice. In general, the wavelet matrix at the kth decomposition level is:

Figure 8. The filter bank representation of the SWT
El-Khamy, Abdel-Malek, & Kamel, 2017.

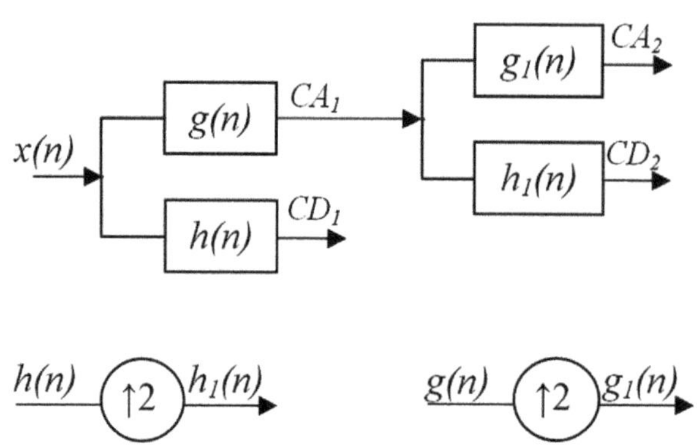

Figure 9. An example comparing the SWT and DWT
Kamel, El-Khamy, & Abdel-Malek, 2014.

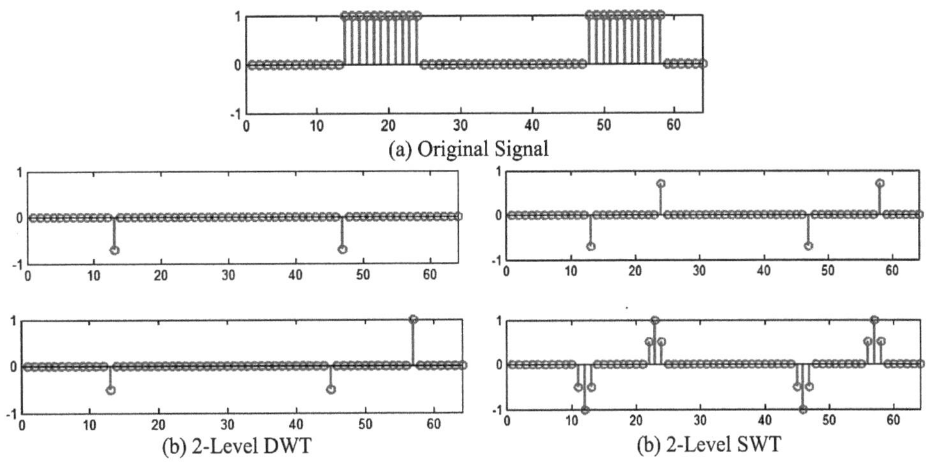

$$W = H_k. \tag{19}$$

Multiscale wavelet sum can also be implemented in case of the SWT as shown in Figure 10. The multiscale matrix in this case is formed by calculating the average sum of the $\mathbf{H}_k$ matrices as follows:

$$W = (H_1 + H_2 + ... + H_k) / k \tag{20}$$

where $\mathbf{W}$ is the inverse of the basis matrix $\boldsymbol{\Psi}$. It is worth noting that these basis matrices may not be exactly the same as their corresponding wavelet transformations, since at every level, the approximation coefficients are the inputs to the high and low pass filters of the following level, but incorporating the low pass filter coefficients into the design of the basis matrix results in a near singular matrix, in which case the reconstruction algorithm fails to find the solution. However, when the results of applying the proposed basis matrices and their corresponding transforms are compared, the coefficients obtained are almost the same and the performance of the system agrees with the theoretically expected results.

Figure 10. Constructing the SWT matrix
El-Khamy, Abdel-Malek, & Kamel, 2017.

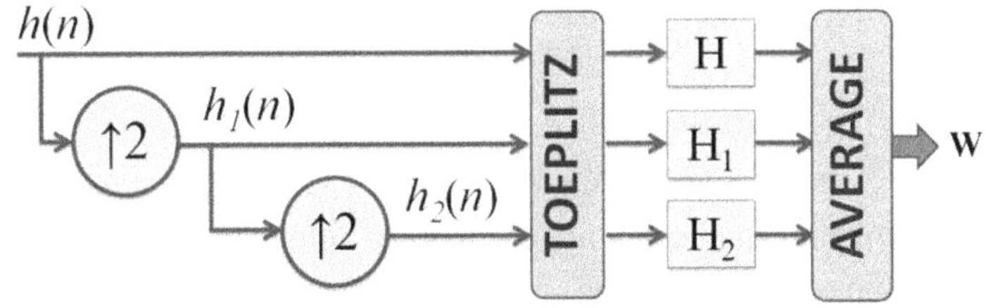

SIMULATION RESULTS

In this section simulation results are presented based on a number of performance parameters for spectrum sensing. The first measure of performance is the subcarrier occupancy error rate (SOER), defined as:

$$SOER = \frac{\sum_{i=0}^{N-1} \left| SOI_a(i) - SOI_d(i) \right|}{N} \tag{21}$$

where SOI(i) is the Subcarrier Occupancy Index of frequency subcarrier i which takes on binary values; 0 if the subcarrier is vacant and 1 if it is occupied; based on the results of energy detection and thresholding. $SOI_a(i)$ and $SOI_d(i)$ are the actual and detected SOI for the ith subcarrier, respectively (Zeng, Liang, & Chia, 2011). This gives an indication of how accurately one can identify spectrum holes, which depends on both the efficiency of the edge detection method and the energy detection step that follows. The SOER is useful in blind detection where the receiver does not know the bandwidth or the distribution of the channels across the spectrum.

In order to assess the accuracy of edge detection, the following parameters are used: the probability of misdetection (PMD), probability of false detection (PFD) and the total error probability (PT), which is somewhat equivalent to the SOER (El-Khamy, El-Mahallawy, & Youssef, 2013). These parameters, unlike the SOER, are unaffected by the energy detection step; they are concerned with the detected edges regardless of which bands are identified as vacant or occupied. The PMD depends on the number of missed edges, and is defined as:

$$PMD = \left(N_{CB} - \tilde{N}_{CB}\right) / N_{CB} \tag{22}$$

where N_{CB} is the total number of actual channel edges and $\tilde{N}_{CB}$ is the number of correctly detected channel edges. The PFD, on the other hand, depends on the number of falsely detected edges; or spurious noise peaks that are incorrectly classified as actual edges, and it is defined as:

$$PFD = \left(N_T - \tilde{N}_{CB}\right) / \left(N - N_{CB}\right) \tag{23}$$

where N_T is the total number of detected peaks (whether false or not) and N is the number of samples of the original signal (as previously defined), while the PT is defined as:

$$PT = \frac{PMD + PFD}{2}. \tag{24}$$

It is worth noting that, when calculating N_{CB} or $\tilde{N}_{CB}$, the frequencies at which the actual edges occur and the frequencies of the detected edges are compared, but since there's always an error, however small, between the exact value of the frequencies corresponding to the actual and detected edges, a small percentage of error is allowed (in most simulations presented in this chapter, an error less than or equal to 5% of the original value is allowed).

For the following simulations, it is assumed that the receiver has no information about the number of channels or the channel width, i.e. blind detection. The received signal is contaminated by additive white Gaussian noise (AWGN) and the performance parameters are evaluated at different signal-to-noise ratio (SNR) values.

Comparing the DWT and SWT

The first set of results test the system performance when edge detection is done by applying the DWT and SWT to the PSD signal without using compressive sensing. Figure 11 shows the SOER, PMD, PFD and PT versus SNR for several different scenarios: single-level, 2-level and 3-level DWT and SWT decomposition as well as multiscale sum for both types of wavelet decomposition; where – for the latter – the detail coefficients resulting from the three decomposition levels are added. The figures clearly show that the SWT is superior to the DWT at any decomposition level and also in case of multiscale sum. This result confirms the theoretical reasoning provided earlier, which predicts that the SWT will perform better than the DWT due to the omission of down sampling in the SWT. It is worth noting that, while the SWT outperforms the DWT, it needs more processing time and memory, because the coefficients of the DWT are down sampled at each scale so the number of samples processed is considerably lower than the SWT.

Comparing the DWT and SWT as Basis Matrices

In this section the DWT and SWT are compared in a compressive sensing framework as shown in Figure 12. The DWT and SWT are represented by the basis matrix or sparsifying basis of compressive sensing. In this case, there is no difference between the SWT and DWT in terms of processing time and memory; they are both represented by a matrix of the same size and the compression ratio K/N (which is equal to 1/8 in this case) is the same.

As in the case of no compression, the SWT matrices give lower error rates. Also, the multiscale sum matrices, labeled as SWT2 and SWT3 for 2-level and 3-level sums, perform better than the matrices used for single level decomposition, labeled as SWTmod2 and SWTmod3.

Since the SWT enhances the system performance, it should be expected that higher compression ratios can be used with the SWT matrices while maintaining the same performance level of a system that uses the DWT. Figure 13 compares the SOER for eight different scenarios: two basis matrices are used; the DWT matrix (single-level decomposition) and the SWT 2-level multiscale sum matrix, and for each of these matrices the compression ratios $K/N = 1/2$, 1/4, 1/8 and 1/16 are tested. Lower compression ratios exhibit lower error rates, which is normal. However the important result here is that a compression ratio of 1/8 using the SWT matrix is almost the same as using a compression ratio of 1/4 with the DWT matrix in terms of SOER. This means the same performance level reached using a DWT matrix can be attained using a SWT multiscale sum matrix with only half the number of samples.

Figure 14 shows four different PSD signals for which the performance of the system, using the 2-level SWT multiscale sum basis matrix, is compared in Figure 15. The SOER shows a different behavior at high SNR values than at low SNR values; for example, signal 3 has only two occupied bands and most of the spectrum is unoccupied, so at low SNR values, noise peaks are higher and may falsely indicate the presence of edges in the unoccupied part of the spectrum, resulting in higher error rates, but at high SNR values, signal 3 has the lowest error rate because the unoccupied part forms most of the signal and

Figure 11. A comparison of system performance in case of DWT, SWT and the multiscale sum cases (no compressive sensing) in terms of (a) SOER, (b) PT, (c) PMD and (d) PFD
El-Khamy, Abdel-Malek, & Kamel, 2017.

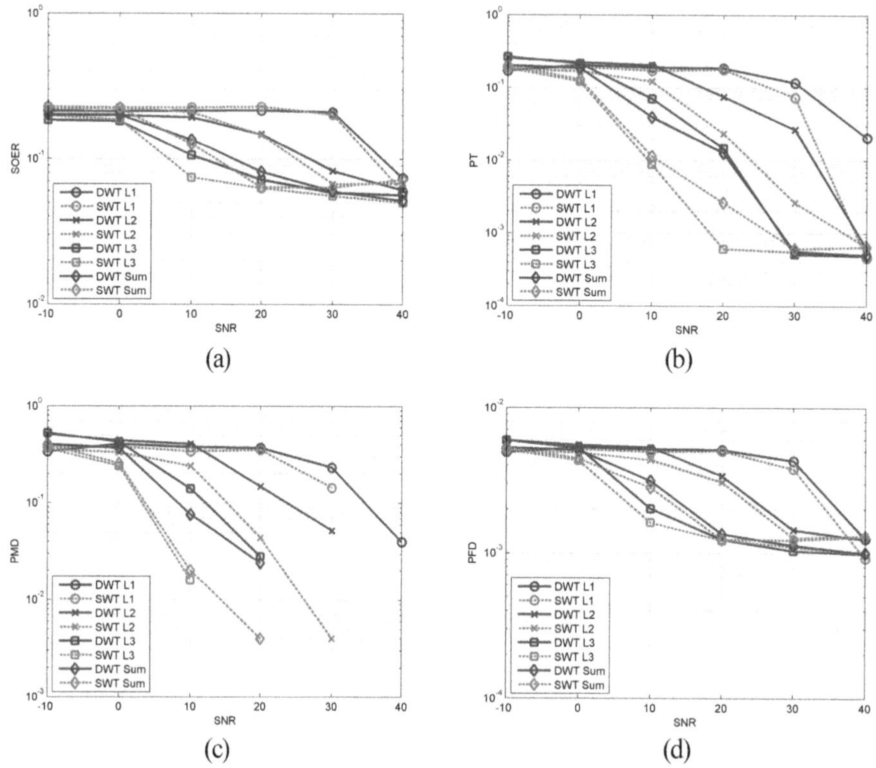

**For a more accurate representation see the electronic version.*

it is highly unlikely that noise peaks will occur there so the SOER drops considerably. This also explains why in Figure 15(d) signal 3 has the highest PFD. Signal 2, which consists of five adjacent occupied bands, results in a SOER that is almost constant over the SNR range. This is mainly due to the fact that the signal is mostly occupied, so even when the location of the edges is not accurately detected, the energy detection step will show that the frequency band between the detected edges is occupied, resulting in a correct estimate of the subcarrier occupancy index, so for signal 2 it is the energy detection step that mainly affects the SOER. Signal 4 has fewer edges than signal 1, which increases the probability of noise peaks being mistaken for actual edges, so the PFD in case of signal 4 is higher than in case of signal 1, unlike the PMD (as shown in Figure 15(c)) which is lower in case of signal 4.

Performance of Reconstruction Methods

Using the auxiliary function method, the system performance using ℓ_p-norm reconstruction for $p = 0.1$, 0.2, 0.5, 0.8 and 0.9 is compared. For the simulation, a permutation measurement matrix and a DWT basis matrix are used for all cases, and the number of iterations is fixed at 10. Figure 16(a) compares the system performance in terms of PT, which clearly shows that as the value of p approaches zero, the error decreases. Also in Figure 16(b), the average relative error is compared, which is defined as:

Figure 12. A comparison of system performance in case of DWT, SWT and the multiscale sum cases in a compressive sensing framework in terms of (a) SOER, (b) PT, (c) PMD and (d) PFD
El-Khamy, Abdel-Malek, & Kamel, 2017.

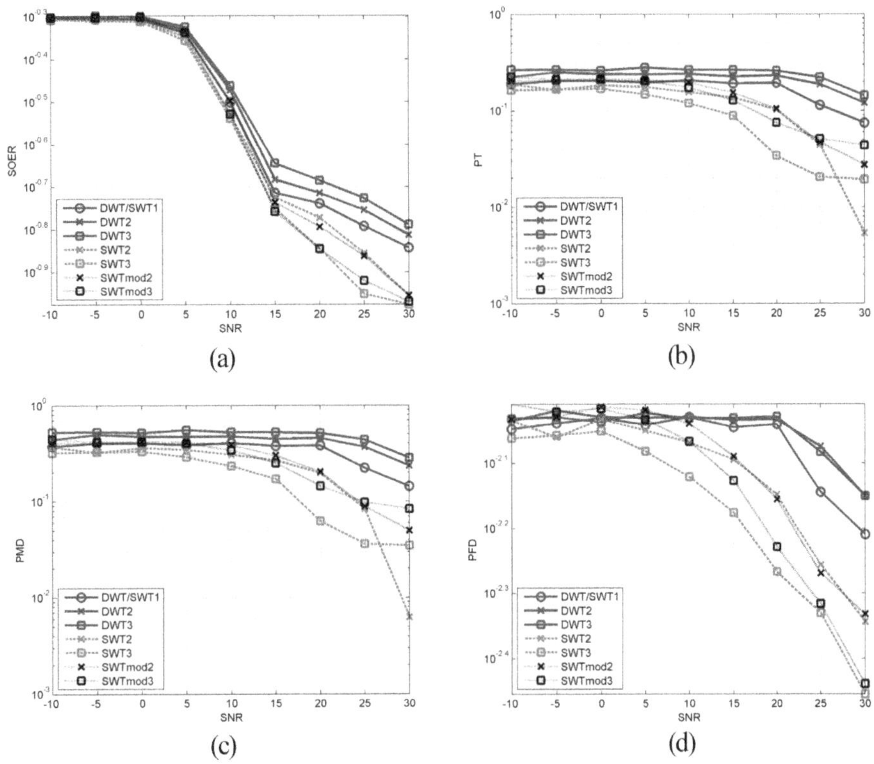

**For a more accurate representation see the electronic version.*

$$\varepsilon_{relative} = \frac{\sum_{i=1}^{N} \left| z_i^{(N_{iter})} - z_i^{(N_{iter}-1)} \right|}{N} \tag{25}$$

where x_i is the ith element of the vector $\mathbf{z}$ and N_{iter} is the number of iterations (which is, in this case, 10). The average relative error computes the error between the solution obtained at the final iteration and the solution at the previous iteration.

The performance of the genetic algorithm as a reconstruction method is also tested in Figure 17. For these simulation results, the initial population consists of 20 vectors formed as described in Figure 5 and a crossover factor of 0.8 is used. Figure 17(a) shows the change of the objective function of the genetic algorithm (using the ℓ_1-norm) with the SNR after 1, 5, 10 and 20 iterations; as the number of iterations increases, the final value of the objective function decreases. In Figure 17(b) the objective function value is plotted against the SNR for different values of p (i.e. different ℓ_p-norms). As the value of p decreases, the final value of the objective function increases, because ℓ_p-norms for smaller values of p represent a stricter measure of sparsity. For example, if a vector of length N has only 2 elements of relatively large values, and the rest of the $(N-2)$ are of negligible value (but not equal to zero), then its ℓ_1-norm will depend on the values of the 2 large elements while its ℓ_0-norm will be equal to N. On the other hand, if the $(N-2)$ elements are equal to zero, its ℓ_0-norm will be equal to 2 while its ℓ_1-norm will be almost

Figure 13. Comparing the DWT and SWT basis matrices for different compression ratios in terms of SOER
El-Khamy, Abdel-Malek, & Kamel, 2017.

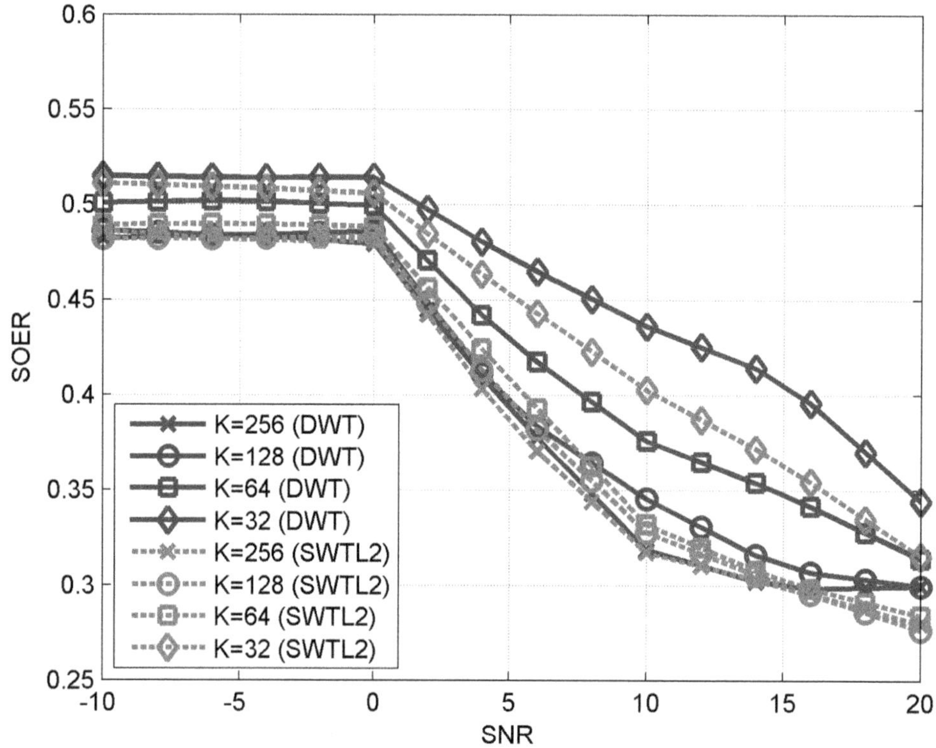

**For a more accurate representation see the electronic version.*

Figure 14. Different PSD signals for testing system performance
El-Khamy, Abdel-Malek, & Kamel, 2017.

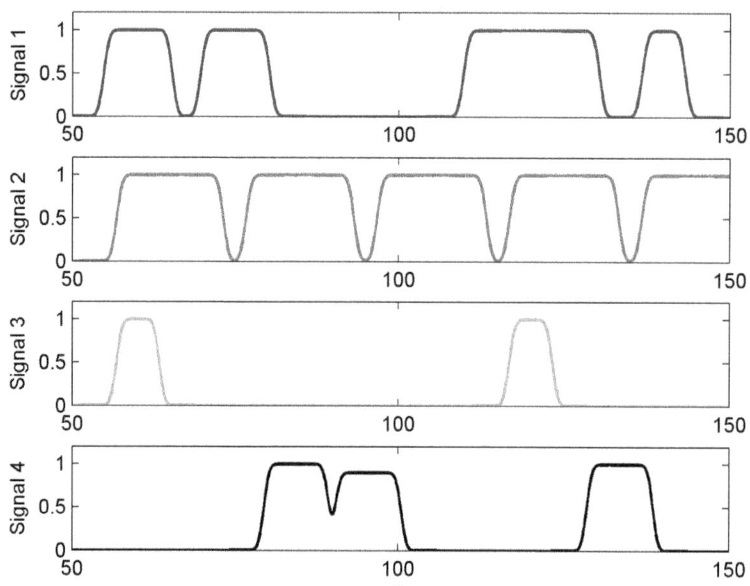

**For a more accurate representation see the electronic version.*

Figure 15. Comparing system performance for different signals in terms of (a) SOER, (b) PT, (c) PMD and (d) PFD
El-Khamy, Abdel-Malek, & Kamel, 2017.

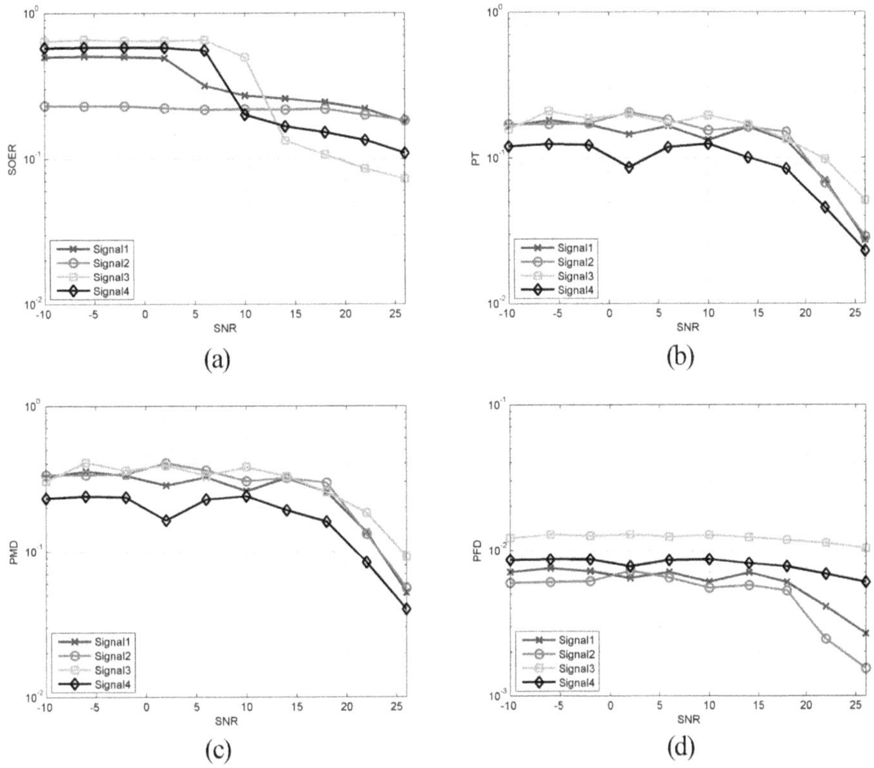

**For a more accurate representation see the electronic version.*

Figure 16. Reconstruction using the auxiliary function algorithm for ℓ_p-norm minimization

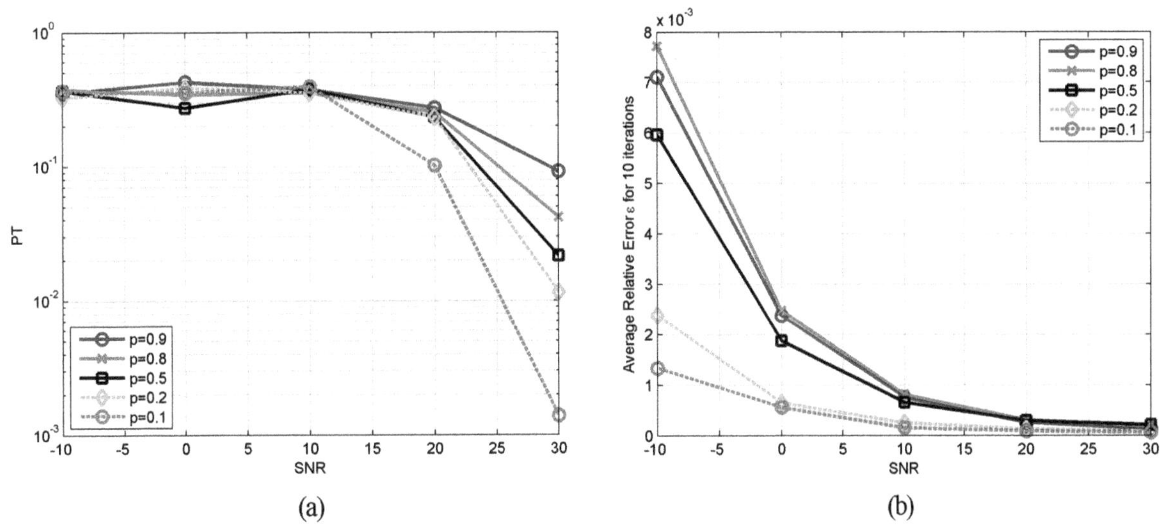

**For a more accurate representation see the electronic version.*

the same. To test the system performance, the PDF is shown in Figure 17(c) for different values of p, where the $\ell_{0.5}$-norm exhibits the lowest error rate, particularly for low SNR values. The SOER is shown in Figure 17(d) for different stopping criteria (number of iterations).

CONCLUSION

In cognitive radio systems, secondary users need efficient and reliable methods to sense the wideband spectrum and identify the location of the vacant bands. Compressive spectrum sensing is one such method; since compressive sensing enables the system to operate at sampling rates lower than the Nyquist rate while maintaining a reliable performance. There are two important design aspects in compressive sensing, namely the design of the sparsifying basis matrix and the reconstruction method. In this chapter, the wavelet transform is used to design the basis matrix for compressive sensing, since the wavelet transform of the signal PSD gives a sparse signal, and this signal indicates the location of the occupied channel edges, which appear in the form of peaks in the detail coefficients of the wavelet transform. It is also shown that the SWT is superior to the DWT and gives better results and more accurate estimates for the location of the channel edges both with and without compressive sensing. Results show that using the SWT, the same performance level of the DWT can be attained at half the sampling rate (or twice the compression ratio). Also, different reconstruction methods and sparsity measures are discussed. The

Figure 17. Performance of the genetic algorithm as a reconstruction algorithm
El-Khamy, Abdel-Malek, & Kamel, 2014.

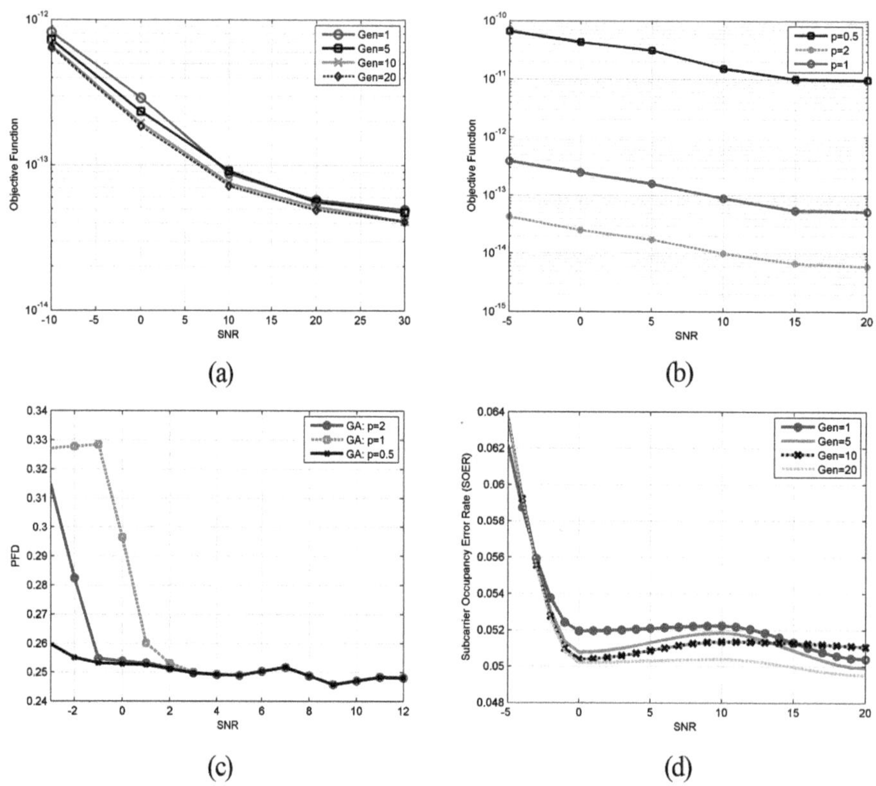

**For a more accurate representation see the electronic version.*

ℓ_1-norm is the most common measure of sparsity since finding the solution of minimum ℓ_1-norm can be reduced to a simple linear program. Theoretically, ℓ_p-norms for $0 < p < 1$ are better measures of signal sparsity, but unlike the ℓ_1-norm, the minimization problem becomes non-convex and therefore cannot be implemented by a linear program. However, using the genetic algorithm, any ℓ_p-norm can be used as an objective function, but its convergence depends on the initial population. The authors also propose the auxiliary function method for ℓ_p-norm minimization, which is a fast algorithm that works for values of p that are less than one, and simulation results have shown that as the value of p decreases, the overall performance is improved and the error rate decreases.

REFERENCES

Adoum, B. A., & Jeoti, V. (2010, May). Cyclostationary feature based multiresolution spectrum sensing approach for DVB-T and wireless microphone signals. In *Computer and Communication Engineering (ICCCE), 2010 International Conference on* (pp. 1-6). IEEE. 10.1109/ICCCE.2010.5556803

Bagwari, A., & Singh, B. (2012, November). Comparative performance evaluation of spectrum sensing techniques for cognitive radio networks. In *Computational Intelligence and Communication Networks (CICN), 2012 Fourth International Conference on* (pp. 98-105). IEEE. 10.1109/CICN.2012.66

Baraniuk, R. G. (2007). Compressive sensing. *IEEE Signal Processing Magazine*, *24*(4), 118–121. doi:10.1109/MSP.2007.4286571

Blair-Stahn, N. (2000). *Random Permutation Matrices*. Academic Press.

Candes, E., & Romberg, J. (2005). *l1-magic: Recovery of sparse signals via convex programming*. Retrieved from www.acm.caltech.edu/l1magic/downloads/l1magic.pdf

Carothers, N. L. (2000). *Real analysis*. Cambridge University Press. doi:10.1017/CBO9780511814228

Divakaran, K., Manikandan, N., & Hari, S. (2011). Wavelet based spectrum sensing techniques for cognitive radio-A survey. *International Journal of Computer Science & Information Technology*, *3*(2), 123–137. doi:10.5121/ijcsit.2011.3209

Donoho, D. L. (2006). Compressed sensing. *IEEE Transactions on Information Theory*, *52*(4), 1289–1306. doi:10.1109/TIT.2006.871582

El-Khamy, S., Abdel-Malek, M., & Kamel, S. (2013, April). C9. A New Wideband Spectrum Sensing Technique based on Compressive Sensing, Wavelet Edge Detection and Parallel Processing. In *Radio Science Conference (NRSC), 2013 30th National* (pp. 202-210). IEEE.

El-Khamy, S. E., Abd-el-Malek, M. B., & Kamel, S. H. (2017). A stationary wavelet transform approach to compressed spectrum sensing in cognitive radio. *International Journal of Communication Systems*, *30*(7), e3140. doi:10.1002/dac.3140

El-Khamy, S. E., Abdel-Malek, M. B., & Kamel, S. H. (2014, April). An improved reconstruction technique for wavelet-based compressive spectrum sensing using genetic algorithm. In *Radio Science Conference (NRSC), 2014 31st National* (pp. 99-106). IEEE. 10.1109/NRSC.2014.6835066

El-Khamy, S. E., El-Mahallawy, M. S., & Youssef, E. N. S. (2013, January). Improved wideband spectrum sensing techniques using wavelet-based edge detection for cognitive radio. In *Computing, Networking and Communications (ICNC), 2013 International Conference on* (pp. 418-423). IEEE. 10.1109/ICCNC.2013.6504120

Gorodnitsky, I. F., & Rao, B. D. (1997). Sparse signal reconstruction from limited data using FOCUSS: A re-weighted minimum norm algorithm. *IEEE Transactions on Signal Processing*, *45*(3), 600–616. doi:10.1109/78.558475

Gray, R. M. (2006). Toeplitz and circulant matrices: A review. *Foundations and Trends® in Communications and Information Theory*, *2*(3), 155-239.

Haggstrom, P. (2016). *The good, the bad and the ugly of kernels: why the Dirichlet kernel is not a good kernel.* Academic Press.

Haykin, S. (2005). Cognitive radio: Brain-empowered wireless communications. *IEEE Journal on Selected Areas in Communications*, *23*(2), 201–220. doi:10.1109/JSAC.2004.839380

Jafarpour, S., Schapire, R. E., & Cevher, V. (2011, May). Compressive sensing meets game theory. In *Acoustics, Speech and Signal Processing (ICASSP), 2011 IEEE International Conference on* (pp. 3660-3663). IEEE. 10.1109/ICASSP.2011.5947144

Kamel, S. H., El-Khamy, S. E., & Abdel-Malek, M. B. (2014, August). An improved compressed wideband spectrum sensing technique based on stationary wavelet transform in Cognitive Radio systems. In *General Assembly and Scientific Symposium (URSI GASS), 2014 XXXIth URSI* (pp. 1-4). IEEE. 10.1109/URSIGASS.2014.6929237

Ma, J., Li, G. Y., & Juang, B. H. (2009). Signal processing in cognitive radio. *Proceedings of the IEEE*, *97*(5), 805–823. doi:10.1109/JPROC.2009.2015707

Merry, R. J. E., & Steinbuch, M. (2005). *Wavelet theory and applications.* literature study, Eindhoven University of Technology, Department of Mechanical Engineering, Control Systems Technology Group.

Migliore, M. D. (2014). A simple introduction to compressed sensing/sparse recovery with applications in antenna measurements. *IEEE Antennas & Propagation Magazine*, *56*(2), 14–26. doi:10.1109/MAP.2014.6837061

Narendar, M., Vinod, A. P., Madhukumar, A. S., & Krishna, A. K. (2010, April). An algorithm for spectrum sensing in cognitive radio using tree-structured filter bank. In *Telecommunications (ICT), 2010 IEEE 17th International Conference on* (pp. 418-424). IEEE. 10.1109/ICTEL.2010.5478782

Nason, G. P., & Silverman, B. W. (1995). The stationary wavelet transform and some statistical applications. *Lecture Notes in Statistics*, 281-281.

Needell, D. (2009, November). Noisy signal recovery via iterative reweighted l1-minimization. In *Signals, Systems and Computers, 2009 Conference Record of the Forty-Third Asilomar Conference on* (pp. 113-117). IEEE. 10.1109/ACSSC.2009.5470154

Polo, Y. L. (2011). *Compressive wideband spectrum sensing for cognitive radio applications.* TU Delft, Electrical Engineering, Mathematics and Computer Science, Telecommunications.

Shannon, C. E. (1949). Communication in the presence of noise. *Proceedings of the IRE*, *37*(1), 10–21. doi:10.1109/JRPROC.1949.232969

Tian, Z., & Giannakis, G. B. (2006, June). A wavelet approach to wideband spectrum sensing for cognitive radios. In *Cognitive Radio Oriented Wireless Networks and Communications, 2006. 1st International Conference on* (pp. 1-5). IEEE. 10.1109/CROWNCOM.2006.363459

Tian, Z., & Giannakis, G. B. (2007, April). Compressed sensing for wideband cognitive radios. In *Acoustics, Speech and Signal Processing, 2007. ICASSP 2007. IEEE International Conference on* (Vol. 4, pp. IV-1357). IEEE. 10.1109/ICASSP.2007.367330

Tropp, J. A., & Gilbert, A. C. (2007). Signal recovery from random measurements via orthogonal matching pursuit. *IEEE Transactions on Information Theory*, *53*(12), 4655–4666. doi:10.1109/TIT.2007.909108

Tsutsu, H., & Morikawa, Y. (2012). An lp norm minimization using auxiliary function for compressed sensing. *Proc. IMECS.*

Wang, J., & Zhang, Q. T. (2009, April). A multitaper spectrum based detector for cognitive radio. In *Wireless Communications and Networking Conference, 2009. WCNC 2009. IEEE* (pp. 1-5). IEEE. 10.1109/WCNC.2009.4917513

Zeng, Y., Liang, Y. C., & Chia, M. W. (2011, May). Edge based wideband sensing for cognitive radio: Algorithm and performance evaluation. In *New Frontiers in Dynamic Spectrum Access Networks (DySPAN), 2011 IEEE Symposium on* (pp. 538-544). IEEE.

Zeng, Y., Liang, Y. C., & Chia, M. W. (2011, May). Edge based wideband sensing for cognitive radio: Algorithm and performance evaluation. In *New Frontiers in Dynamic Spectrum Access Networks (DySPAN), 2011 IEEE Symposium on* (pp. 538-544). IEEE.

KEY TERMS AND DEFINITIONS

Compressive Sensing: A signal processing technique that allows the reconstruction of signals from samples obtained by sampling the signal at a sub-Nyquist rate.

Genetic Algorithm: An evolutionary optimization algorithm based on the principles of genetics and the survival-of-the-fittest law of nature. Starting with a population of solutions, the algorithm applies crossover and mutation to the members of the population that best fit the objective function in order to obtain better fitting solutions.

ℓ_p-norm: A function that measures the length of a vector and can be considered as a generalization of the Euclidean norm.

Primary Users: Users licensed to use certain frequency bands and hence have higher priority over these bands.

Secondary User: Unlicensed cognitive radio users who attempt to opportunistically access the spectrum while avoiding interference with primary users.

Spectrum Hole: A frequency band that is momentarily unoccupied by a primary user. Secondary users use spectrum sensing to find spectrum holes.

Stationary Wavelet Transform: A modified version of the wavelet transform where the wavelet coefficients are not decimated at every stage.

Sub-Nyquist Rate: A sampling rate lower than twice the value of the highest frequency component of a signal (lower than the Nyquist rate).

Wavelet Transform: A transform characterized by its ability to provide time-frequency resolution and is considered a powerful tool for analyzing local singularities in a signal, which makes it suitable for edge detection.

Chapter 13
Improved Spectrum Sensing Based on Polar Codes for Cognitive Radio Networks

Idriss Chana
Moulay Ismail University, Morocco

Reda Benkhouya
Ibn Tofail University, Morocco

Abdallah Rhattoy
Moulay Ismail University, Morocco

Youssef Hadi
Ibn Tofail University, Morocco

ABSTRACT

One challenge of a sensing technique is reducing sensing time while ensuring good effective data rate. In fact, once compressive sensing based on sub-Nyquist sampling is adopted, sensing time can be reduced by saving number of samples. This increases the probability of missed detection which causes collisions with primary service and worsens channel imperfections. In such case erasures occur in addition to errors. In this chapter, the authors propose a new technique to correct erasures while keeping sensing time at a desired level. Based on polar code and low complexity decoding algorithm, the proposed technique exhibits for high code rates better performance in terms of bit error rate (BER) compared to two existing techniques based on other codes, namely low-density parity check (LDPC) and BCH.

INTRODUCTION

Electromagnetic spectrum is a scarce natural resource that needs to be managed. Cognitive radio (CR) is believed to ease congestion and overcome underutilization of the usually called spectrum. However, designing CR networks faces some challenging issues, such as spectrum sensing (SS). SS is a process of identifying the spectrum status for self-usage. The critical function of SS is required to identify spectrum

DOI: 10.4018/978-1-5225-5354-0.ch013

status and to determine if the spectrum allocation satisfies the quality of service (QoS) requirements of all users.

Generally, detection performance is characterized by false alarm and missed detection probabilities. False alarm occurs when the SS inaccurately detects primary users. Missed detection occurs when spectrum holes are detected. Moreover, in case of missed detection, this may cause collision between the primary and the secondary transmission.

SS techniques are indeed not perfect. When SS fails, it not only causes collisions, but also increases the error rate. Such imperfections can be handled either by using channel coding techniques, or increasing the sensing time. Both of the methods result in the reduction of effective data rate for secondary users. One can say that increasing sensing time enhances detection, whereas time resource is limited and shared by SS and data transmission. Since the sensing time determines network responsiveness and performance, thus it is necessary to reduce it as minimum as possible. However, reducing sensing time increases collision probability, which causes erasures and data loss along the secondary user bit stream. Thus, to counter such effects, the secondary receiver can employ channel coding technique while maintaining a desired sensing time.

To optimize the performance of the secondary user (SU) without incurring undesirable collisions, one may use P_m as the constraint for ensuring protection of primary users (PU) while setting P_f as the optimization target to ensure maximization of spectrum reuse by secondary users (Lin, Liu, & Hsieh, 2013). Such problem can be formulated as follows:

Minimize Pf, conditional on the constraint of $P_m \leq \varepsilon$ (1)

where ε is the desired threshold for PU protection.

In this chapter, the compressive sensing (CS) is used for detecting the presence of the PU. Authors consider the use of a CS thanks to its applicability to a wideband spectrum, and mathematical amenity. The models in literature have not considered both number of samples and sensing time in CS. These factors are the primary constraints in this chapter. A priori, the sample number M is determined as:

$$M = \tau f_s \tag{2}$$

where f_s is a sampling frequency and τ is sensing time. Obviously from the Equation 2, increasing parameter M increases sensing time, which affects network responsiveness and performance. Nevertheless, decreasing M worsens sensing imperfections.

At given SNR, authors showed in (Benkhouya, Chana, & Hadi, 2016) that there is an appropriate M_e such that no more samples M are required to verify the Equation 1 above. If M_0 is denoted the value of M for which $P_m = 0$, then $\Delta M = M_0 - M_e$ is the number of saved samples. Authors proved that for the CS deployed, there is a time reduction corresponding to $\Delta\tau = \Delta M / f_s$. Since the sensing time is now reduced, this causes erasures and channel coding techniques should be adopted to tackle these defects.

The adoption of channel coding has been studied by Haddadi et al. (2013) using rate-compatible (RC) LDPC codes. And obtained results have proven the effectiveness of the solution. Authors in (Benkhouya,

Chana, & Hadi, 2016) have proposed a method based on BCH codes. Nevertheless, the literature on the correction of the sensing errors is still limited. In this chapter, a new technique based on polar codes is proposed to correct errors and erasures, taking into consideration the probability of erasures 10^{-4}, required by the appropriate M_e already discussed.

Hence, authors complete here their study of (Benkhouya, Chana, & Hadi, 2016) to compensate for the decrease in sensing time. Computer simulations were carried out to evaluate the performance of the proposed technique. Since both BCH codes and polar codes have the soft-in hard-out (SIHO) decoding approach, it is obvious to compare them for the same code length and code rate. Authors plotted curves of bit error rate (BER) versus signal-to-noise (SNR), for the code lengths of $N = 128$ and $N = 127$ for polar code and BCH code respectively for various code rate, namely $R = 0.83$, $R = 0.78$, and $R = 0.5$ as medium code rate.

The RC-LDPC codes have soft-in soft-out (SISO) decoding technique, and were constructed in (Haddadi, Saeedi, & Navaie, 2013) with the block length of 10000 bits. Nevertheless, it is now judicious to increase the code length of polar codes which has a SIHO decoder. The main aim is to make a fair comparison even though the decoding approaches are different. Let $N = 1024$ be the higher code length of polar codes. Thus, authors plotted new curves of the BER performance versus SNR, for the above code rates $R = 0.83$, $R = 0.78$, and $R = 0.5$ as medium code rate.

While the proposed method outperforms the BCH based method for short code length for it exhibits good performance in terms of BER for long code length, in comparison with RC-LDPC based technique for high code rate.

Interestingly enough, the proposed method has lower time complexity compared to both methods described in (Haddadi, Saeedi, & Navaie, 2013) and (Benkhouya, Chana, & Hadi, 2016). As a result, the proposed method is an efficient technique to improve the spectrum sensing when adjusting sensing time at lower level.

COMPRESSIVE SENSING FOR WIDEBAND COGNITIVE RADIO

Even with the remarkable expansion of wireless applications, the wideband still offers new opportunities which need to be identified in a quick manner. Although the Nyquist rate sampling appears to be the best conciliation for wideband sensing however, it requires a high-speed signal processors and large number of samples. CS has been successfully applied to resolve the sampling issue, by leveraging the sparsity of the spectrum signal. Due to the inherent sparsity of spectral utilization, CS is hoped to reduce the sampling rates obviously for wideband sensing. Most of current works on wideband sensing are focused on CS, and many papers have been published about CS theory and applications. In the following, authors refer to (Rashidi, Haghighi, Owrang, & Viberg, 2011) and references therein, where the proposed CS method does not needs to know the usage information of the whole band, or even to reconstruct the wideband signal.

Figure 1 illustrates the model under consideration. The analog received signal is sampled by multicoset sampler at a sample rate lower than the Nyquist rate. The sampling reduction ratio is affected by the channel occupancy and multicoset sampling parameters. The output of the multicoset sampler is partially shifted using a multi-rate system, passing through the interpolation, delaying and down-sampling stages. Next, the sample correlation matrix is computed from the finite number of obtained data. Finally, the correlation matrix is used to discover the position of the active channels by subspace methods.

Figure 1. The adopted wideband spectrum sensing model

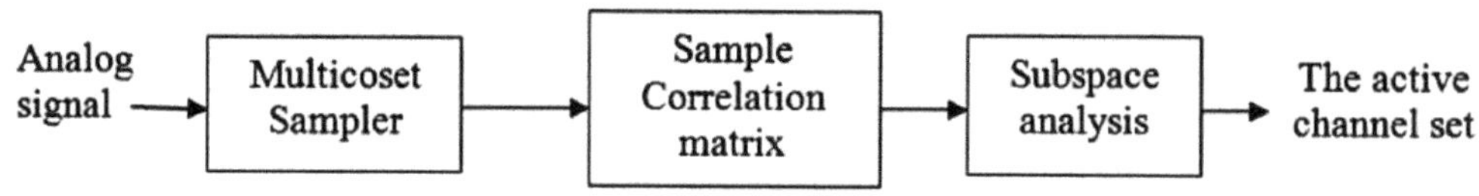

The received signal is assumed to be an analog wideband sparse spectrum signal, corrupted by additive white Gaussian noise (AWGN), and band-limited to $[0, B_{max}]$. Depending on the application, the entire frequency band is segmented into L narrowband channels, each of them with bandwidth B, where $B_{max} = L \times B$. Since the distribution of the signal is unknown, the real correlation matrix R of observation vector cannot be achieved.

The multicoset sampler provides randomly p data sequences. The $p \times p$ sample correlation matrix $\hat{R}$ is computed from M samples of the partially shifted sequence, under suitable assumptions $\hat{R} \to R$ when $M \to \infty$. As stated in Equation 2, M is a function of sensing time. This means that the computed correlation matrix needs an infinite sensing time to be estimated correctly. Therefore the estimation problem becomes a large-scale computationally intractable problem.

IMPERFECTIONS OF SPECTRUM SENSING

Minimizing errors in spectrum sensing is highly desired, since it both avoids collision with primary user and enhances usage level of vacant spectrum. IEEE 802.22 defines requirements for spectrum sensing performances where transmitting secondary user should stop its transmission within 2 seconds after the start of primary user transmission with false alarm probability P_f less than 0.1. The goal is to minimize the probability of missed detection P_m for a fixed level of false alarms. Moreover, maximizing the SU utility requires optimizing its probability of false alarm with respect to collision cost (Aulakh & Vig, 2014).

High false alarm is one of the problems to overcome in order to allow an application of spectrum sensing. Current researches mainly focus on the effect of false alarm on sensing time and on the effect of unnecessary spectrum handoff due to false alarm. By choosing a proper sensing time, the SU may make an attractive tradeoff between the false alarm probability and missed detection probability. When a false alarm arises in a spectrum sensing, the secondary user should retrieve another unused channel and carry out handoff to it. False alarms prevent wrongly spectrum holes from being accessed by secondary users, and hence result in spectrum underutilization (Ding, Wang, Wu, Zhang, Song, Zou, Yao, & Feng, 2014). Therefore wider spectrum band is progressively sought, leading to an amplified phenomenon of underutilization.

Spectrum sensing during transmission should satisfy the required missed detection probability. To ensure a desired QoS for the primary service, a maximum value for the probability of missed detection is enforced to the secondary service (Haddadi, Saeedi, & Navaie, 2013). But still a challenge is the tradeoff between the goals of minimizing P_m and minimizing P_f (Lin, Liu, & Hsieh, 2013). Figure 2 shows an example of P_m versus P_f. Such curve is called the receiver operating curve (ROC).

Figure 2. The ROC representing P_m vs. P_f

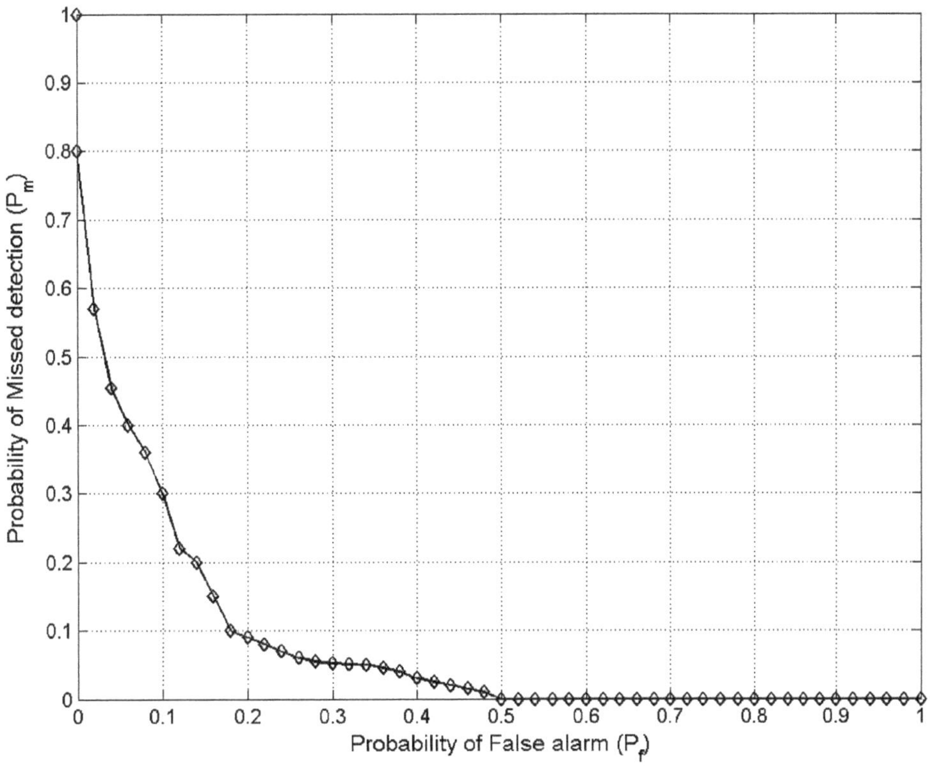

It is critical to reduce spectrum sensing time while satisfying the required sensing error probabilities. This leads to reduction of the energy consumption needed by the SU for sensing process, and also increase of data transmission rate (Lee & Jang, 2009).

In the next section, LDPC codes and BCH codes are reviewed in order to introduce some related works. Polar codes are presented independently in the section VI where the authors propose the new method.

PRELIMINARIES

LDPC Codes

LDPC codes are a family of linear block codes first introduced in 1960 by Robert Gallager in his thesis. The proposal was later published in (Gallager, 1963).

LDPC codes are block structured and linear divisible codes which are characterized by sparse parity-check matrix containing mostly zeros and only a small number of ones. A binary (N, J, K) LDPC code has block length N and a parity-check matrix with exactly J ones in each column and K ones in each row, assuming $J \geq 3$ and $K > J$. When the finite field is $GF(2)$, encoding complexity of LDPC decreases from N^2 to $N \log N$.

LDPC codes can perform close to the Shannon limit using iterative decoding, namely, the belief propagation (BP) or sum-product (SP) algorithms. The BP algorithm is more a striking choice when log-likelihood ratios (LLR) are used as messages between symbol and parity-check nodes. In the LLR based BP algorithm, the most computationally complex part is the check-node updates (Li & Zhang, 2008).

BP is based on iterative exchange of information between variable and function nodes in a graphical model (Bavarian & Cavers, 2007). Let N denote the block length. The BP needs to store $N\left(1+\log N\right)$ LLRs, $2N\log N$ additions and $2N\log N$ comparisons when $N \geq 64$ (Lin, Xiong, & Yan, 2015).

On time-varying channels, it is desirable to adjust the forward error correction (FEC) code rate according to the channel state information (CSI) (Zhang & Huang, 2009). Rate design for channel coding is indispensable to rate adaptive communications. Generally stated, low-rate codes can guarantee the transmission reliability under poor channel conditions, whereas high-rate codes can improve the transmission efficiency under good channel conditions. RC techniques can provide a way of low-cost implementation and capacity-approaching performances for practical rate-adaptive communications (Zhou, Huang, Zhao, & He, 2017).

In addition to their excellent performance and low decoding complexity, a major advantage of using RC codes is to reduce the storage memory. Moreover, any arbitrary rate can be reached by puncturing the parent code and the constructed code performs well.

Using RC techniques, a class of LDPC codes can be designed based on a parent LDPC code and decoded by only a single decoder for the parent code. RC techniques include puncturing, shortening, and protograph extension. Puncturing is the most popular technique for constructing RC-LDPC codes with high-rate codes as subsets of lower-rate codes (Zhou, Huang, Zhao, & He, 2017). RC-LDPC codes are obtained by gradually designing a puncturing pattern for a low-rate parent code (Babich, Noschese, & Vatta, 2016).

Since the use of codes with different rates is needed, Haddadi et al. (2013) proposed to employ those RC-LDPC codes. They consider an irregular LDPC parity check matrix $N = 2^m - 1$ with degree distribution polynomials $\lambda\left(x\right)$ and $\rho\left(x\right)$, defined as follows

$$\lambda\left(x\right) = \sum_{i=2}^{d_v \max} \lambda_i x^{i-1},\ \rho\left(x\right) = \sum_{i=2}^{d_c \max} \rho_j x^{j-1} \tag{3}$$

where λ_i and ρ_j denote the fraction of edges connected to variable nodes and check nodes of degree i, respectively. $d_v \max$ and $d_c \max$ denotes the maximum bit node and maximum check node degree respectively.

Due to the technical restrictions of time and the complexity of decoder design, LDPC codes were neglected for decades. Nowadays, LDPC codes are employed by several standards such as IEEE 802.16e (WiMAX), IEEE 802.3an (10GBASE-T), and IEEE 802.15.3c (WPAN) for its excellent performance and data rate (X. Lee, 2014). LDPC codes are also used for Wi-Fi 802.11 standard as an optional part of IEEE 802.11n and 802.11ac (Yang & Ku, 2008).

BCH Codes

Bose, Ray-Chaudhuri, Hocquenghem (BCH) codes, in turn, are among the best block codes, and form a special class of cyclic codes, because it detects and corrects a predetermined number of independent errors. As BCH codes are cyclic, they are best defined with the aid of finite-field theory (Peterson, 1972). The BCH code exists for any values $m \geq 3$ with code length $N = 2^m - 1$ with correcting capacity of $t = \left(d-1\right)/2$ where d is the minimum distance.

These BCH codes are called primitive because they are built using a primitive element of $GF\left(2^m\right)$. Thus, the BCH code length cannot be chosen randomly but it depends on the parameter m. For every K input bits, the BCH encoder adds $\left(N-K\right)$ redundant bits to generate codeword, leading to a code rate K / N. An N-bit BCH code is represented in terms of $BCH\left(N, K, t\right)$, where N is the length of the BCH code and K is the length of the message.

There are two methods for BCH encoding. The first one is based on matrix multiplication and has complexity of $\mathcal{O}\left(NK\right)$, the second one is polynomial and has complexity of $\mathcal{O}\left(\log\left(N-K\right)\log N\right)$.

The algebraic decoding methods for BCH code require solving the high degree equation in the Galois field like Berlekamp-Massey (BM) and the Euclidean algorithms. These algorithms have the high delay processing when n and t increase. Such methods are called hard in hard out (HIHO) decoders, and correct all combinations of t or less error.

Chase proposed a soft-decision decoding algorithm that first flips η least reliable bits before a conventional hard-decision decoding. By using soft information, Chase II algorithm can correct up to $t+\eta$ errors.

BCH codes are implemented in many applications, including but not limited to optical transport network, and digital video broadcasting. All the DVB standards use a BCH code outer code to eliminate residual errors after LDPC decoding (Wells, 2011).

RELATED WORKS

RC-LDPC Based Technique for Erasure Correction

Haddadi et al. (2013) have proposed to employ channel coding techniques where they used RC-LDPC codes. They adopted a local sensing technique, namely the energy detection which is not suited for the wideband spectrum. They claimed after simulation results, that effective rate of data transmission for the coded case can be significantly more than that of the case without channel coding while exhibiting a considerably better performance.

In (Haddadi, Saeedi, & Navaie, 2013) an arbitrary rate is expected, thus Haddadi et al. (2013) started from a parent code and obtained the desired rate by puncturing. They considered a rate one-half parent code with degree distributions:

$$\lambda(x) = 0.327x + 0.235x^2 + 0.438x^7$$
$$\rho(x) = 0.229x^5 + 0.771x^6 \quad (4)$$

The code was intended particularly for the proposed cognitive channel model with a combination of Rayleigh fading and erasure channel. The simulation results showed that the designed ensemble when used over the considered channel outperforms rate one-half ensembles with similar constituent degrees which were already optimized for the Gaussian channel. The parity check matrices are built based on the progressive edge growth (PEG) method. The block length is set to 10000 bits and the maximum number of BP iterations is limited to 50.

Haddadi et al. (2013) have considered both random and intentional puncturing starting from the parent code whose rate is 0.50. The performance of the parent code is shown by solid star curve in Figure 3. Using random puncturing, when the rate is increased to 0.79, the corresponding performance is given by the solid circle curve. A slight increase in the code rate to 0.82 comes up with the solid square curve, which gives the worst performance. Considerable performance improvement can be achieved if authors use intentional puncturing.

The performance of almost all coded systems is better than the uncoded one with some exception at code rate 0.82. When the code rate is reduced, the BER performance improves quickly. In fact, the code rate of 0.5 leads early to a BER less than 10^{-1} which is not the case for all other systems.

As a conclusion, the above results mean that it is judicious to explore the case of high code rates, in order to improve the performance of such case. In other words, the RC-LDPC based technique should be challenged.

BCH Based Technique for Erasure Correction

In the same perspective as (Haddadi, Saeedi, & Navaie, 2013), authors have proposed in (Benkhouya, Chana, & Hadi, 2016) a technique based on BCH codes to correct erasures caused by sensing time reduction. This technique was intended for compressive sensing instead of energy detection adopted by Haddadi et al. (2013).

To decode BCH codes, authors proposed a decoding scheme based on conventional Chase algorithm described in following paragraph.

1. Chase Algorithm

The Chase algorithm is a sub-optimal decoding method that uses a set of most likely error patterns to correct only errors due to channel transmission (not erasures). These error patterns are chosen based on the reliability of the received symbols. Each error pattern is added to the hard-decision received word and decoded using a hard-decision decoder. A decoded codeword is scored by computing its metric with respect to the received sequence (soft decision). The codeword with maximum metric is chosen as the most expected. Thanks to its reduced complexity and good performance, the most popular Chase Algorithm is the so-called Type-II to be described in the Algorithm 1.

Figure 3. Performance comparison of the uncoded system and the coded one for rates 0.82, 0.79 and 0.5 with random and intentional puncturing

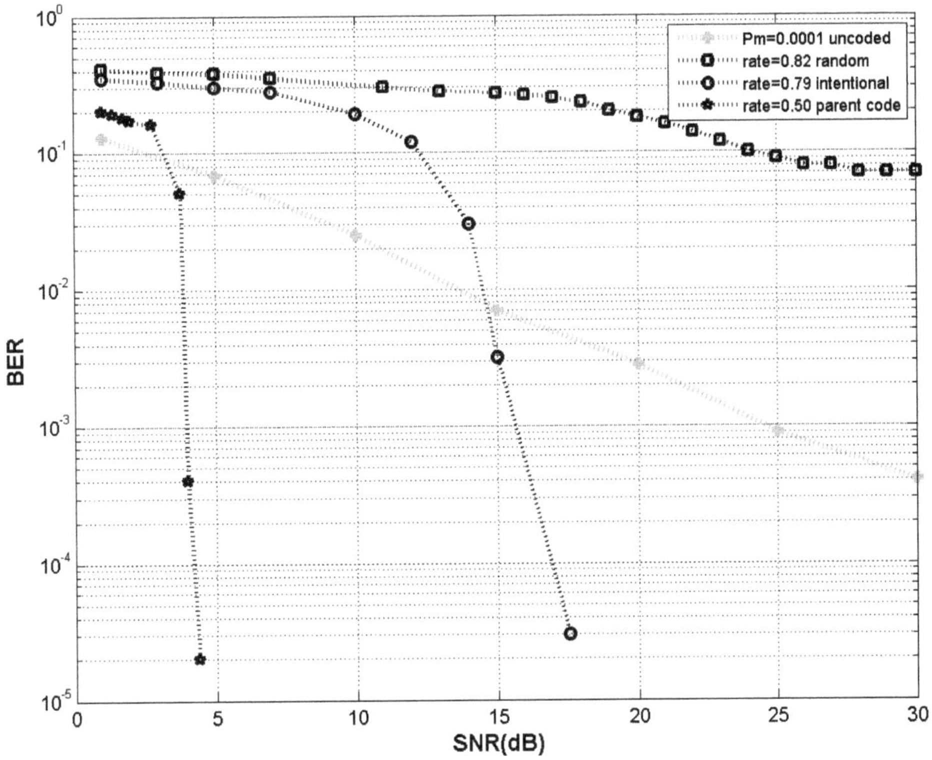

2. Chase Based Algorithm for Erasure Correction

Authors aimed in (Benkhouya, 2016) to correct both errors and erasures caused by channel transmission and collisions respectively. To this end, they proposed a new decoding scheme based on Chase Type-II algorithm. Before presenting the algorithm, some notations used later are listed as follows:

Let $BCH(N,K,t)$ be a block code capable of correcting any combination of $t=(d-1)/2$ or less random bit errors, and $R\{r_0,\ldots,r_{N-1}\}$ be the received word from the output of a combination of Rayleigh fading and erasure channel, where binary antipodal signaling (BPSK) is used. At the receiver side a hard decision $H\{h_0,\ldots,h_{N-1}\}$ is made by the demodulator of the received word $R\{r_0,\ldots,r_{N-1}\}$ where:

$$h_i=\begin{cases}0, r_i>0\\1, otherwise\end{cases}, \ 0\le i\le N-1 \tag{5}$$

To handle with erasures, Benkhouya et al. (2016) proposed a metric called the Euclidian Erasure Distance (EED) and denoted by $(.)$. It can be defined as a classical Euclidian distance with no contribution of the position of erasures.

Algorithm 1. Chase Type-II decoding

Step 1: Generate 2^t error patterns $\{e_i\}$ among the t least reliable positions. Precisely, positions $\{I_1,\ldots,I_t\}$, for which the reliabilities (amplitudes) are the smallest

Step 2: Add $\overline{e}_i$ to the hard-decision received word $\overline{z}_i = \overline{e}_i \oplus \overline{z}_0$, where $\overline{z}_0$ is the hard version of the received vector

Step 3: Use $\overline{z}_i$ as input to a hard-decision decoder, generating a codeword $\overline{v}_i$, $i = 1, 2, \ldots 2^t - 1$

Step 4: Calculate the metric according to $v_0 = \sum_{j=1}^{n} (-1)^{v_{0j}} \times r_j$ and if maximum, codeword $\overline{v}_i$ stored as the most likely

Let $R\{r_0,\ldots,r_{N-1}\}$ be the received vector with set of erasures ϵ. Let $\mathrm{D}\{d_0,\ldots,d_{N-1}\}$ be the hard vector where $d_i \in \{0,1\}$.

$$(D,R) = \sum_{\substack{i=0 \\ i\notin\epsilon}}^{n-1} \sqrt{\left(r_i - (2d_i - 1)\right)} \tag{6}$$

The new decoding algorithm is applied as described in Algorithm 2.

To guarantee the QoS of secondary service, the maximum acceptable probability of erasure is considered to be $P_m = 10^{-4}$. Which is equivalent to creating in the worst case a channel with probability of erasure 10^{-4}. This allows tolerating collisions by reducing sensing time.

The study consists on evaluating the technique using $BCH(127, K, t)$ codes with various code rates, namely $R = 0.83$ for $BCH(127,106,3)$, $R = 0.78$ for $BCH(127,99,4)$ as highest code rate. $R = 0.5$ for $BCH(127,64,10)$ as medium code rate.

Two types of channel transmission are considered, namely AWGN and Rayleigh fading with erasures channel, by using the BPSK modulation.

Figure 4 illustrates the BER performances of AWGN and Rayleigh fading combined to erasures channel. The corresponding curves are represented by dashed and solid lines respectively. It is shown that the coded system exhibits better results compared to the uncoded one, and this for all BCH studied codes.

Figure 5 shows the results of Benkhouya et al. (2016) compared to those obtained by Haddadi et al. (2013) who used the RC-LDPC codes in Rayleigh fading with erasures channel. The results of Haddadi et al. (2013) corresponds to the so-called parent code with rate

$R = 0.50$, intentional puncturing with rate $R = 0.79$ and random puncturing with rate $R = 0.82$. BCH codes and RC-LDPC codes are illustrated by solid line curves and dotted line curves respectively.

Algorithm 2. Erasure correction using Chase and EED

Step 1: Identify the erasure positions on the received vector R

Step 2: Substitute each erasure position detected on R with -1 to obtain a new vector R_1

Step 3: Perform Chase algorithm for R_1 to obtain first decision D_1

Step 4: Substitute each erasure position detected on R with +1 to obtain a new vector R_2

Step 5: Perform Chase algorithm on R_2 to obtain second decision D_2

Step 6: For final decision D_f, use Equation 6 to compute the metric for both D_1 and D_2 according to R

Step 7: if $(D_1, R) > (D_2, R)$

then $D_f = D_2$

else $D_f = D_1$

Step 8: end if

Figure 4. Performance of the BCH coded system under AWGN and Rayleigh with $\boldsymbol{P}_m = 10^{-4}$

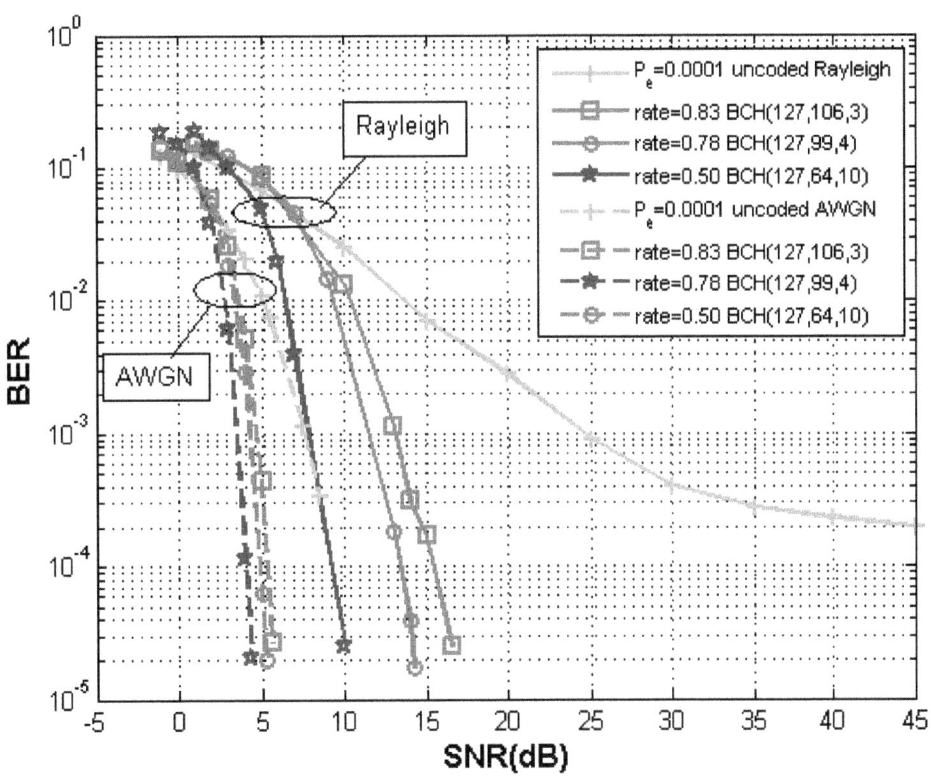

**For a more accurate representation see the electronic version.*

For all the BCH codes mentioned so far except the one with rate 0.50, the performance is better than that of RC- LDPC codes.

In (Benkhouya, Chana, & Hadi, 2016), authors kept the sensing time low while combining compressive sensing with channel coding in a communication system. Authors started by studying the effect of the number of samples, which means the effect of the sensing time on probability of missed detection. Authors showed that reducing sensing time increases the probability of missed detection and causes erasures.

To alleviate that, authors proposed a new technique based on BCH codes and Chase Type-II algorithm. Authors showed that they can reduce sensing time to desired value, while compensating sensing imperfections caused by such reduction. In addition, the use of highest code rates allows improving effective date rates. Using simulations authors showed that, compared to the corresponding algorithms which use RC-LDPC codes, the proposed method shows substantial improvement in performance for high rates under the same traffic conditions.

THE NEW METHOD BASED ON POLAR CODES

Polar code has attracted more attentions since it was firstly introduced by Arikan in 2008. Authors propose in this section to employ polar codes in order to improve the compressive sensing. Unlike RC-

Figure 5. Performance comparison between BCH coded system and that of Haddadi et al. (2013) under Rayleigh fading for rates 0.83, 0.78 and 0.5, with $P_m = 10^{-4}$

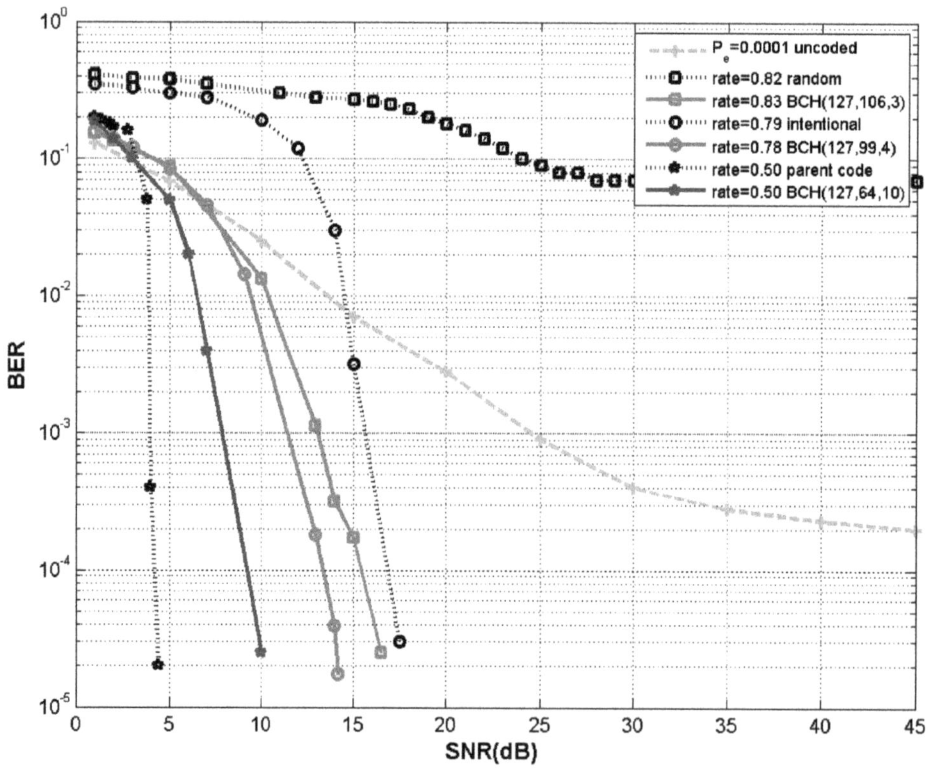

**For a more accurate representation see the electronic version.*

LDPC, BCH codes and maybe other types of codes, polar codes are known for their encoding/decoding simplicity and low complexity. To complete encoding and decoding scheme, authors propose new algorithm for erasure correction.

With the same approach as the related work in (Benkhouya, Chana, & Hadi, 2016), authors highlights in this section the relation between number of samples and sensing time before presenting the new coding scheme.

Number of Samples vs. Sensing Time

The compressive sensing is based on sub-Nyquist sampling, whose parameter is the number of samples M. Since the sensing technique is characterized by the probability of missed detection P_m, which is imposed to secondary service, it is important to focus on the Figure 6 that illustrates P_m versus M for various SNR.

The probability of false alarm P_f, in turn, determines the transmission data rate of secondary service. In fact, as P_f decreases there are more opportunities to transmit data. Figure 7 illustrates the P_f versus M for various SNR.

Recall that to ensure the desired QoS of primary service; it is necessary to respect the Equation 1:

$$Minimize\ P_f, subject\ to\ the\ requirement\ of\ P_m \leq 10^{-4}$$

where the protection threshold is $\varepsilon = 10^{-4}$.

Authors know that for $P_m = 0$, the sensing time becomes $\tau = M_0 / f_s$, where f_s the sampling frequency.

On the other hand, from Figure 6 and Figure 7, authors consider an appropriate $M_e \leq M_0$ from which no more samples are needed to verify the Equation 1, such that $P_f = 0$ and $P_m \leq 10^{-4}$. Thus, time reduction corresponds to $\Delta\tau = \frac{\Delta M}{f_s}$, where $\Delta M = M_0 - M_e$

Table 1 presents from Figure 6, the different values of M_0 and M_e for each SNR.

Figure 6. Probability of missed detection versus M and SNR for the simulated system, under AWGN

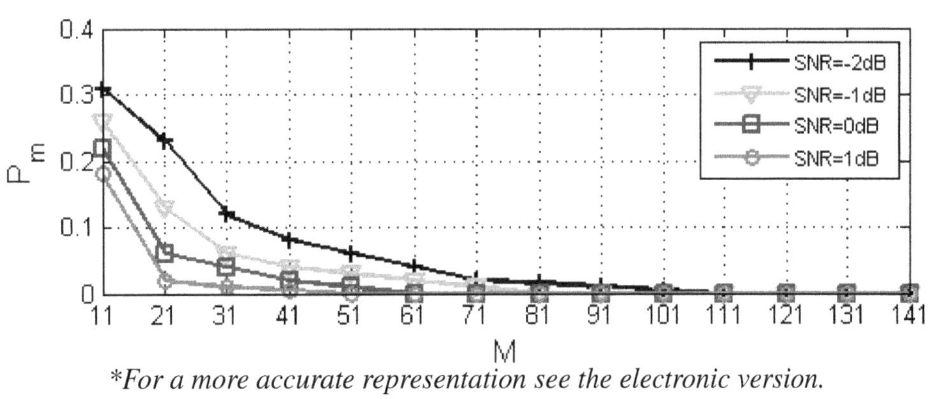

For a more accurate representation see the electronic version.

Figure 7. Probability of false alarm versus M and SNR for the simulated system, under AWGN

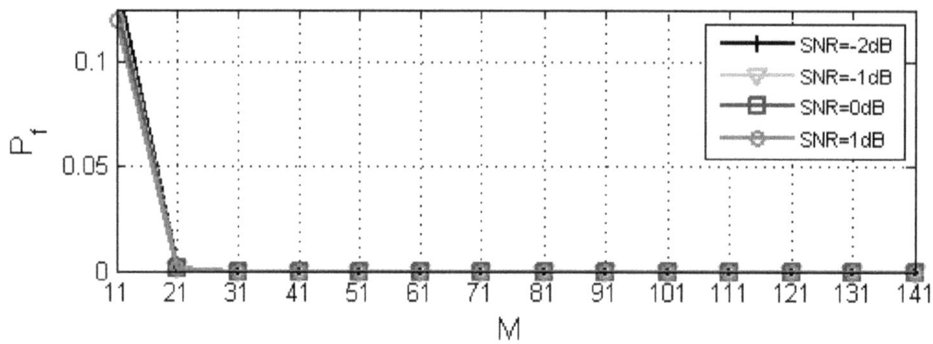

**For a more accurate representation see the electronic version.*

Table 1. M vs. SNR for $\boldsymbol{P}_m = 10^{-4}$ *and* $\boldsymbol{P}_m = 0$

SNR	-2	-1	0	1
M_e ($P_m = 10^{-4}$)	101	71	51	41
M_0 ($P_m = 0$)	111	81	61	51

Since the sensing time is now reduced, there is a need to deal with coding and decoding scheme. For this, authors propose new decoding scheme considering that the probability of erasures is $P_m = 10^{-4}$.

The Proposed Coding Scheme

1. Polar Codes

Let (N, K) be a linear bock code, where $N = 2^n$ is the code length, n is an arbitrary integer, K is the code dimension, the code rate of such coding scheme is defined by $R = K / N$, $0 \le K \le N$.

A binary polar code is generally specified by a triple $(N, K, \mathcal{F})$, where $\mathcal{F} \subseteq \mathbb{N}$, $|\mathcal{F}| = N - K$ is the set of the frozen bit indices. The remaining K elements are the information bits indices. For simplicity, let denote such code $P(K, N)$ in the rest of this chapter.

The construction itself was first described by (Stolte, 2002), and later independently by (Arikan, 2009).

Let $F^{\otimes n}$ denotes the n-th tensor power of F and it is calculated by applying the Kronecker product recursively according to $F^{\otimes n} = F^{\otimes(n-1)} \otimes F$.

Then for a vector of information bits u of length K, a codeword is generated as

$$x = G.u \tag{7}$$

where $G \triangleq \left(F^{\otimes n}\right)_{\mathcal{F}^C}$ is the generator matrix of polar code which picks a specific subset of K rows of the $N \times N$ matrix, and $\mathcal{F}^C \triangleq \left\{0,1,\cdots,N-1\right\} \setminus \mathcal{F}$ corresponds to the set of non-frozen bit indices.

Implicitly, with respect to classical $N \times N$ matrix based encoding, the frozen bits $\mathcal{F}$ are set to zero and the authors follow this convention throughout the paper. The authors use the notation W to denote a discrete memoryless channel $W : X \rightarrow Y$ with input alphabet X, output alphabet Y, and channel transition probability $W(y \mid x)$ that $y \in Y$ is received given that $x \in X$ $\Delta M = M_0 - M_e$ is transmitted.

2. Construction of Polar Code

The selection of the list of frozen bit $\mathcal{F}$ is referred to as code construction. According to polarization phenomenon, two groups of bits are established from the many independent copies of channels.

The noiseless or pure-noise of bit-channels are determined by the Bhattacharyya parameter. The Bhattacharyya parameter is defined as the parameter to measure the reliability of channel since it is an upper bound on the probability of maximum-likelihood (ML) decision error for an uncoded transmission over a channel.

The earliest polar coding is based on the evolution of simple bounds on the Bhattacharyya parameters of bit channels. Due to its simplicity, this construction has been widely used.

Let us observe that the Bhattacharrya parameters approach 0 or 1 exponentially with the block length N. For general binary memoryless channels, the Bhattacharrya parameters are exponentially hard to compute as block length increases. To calculate the Bhattacharyya parameter, the next key element is the initial value.

The original recursive algorithm requires an initial value and this was proposed by Arikan as 0.5, corresponding to the worst BER. This value is actually the Bhattacharyya parameter of the underlying binary input AWGN (BI-AWGN) channel; therefore it may be replaced with $\exp\left\{-RE_b / N_0\right\}$ due to the non-universality of polar codes.

There is a design condition based construction, denoting the use of a unique code designed by running the code construction at design state and keeping the code unchanged for all channel conditions. One can read (Benkhouya, Chana, & Hadi, 2017) to get a glimpse of this.

Polar codes are non-universal by definition. Hence, in the proposed system, the code construction is repeated at every value of channel state in channel range. This way, the polar code changes at every time channel changes.

Figure 8 shows the way to construct a polar code in this paper, which is depicted in (Benkhouya, Chana, & Hadi, 2017; Qin, Guo, Bhatia, Fàbregas, & Siegel, 2017).

3. Polar Decoder

The decoding algorithm of the proposed system is based on the successive cancellation (SC) algorithm. As is the case with polar encoding, SC decoding has the same complexity of $\mathcal{O}\left(N \log N\right)$, for code length N. The SC decoding algorithm is often defined as a path search procedure in the code tree.

Recently, a leading global telecom provider declared it has accomplished downlink speeds of tens gigabits using polar code, a new advance in channel coding. During a field test the provider was able to

Figure 8. Construction of polar code

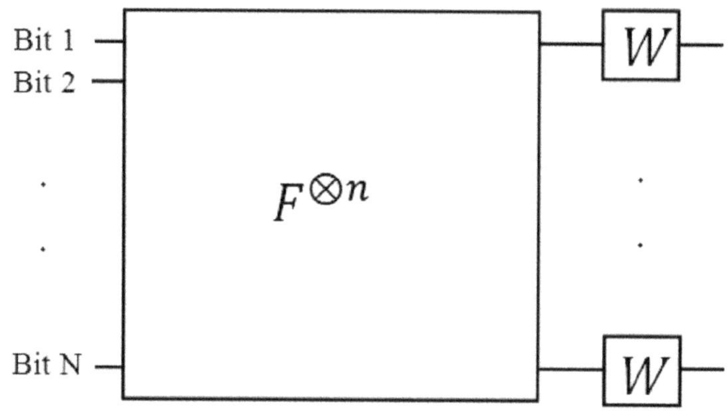

prove that polar code can meet the typical use cases of the 5G definition as stated by IEEE 802.11ac. This technology provides speeds hundred times higher than the current 4G LTE, and enables network expansion to billions of connections (Bushell-Embling, 2016). The

5G technology has been improved for better coverage and efficiency of spectrum and signaling. These improvements promise to make possible changes like the Internet of Things.

4. Proposed Algorithm for Erasure Correction

The authors aim to correct erasures caused by collisions. For this, they propose a new decoding scheme. Firstly, they define some useful notations as follows:

$H = \{h_i\}, i = 1,\ldots,N$ which returns the hard decision of a vector $V = \{v_i\}, i = 1,\ldots,N$, such that

$$h_i = \begin{cases} -1, v_i \leq 0 \\ 1, v_i > 0 \end{cases} \tag{8}$$

Authors define $rabs = |v_i|$, where $|.|$ returns the absolute value, synonymous of the reliability of the $\{v_i\}$.

Let $\delta(.,.)$ denotes the Euclidian distance between two vectors in the Euclidian space. Let $R = \{r_i\}, i = 1,\ldots,N$ be the received vector from an erasure channel.

Let define the SC decoder as function $SC : Y \to X$, where $X \in \{0,1\}^K$ and $Y \in \{-1,1\}^N$. The function is denoted $SC(H)$ when applied to vector H.

The native decoder SC of polar code cannot performs well when transmission is over an erasure channel. Authors believe that the conventional SC decoder should be enhanced to overcome its shortcomings. For this, they develop the new algorithm described below in Algorithm 3:

Algorithm 3. Correction of erasures using SC decoder

Input: probability of miss detection P_m, number of erasures cpt, parameter of correction p, received vector R

Output: final decision D

Step 1: Detect number of erasures cpt

Step 2: Replace the erasure position with 0 to obtain R'

Step 3: Calculate the two vectors H and $rabs$ of length N

$H = h\left(R'\right)$

$rabs = \left|R'\right|$

Step 4: Arrange $rabs$ in ascending order reliability

Step 5: Create the error patterns among the p least reliable positions

Step 6: Combine H with error patterns to obtain candidates C_j, $j = 1,\ldots,N$ and apply SC to decode each one to T_j

Step 7: Apply re-encoding on each result to obtain a list of decisions d_j

Step 8: Calculate all $\delta_j\left(R',d_j\right)$ and save d_j

Step 9: The final decision is $D = \left\{SC\left(d_i\right) \mid \delta_i\left(R',d_i\right) = \min\left\{\delta_j\left(R',d_j\right), j \in \left\{1,\ldots,2^p\right\}\right\}\right\}$

where $i \in \left\{1,\ldots,N\right\}$

a. Description of Algorithm Steps

In this paragraph authors would like to explain their algorithm, to make it more understandable, and this for each step.

Step 1: Since the channel is with erasures, the algorithm starts by determining the number (cpt) and positions of erasures.

Step 2: This step consist on attributing a value for all erased symbols.

Step 3: Calculate H the hard decision of the vector R' derived from R. Calculate absolute values for the so-called derived vector R' to obtain $rabs$. Note that the vector $rabs$ still contains some zeros which correspond to erasure positions.

Step 4: In this step, it is obvious that zeros corresponding to erasures will be in the first positions followed by the remaining symbols. This way, it is certain that erasures will be always with the least reliable positions. Those positions will be objects of the 2^p error patterns, i.e. all combinations of $\left\{-1,1\right\}$ among p positions.

Step 5: In this step, three cases are distinguished.

The first one is when parameter p is equal to the number of erasures ($p = cpt$). Hence, all error patterns belong to erasures positions.

The second case is when $p > cpt$. Error patterns cover in addition to erasures, some not erased symbols corresponding to least reliable ones.

The last case is when $p < cpt$ which is considered as rare. In this case, not all erasures are covered by error patterns. However, they were already fixed as -1 when the hard decision was applied (step 3).

Step 6: The result of such combining is a set of 2^p candidates of length N. This set will be decoded by SC to obtain the test sequences with length K. Authors intend to apply slight change on input of the original SC decoder. This way, instead of handling directly with the channel outputs as real domain, the SC decoder will deal with $\left\{-1,1\right\}^N$ as input.

Step 7: To allow composition with the received vector R', the test sequence is re-encoded to restore the original length N of the codeword, and this for each test sequence.

Step 8: Calculate Euclidian distance as metric from received vector R' of each test sequence.

Step 9: Finally the output of decoder D is the d-th test sequence which corresponds to the smallest Euclidian distance such that $D = \left\{T_d \mid \delta_d = \min\left\{\delta_j, j = 1,\ldots,2^p\right\}\right\}$.

b. Example of the Algorithm Execution

Let $p = 2$ be the parameter of correction. Let consider a polar code $P\left(8,4\right)$. The information to encode is $I = \left(1,0,1,1\right)$. The codeword to transmit over a Rayleigh channel with erasures is $X = \left(1,0,0,1,0,1,1,1\right)$. The received word is $R = \left(0.7,-0.8,\Theta,2.1,-0.2,0.4,-0.1,1.6\right)$ where Θ corresponds to the erased symbols. In this case, there is one erased symbol (step 1).

Now let replace erased symbol by zero. The derived vector is then (step 2):

$$R' = \left(0.7,-0.8,0,2.1,-0.9,0.4,-0.1,1.6\right)$$

The hard decision and absolute values of the vector R' is calculated to obtain H and $rabs$ (step 3):

$$H = \left(1,1,-1,1,-1,1,-1,1\right)$$

$$rabs = \left(0,0.1,0.4,0.7,0.8,0.9,1.6,2.1\right)$$

The ascending order of reliabilities is arranged by positions (Step 4):

$$rabs_3 < rabs_7 < rabs_6 < rabs_4 < rabs_1 < rabs_8 < rabs_2 < rabs_5 < rabs_4$$

which yields to positions $\left(3,7,6,4,1,8,2,5,4\right)$.

Since parameter of correction is set on $p = 2$, authors come up with the second case. The 2 positions that contribute in test configuration are 3 and 7. While the position 3 corresponds to the erasure, the position 7 corresponds to the least reliable one (step 5).

According to the parameter $p = 2$, the SC decoder will be performed on 4 candidates that are derived as a combination between H and symbols $\{1, -1\}$ in position 3 and 7 (in bold). Thus, the candidates are:

$$C_1 = (1, -1, 1, 1, -1, 1, 1, 1)$$

$$C_2 = (1, -1, 1, 1, -1, 1, -1, 1)$$

$$C_3 = (1, -1, -1, 1, -1, 1, 1, 1)$$

$$C_4 = (1, -1, -1, 1, -1, 1, -1, 1)$$

Next step is applying the SC decoder to the four candidates and obtain test sequences as follows (Step 6):

$$T_1 = SC(C_1) = (1, 0, 0, 1)$$

$$T_2 = SC(C_2) = (1, 0, 0, 0)$$

$$T_3 = SC(C_3) = (1, 0, 1, 1)$$

$$T_4 = SC(C_4) = (0, 0, 1, 1)$$

The Step 7 of the algorithm consist on re-encoding each to obtain 4 decisions d_1, d_2, d_3, d_4 and their metrics $\delta(R', d_i)$, $i \in \{1, 2, 3, 4\}$:

$$\{d_1, \delta(R', d_1)\} = \{(1, 0, 0, 1, 0, 1, 1, 1), 20.04\}$$

$$\{d_2, \delta(R', d_2)\} = \{(1, 1, 1, 1, 0, 0, 1, 1), 15.76\}$$

$$\{d_3, \delta(R', d_3)\} = \{(1, 0, 0, 1, 0, 1, 1, 1), 6.68\}$$

$$\{d_4, \delta(R', d_4)\} = \{(0,0,1,1,0,1,0,1), 26.72\}$$

As a result, the final decision is d_3 since it is achieving the minimum Euclidian distance from R'.

SIMULATION RESULTS

This section reports computer simulations that have been conducted to evaluate the performance of the decoding algorithm, and demonstrate the effectiveness of the proposed technique. Considering the used polar code, authors focused on $P(128, K)$ with two levels, high and medium code rate. The code rates of the dedicated codes are $R = 0.83$ for $P(128,107)$, $R = 0.78$ for $P(128,100)$, and $R = 0.5$ for $P(128,64)$. Table 2 shows the simulation parameters.

The transmission is over AWGN/Rayleigh fading and erasures channel, by using the BPSK modulation. In that event, the simulation will stop when achieve at least 10000 blocks and 100 errors. The maximum acceptable probability of erasure P_m of the secondary service is considered to be 10^{-4}. The parameter of correction p is set to 3.

For all next figures, the dashed curve with plus sign illustrates the BER performance versus SNR for the uncoded case.

Figures 9 and 10 correspond to the case of AWGN combined to erasure channel. In both figures, it is shown that the coded system exhibits very interesting results compared to the uncoded one, and this for all studied code rates $R = 0.83$, $R = 0.78$ and both code lengths $N = 128$ and $N = 1024$.

The same behavior is observed in the case of Rayleigh fading combined to erasure channel, as depicted in Figures 11 and 12. The coded system improves significantly the performance in terms of BER compared to the uncoded system from a certain SNR.

For example, for the code rate $R = 0.83$, the proposed coded system starts outperforming the uncoded system when SNR is greater than $6.5dB$ and $12.4dB$ respectively for $N = 128$ and $N = 1024$. This means that given code rate R, increasing code length N changes the operational SNR domain. The same behavior can be observed for other code rates.

On the other hand, the coded system delivers a good performance for both code lengths $N = 128$ and $N = 1024$.

Table 2. Simulation parameters

Minimum Block Samples	10000
Number of Residual Errors	100
Modulation	BPSK
Channel	AWGN/Rayleigh fading Erasures
Probability of Erasures P_m	0.0001
Parameter of Correction	3

Figure 9. Performance of the proposed coded system under AWGN for rates 0.83, 0.78 and 0.5, with N=128 and $\boldsymbol{P}_m = 10^{-4}$

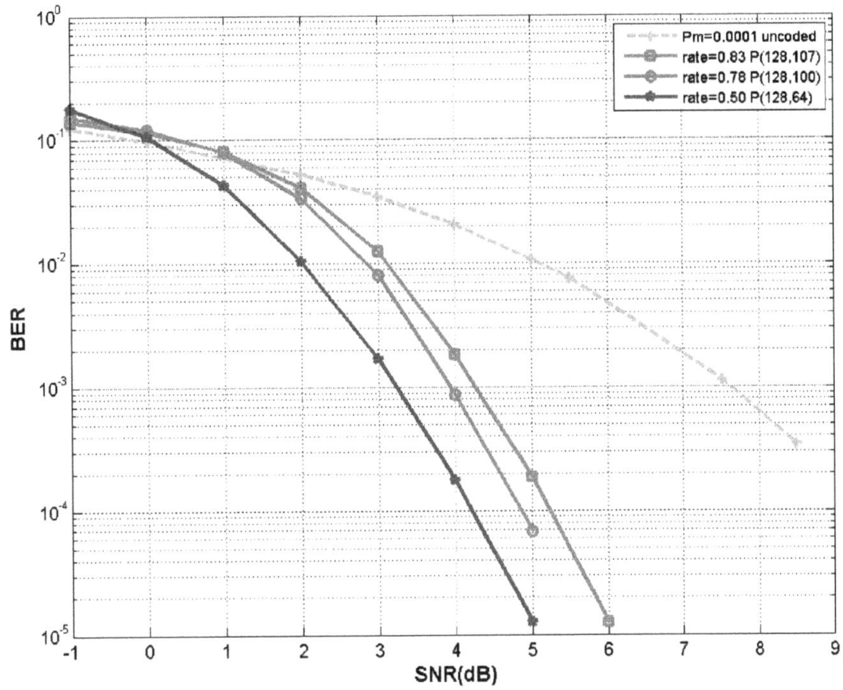

**For a more accurate representation see the electronic version.*

Figure 10. Performance of the proposed coded system under AWGN for rates 0.83, 0.78 and 0.5, with N=1024 and $P_m = 10^{-4}$

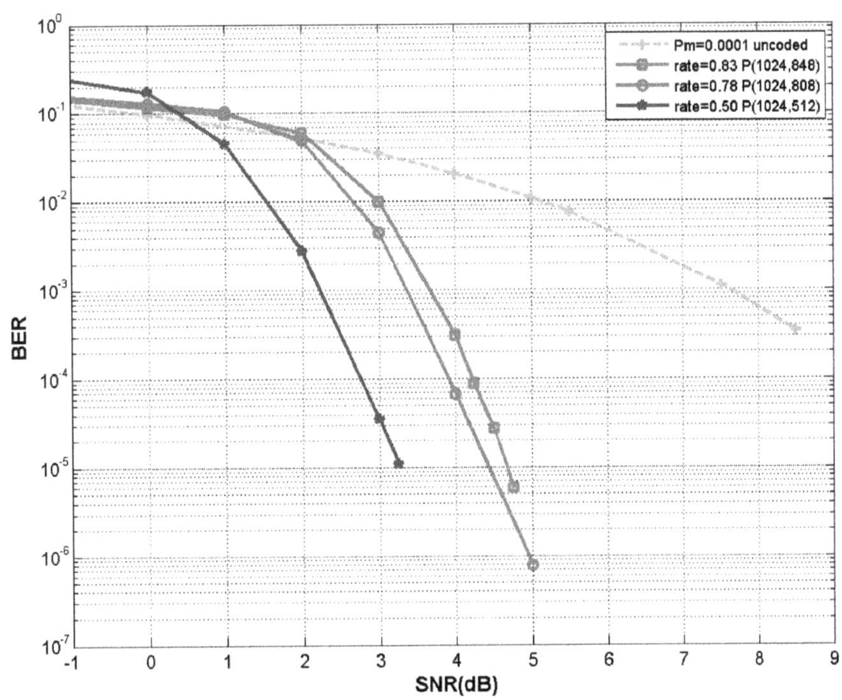

**For a more accurate representation see the electronic version.*

Figure 11. Performance of the proposed coded system under Rayleigh fading for rates 0.83, 0.78 and 0.5, with N=128 and $P_m = 10^{-4}$

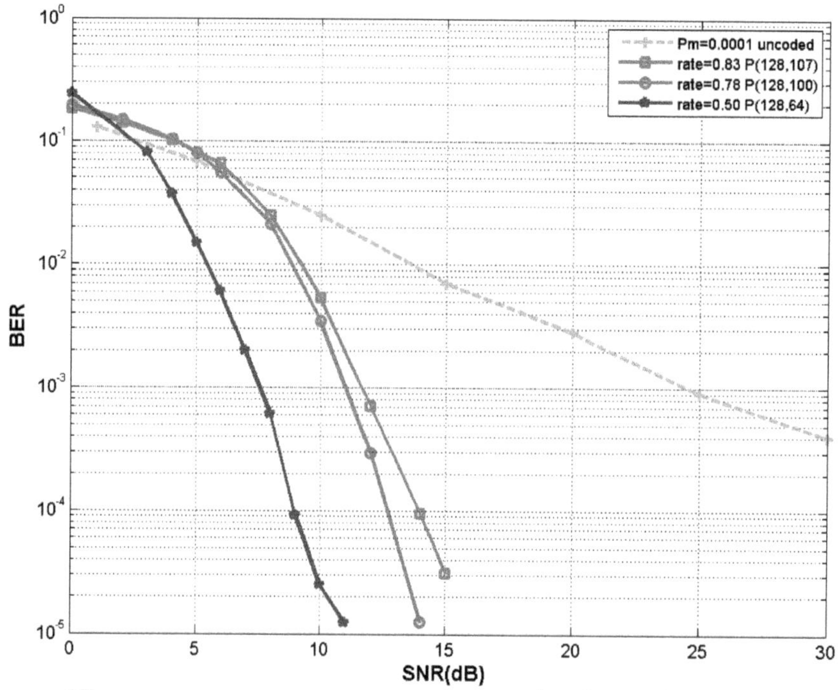

**For a more accurate representation see the electronic version.*

Figure 12. Performance of the proposed coded system under Rayleigh fading for rates 0.83, 0.78 and 0.5, with N=1024 and $P_m = 10^{-4}$

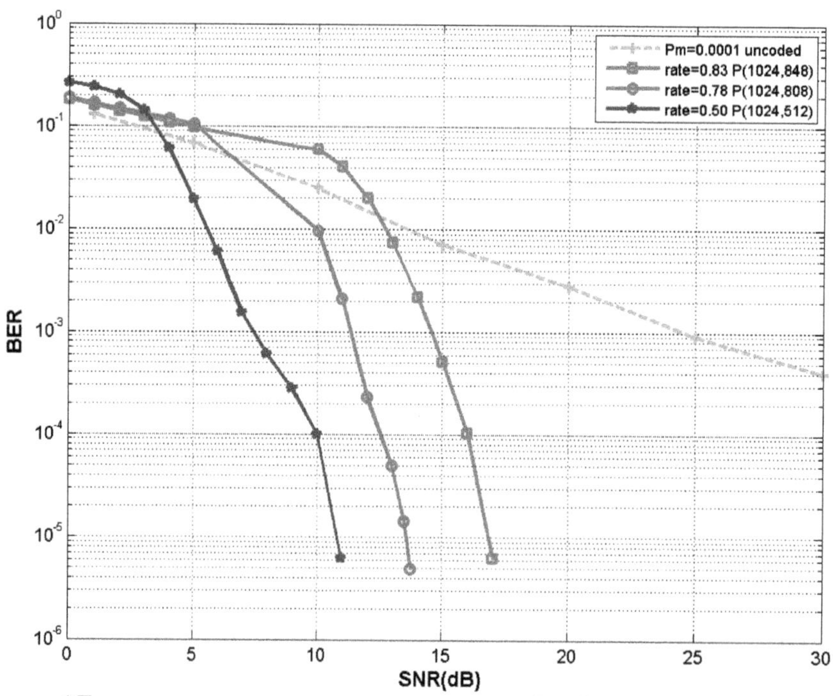

**For a more accurate representation see the electronic version.*

In Figure 13 the results are compared to those obtained by Benkhouya et al. (2016) who used the BCH codes, with rate $R = 0.50$, $R = 0.78$ and rate R=0.83. The corresponding performances of the proposed coded system are shown by solid line.

If the rate is started with 0.50 using $P(128,64)$, authors come up with the solid star curve which performs relatively better than the $BCH(127,64,10)$.

By slight increase in the code rate to 0.78 and 0.83, authors can improve the performance. For all the polar coded cases mentioned so far, the performance is better than that of BCH codes, especially for the code rate $R = 0.83$.

In Figure 14 the results are compared to those obtained by Haddadi et al. (2013) who used the RC-LDPC codes, the so-called parent code with rate of $R = 0.50$, intentional puncturing with rate of $R = 0.79$ and random puncturing with rate of $R = 0.82$. The corresponding performances of the proposed coded system are shown by solid lines.

If the rate is started to 0.83 using $P(1024, 848)$, authors come up with the solid square curve which performs better than the random puncturing.

By slight decrease in the code rate to 0.78, the corresponding performance is shown by solid circle curve. The performance of the proposed coded system remains better than intentional puncturing by 5dB at a BER of 10^{-4}. However, for the last code rate $R = 0.50$, the parent code outperforms the proposed coded system by about 6dB a BER of 10^{-4}.

Figure 13. Performance comparison between the proposed coded system and that of Benkhouya et al. (2016) under Rayleigh fading for rates 0.83, 0.78 and 0.5, with $N = 128$ *and* $P_m = 10^{-4}$

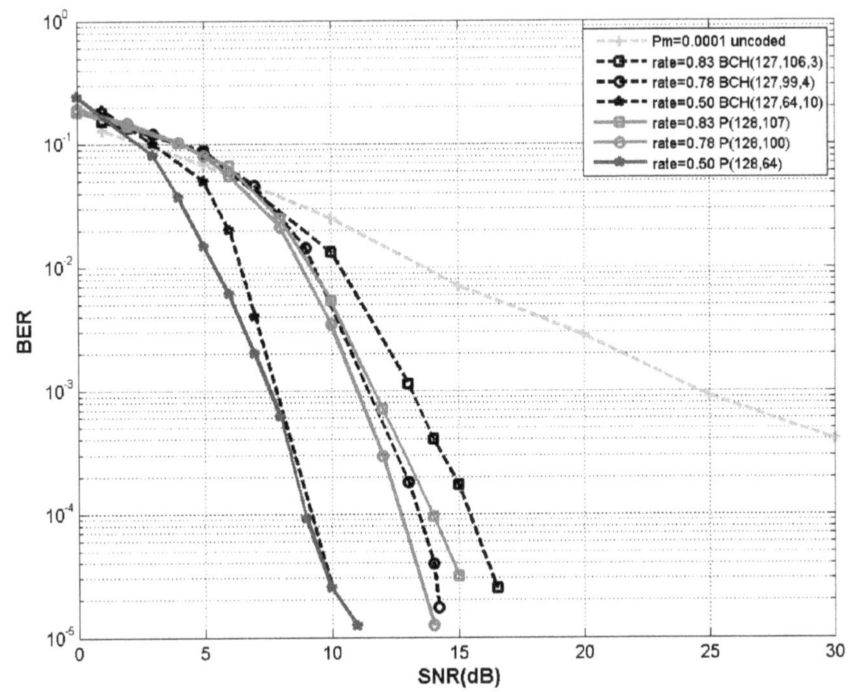

**For a more accurate representation see the electronic version.*

Figure 14. Performance comparison between the proposed coded system and that of Haddadi et al. (2013) under Rayleigh fading for rates 0.83, 0.78 and 0.5, with $N = 1024$ and $P_m = 10^{-4}$

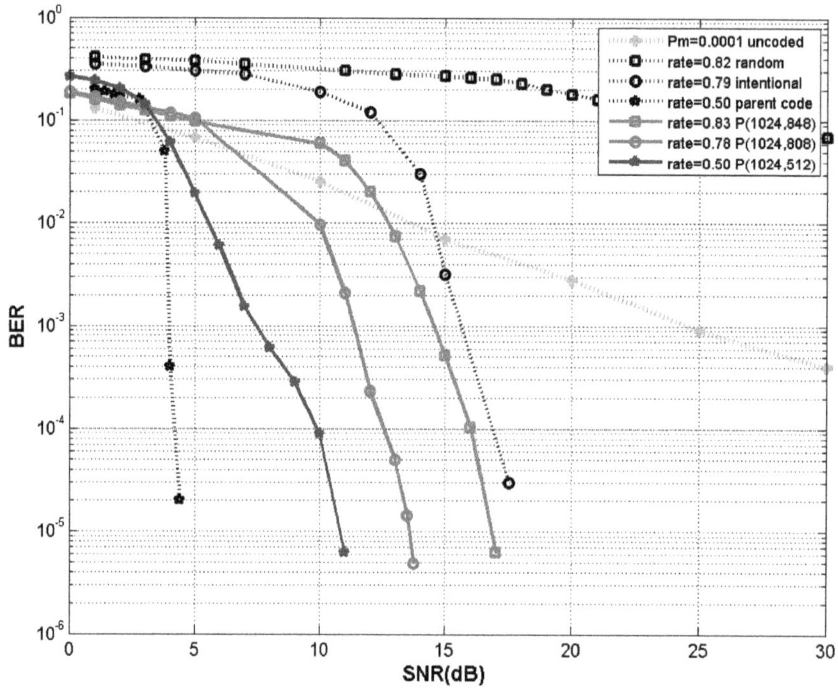

**For a more accurate representation see the electronic version.*

In Table 3 authors compare LDPC code, BCH code and polar code in terms of computational cost. The block length is N, $iter = 50$ is the number of iterations to decode RC-LDPC code with BP. As shown in the table, when $n \geq 6$ ($N \geq 64$), the decoding complexity of polar code is the smallest among all three codes. It is divided by $4M$ and $2^t N$ when compared to that of LDPC code and BCH code respectively.

For the case of code rate $R = 0.50$, the parent code outperforms the proposed coded system, but at the expense of the decoding complexity. Moreover, the proposed algorithm of decoding is of a SIHO type unlike the BP decoder which is a SISO one. Since the proposed SIHO decoder is now performance-proven, a SISO version could easily do better.

Table 3. Comparison of the three codes by their decoding complexities

Code	Code Length N	Decoder	Decoder Type	Decoding Complexity
LDPC	10000	Belief propagation (BP)	SISO	$iter \times (4N \log N)$ (with $N \geq 64$)
BCH	127	Modified Chase II	SIHO	$2^t N^2 \log N$
Polar	128, 1024	SC for erasures	SIHO	$N \log N$

CONCLUSION

This chapter develops a technique dedicated to countering the effect of decreasing sensing time for wideband cognitive radio. Authors started by proving the relation between the number of samples and the sensing time. Then they demonstrated that it is possible to save more number of samples in the adopted compressive sensing and this reduces sensing time. Once the sensing time is reduced the probability of collision increases, which consequently causes erasures and worsens channel imperfections.

Authors proposed a coding scheme based on new decoding algorithm for polar code. Using the proposed scheme, authors showed that they can compensate channel imperfections caused by sensing time reduction.

To study and evaluate the proposed method, medium and high code rates were considered. Using simulations authors showed that, compared to two state-of-the-art techniques based on RC-LDPC codes and BCH codes, the proposed technique shows significant improvement in performance for high code rates under the same traffic conditions, but with less decoding complexity. The punctured codes showed good error performance at medium code rate. Yet, the complicated puncturing algorithm and large block size are not practical for hardware implementation. Finally, the proposed technique is the legitimate candidate for wideband cognitive radio operating with high effective data rate. For future work, authors expect to design the SISO version of the proposed scheme, which will save more samples and correct even more erasures.

REFERENCES

Arikan, E. (2009). Channel Polarization: A Method for Constructing Capacity-Achieving Codes for Symmetric Binary-Input Memoryless Channels. *IEEE Transactions on Information Theory*, *55*(7), 3051–3073. doi:10.1109/TIT.2009.2021379

Aulakh, I. K., & Vig, R. (2014). Optimization of SU's probability of false alarm for dynamic spectrum access in cognitive radio. In *International Conference on Computing for Sustainable Global Development* (pp. 710–715). New Delhi, India: Academic Press. 10.1109/IndiaCom.2014.6828053

Babich, F., Noschese, M., & Vatta, F. (2016). Analysis and Design of Rate Compatible LDPC Codes. In *IEEE 27th Annual International Symposium on Personal, Indoor, and Mobile Radio Communications* (pp. 1-6). Valencia, Spain: IEEE. 10.1109/PIMRC.2016.7794684

Bavarian, S., & Cavers, J. K. (2007). Reduced Complexity Belief Propagation Algorithm Based on Iterative Groupwise Multiuser Detection. In *Canadian Conference on Electrical and Computer Engineering* (pp. 466-468). Vancouver, BC, Canada: Academic Press. 10.1109/CCECE.2007.123

Benkhouya, R., Chana, I., & Hadi, Y. (2016). Correction of sensing imperfections using channel coding in cognitive radio networks while keeping sensing time low. *International Conference on Advanced Communication Systems and Information Security* (pp. 1-5). Marrakech, Morocco: Academic Press. 10.1109/ACOSIS.2016.7843938

Benkhouya, R., Chana, I., & Hadi, Y. (2017). Investigation on BER sensitivity to design SNR of Bhattacharrya bounds based construction. *Journal of Theoretical & Applied Information Technology, 95*(7), 1398–1404.

Berlekamp, E. R. (1968). *Algebraic coding theory*. Dunfermline, UK: McGraw-Hill Education.

Bushell-Embling, D. (2016). *Huawei achieves 27Gbps 5G speeds with Polar Code*. Retrieved July 26, 2017, from https://www.telecomasia.net

Chen, Y., & Tang, Z. (2012). Effect of spectrum sensing errors on the performance of OFDM-based cognitive radio transmission. *IEEE Transactions on Wireless Communications, 11*(6), 2342–2350. doi:10.1109/TWC.2012.040412.111744

Ding, D., Wang, J., Wu, Q., Zhang, L., Song, F., Zou, Y., ... Feng, S. (2014). Robust Spectrum Sensing With Crowd Sensors. *IEEE Transactions on Communications, 62*(9), 3129–3143. doi:10.1109/TCOMM.2014.2346775

Gallager, R. G. (1963). *Low-density parity-check codes*. MIT Press.

Haddadi, S., Saeedi, H., & Navaie, K. (2013). Channel coding adoption versus increasing sensing time in secondary service to manage the effect of imperfect spectrum sensing in cognitive radio networks. In *Iran Workshop on Communication and Information Theory* (pp. 1–5). Tehran, Iran: Academic Press. 10.1109/IWCIT.2013.6555757

Lee, D. J., & Jang, M. S. (2009). Optimal spectrum sensing time considering spectrum handoff due to false alarm in cognitive radio networks. *IEEE Communications Letters, 13*(12), 899–901. doi:10.1109/LCOMM.2009.12.091448

Lee, X. R., Chen, C. L., Chang, H. C., & Lee, C. Y. (2014). A 7.92 Gb/s 437.2 mW Stochastic LDPC Decoder Chip for IEEE 802.15.3c Applications. In *IEEE Transactions on Circuits and Systems I* (Vol. 62, pp. 507–5016). IEEE Circuits and Systems Society.

Li, J., & Zhang, D. (2008). Reduced-complexity belief propagation decoding for low-density parity-check codes. *Electronics Letters*.

Lin, J., Xiong, C., & Yan, Z. (2015). Reduced Complexity Belief Propagation Decoders for Polar Codes. In *IEEE Workshop on Signal Processing Systems* (pp. 1-6). Hangzhou, China: IEEE. 10.1109/SiPS.2015.7344984

Lin, Y. E., Liu, K. H., & Hsieh, H. Y. (2013). On Using Interference-Aware Spectrum Sensing for Dynamic Spectrum Access in Cognitive Radio Networks. *IEEE Transactions on Mobile Computing, 12*(3), 461–474. doi:10.1109/TMC.2012.16

Peterson, W. W., & Weldon, E. G. (1972). *Error-correcting codes*. MIT Press.

Qin, M., Guo, J., Bhatia, A., Fàbregas, A. G. I., & Siegel, P. H. (2017). Polar code constructions based on LLR evolution. *IEEE Communications Letters, 21*(6), 1221–1224. doi:10.1109/LCOMM.2017.2656126

Rashidi, M., Haghighi, K., Owrang, A., & Viberg, M. (2011). A wideband spectrum sensing method for cognitive radio using sub-Nyquist sampling. In *Digital Signal Processing Workshop and IEEE Signal Processing Education Workshop* (pp. 30–35). Sedona, AZ: IEEE. doi:10.1109/DSP-SPE.2011.5739182

Stolte, N. (2002). *Rekursive codes mit der Plotkin-Konstruktion und ihre Decodierung* (Unpublished doctoral dissertation). Technische Universität Darmstadt, Germany.

Wells, N. (2011). *DVB-T2 in relation to the DVB-x2 Family of Standards*. Paper presented at the ATSC Symposium on Next Generation Broadcast Television, Rancho Las Palmas, Rancho Mirage, CA.

Yang, C. Y., & Ku, M. K. (2008). LDPC coded OFDM modulation for high spectral efficiency transmission. *4th European Conference on Circuits and Systems for Communications*. 10.1109/ECCSC.2008.4611693

Zhang, B., Shen, H., Yin, B., Lu, L., Chen, D., Wang, T., … Kishiyama, Y. (2016). A 5G Trial of Polar Code. IEEE Globecom Workshops, 1-6.

Zhang, K., & Huang, X. (2009). A Low-Complexity Rate-Compatible LDPC Decoder. In C*onference Record of the Forty-Third Asilomar Conference on Signals, Systems and Computers* (pp.749-753). Pacific Grove, CA: Academic Press.

Zhou, L., Huang, W. C., Zhao, R., & He, Y. C. (2017). A Non-Greedy Puncturing Method for Rate-Compatible LDPC Codes. *Journal of Communications and Networks (Seoul)*, *19*(1), 32–40.

KEY TERMS AND DEFINITIONS

AWGN: Additive white Gaussian noise is a basic noise model which is often used as a channel model.

BCH: The BCH codes or Bose-Chaudhuri-Hocquenghem codes, part of cyclic codes, are constructed using polynomials over a finite field. Binary BCH codes were first discovered by Alexis Hocquenghem in 1959.

Cognitive Radio: Is an emerging technology to deal with spectrum scarcity and also a way of improving spectral utilization efficiency. Cognitive radio enables the use of radio spectrum that is underutilized due to the fixed allocation of frequency bands.

GF: Galois field, so-named in honor of Evariste Galois, is a field that contains a finite number of elements.

LDPC: Low-density parity-check code, is a linear error correcting code. LDPC codes are also known as Gallager codes, in honor of Robert G. Gallager.

Multicoset: is a sampling technique which is more suitable for practical implementation in comparison to more sophisticated sub-Nyquist systems. Based on the prior information on the number of narrowband channels, multicoset sampling is applied to each secondary user by taking non-uniform samples at the time instants $t = \left(mL + c_i\right)T$, where $i = 1,\ldots,p$, $m \in \mathbb{Z}$ and $1/T = f_s$ is the Nyquist sampling rate. The set $\mathsf{C} = \left\{c_i\right\}_{i=1}^{p}$, which comprises of p distinct integers chosen from $\left\{0,1,\ldots,L-1\right\}$ is referred as a $\left(L,p\right)$ sampling pattern.

Polar Code: Is a novel channel coding scheme that provably achieves the capacity of memoryless channels with low complexity encoding and decoding algorithms.

Rayleigh: Can be modeled by generating the real and imaginary parts of a complex number according to independent normal Gaussian variables.

Spectrum Sensing: Implies a continuous survey of the status of the channels. Spectrum sensing plays a key role in detecting the holes or vacant channels.

Section 4

Role of Artificial Intelligence in CRNs

Chapter 14
Role of Artificial Intelligence in Cognitive Radio Networks

Shikha Singhal
Jaypee Institute of Information Technology, India

Shashank Gupta
Jaypee Institute of Information Technology, India

Adwitiya Sinha
Jaypee Institute of Information Technology, India

ABSTRACT

The role of artificial intelligence techniques and its impact in context of cognitive radio networks has become immeasurable. Artificial intelligence redefines and empowers the decision making and logical capability of computing machines through the evolutionary process of leaning, adapting, and upgrading its knowledge bank accordingly. Significant functionalities of artificial intelligence include sensing, collaborating, learning, evolving, training, dataset, and performing tasks. Cognitive radio enables learning and evolving through contextual data perceived from its immediate surrounding. Cognitive science aims at acquiring knowledge by observing and recording externalities of environment. It allows self-programming and self-learning with added intelligence and enhanced communicational capabilities over wireless medium. Equipped with cognitive technology, the vision of artificial intelligence gets broadened towards optimizing usage of radio spectrum by accessing spectrum availability, thereby reducing channel interferences while communication among licensed and non-licensed users.

INTRODUCTION

Artificial Intelligence (AI) refers to intellectual capability exhibited by computing systems through leaning, adapting and updating knowledge base. More essentially, a device that senses its environment and contextually performs actions based on the perceived information basically forms an artificial intelligence system. The main goals of AI include sensing, learning, training and performing tasks. In fact, an AI System is extended with enhancements in such a manner that allows it to perform tasks with human-like

DOI: 10.4018/978-1-5225-5354-0.ch014

intelligence. Recently, the role of AI in Cognitive Radio Networks (CRNs) has become impeccable & inevitable in executing basic processes, involving sensing, learning, training, and performing overall functions dynamically and intelligently, similar to operations of a human brain (He, Bae, Newman, Gaeddert, Kim, Menon… & Tranter, 2010).

Cognitive Radio (CR) is a technology capable of learning and adapting the surrounding environment with help of sensed environmental parameters. The term cognition means gaining of knowledge by experiencing the externalities of the environment. CR exhibits a self-programming ability that allows learning intelligently and autonomously over the wireless domain. Main objective of CR device is to optimize the use of radio frequency spectrum by sensing free spectrum availability and minimize interference between the primary and secondary users, i.e. managing permanent reliable communication dynamically. CR has an ability to access spectrum and allot available spectrum to secondary user animatedly. In order to equip spectrum management with power of dynamism, CR must rely on artificial intelligence. The learning problems in CRN mostly include decision making and feature classification problems. Feature classification problem occur at the time of spectrum sensing while decision making problem is to determine the spectrum sensing rules along with policies. To overcome these challenges, different types of learning paradigms are available, such as supervised learning and unsupervised learning for classification problem and reinforcement learning for decision making problem. In addition, AI technique extends learning paradigms over Artificial Neural Network (ANN), K-means, Support Vector Machine (SVM), Hidden Markov, and Game Theory. The main factors that need to be focused when designing algorithms involve learning in partially perceivable environments, multi-agent learning and autonomous learning in unknown radio frequency environments. The cognitive technology aims to perform its task optimally with increased robustness in CR-based applications. Normally, in case when no prior information of the radio parameters (i.e. signal to noise ratio, bandwidth and bit error rate, vulnerable rate), AI-based techniques tend to behave more optimally over Very High Frequency (VHF) and Ultra High Frequency (UHF) FM and TV bands.

ARTIFICIAL INTELLIGENCE

Artificial Intelligence is fundamentally a computer, software or computer controlled machine which thinks intelligently same as like human brain works. These computers do not use a fixed programmed to solve real-world problems. They are based on learning, training, reasoning, planning, and decision making aspects. The foundation of AI is based on how human brain works intelligently towards solving the problem i.e., how human brain thinks, sense, learn, decide, take actions to try to address any issue. On the basis of these studies, researchers developed some intelligent software so as to reduce manual efforts. The engineering science makes intelligent machine possible which behaves like as human brain (Russell, Norvig, Canny, Malik, & Edwards, 1995; He, Bae, Newman, Gaeddert, Kim, Menon, Morales-Tirado, Neel, Zhao, Reed, & Tranter, 2010).

AI assimilates science and engineering concepts through which we can develop a machine which treats any situation similar to the way human brain would respond, and produces the result intelligently with help of sensed and learned information (database). Technically, artificial intelligence refers to the ability of a system to calculate, evaluate a correct reason, perceive the current environment, analyse all possible relationships, learnt from experience, store and retrieve information from database, provide solution, problem solving, comprehend new ideas, use natural language, classify, generalize, and able to adapt new situations/ environment. In other words, AI is employed to search for best solution to any problem while interacting with real world challenges, thereby extending ability to handle uncertainties in unexpected problems and learn from current working environment to adapt it. The major goals of AI include:

- Knowledge representation
- Reasoning, problem solving, planning, learning
- Designing Expert Systems that manifest intelligent behaviour i.e., sense, learn, demonstrate, elucidate and give advice its users
- Implementing machines based on human intelligence which has the ability to analyse, think, learn and take decision like human

Some AI researchers provide several specialized programming language to develop artificial intelligence system, include LISP, Prolog, IPL, STRIPS etc. AI-based methods tend to utilize knowledge which is unimaginable, hence explores different properties to represent it a better comprehendible way. AI techniques help to refine the different aspects of knowledge according to situation, provide useful information, and modifies it to remove errors. Represented knowledge can be used in many situations but it is not complete or accurate. In that way AI technique are immensely productive. Some predominant AI techniques involve Natural Language Processing, Neural Networks, Machine Learning, Fuzzy Logic, etc. These techniques help us to solve complex problems that cannot be tackled by traditional programming code. The use of AI technique is escalating rapidly (He, Bae, Newman, Gaeddert, Kim, Menon… & Tranter, 2010; He, Bae, Newman, Gaeddert, Kim, Menon, Morales-Tirado, Neel, Zhao, Reed, & Tranter, 2010). AI is vast domain supporting applications that include consumer marketing, identification technologies, game playing, robotics and navigation, military purpose, medical equipments, etc.

COGNITIVE RADIO NETWORK

The ever-increasing demand for wireless communication originates efficient dynamic radio spectrum utilization challenge. The radio spectrum is very expensive resource which is required to be utilized in best possible way. Wireless network provisions to assign fixed spectrum to authorized users. Resultantly, major portion of the spectrum is available only for authorized users. This type of assignment arises spectrum scarcity and under-utilization challenges. To deal with these challenges, researchers emerge a new key technology which enables the effective use of spectrum dynamically, called Cognitive Radio (CR). It helps to tackle the problem of utilizing radio spectrum efficiently. The main profitable features originated from cognitive radio lead to enhance the spectrum utilization and escalate communication quality. These enchanting features deals with resource constrained challenge in multi-hop wireless sensor network and fulfils all the major requirements in wireless communications. Cognitive radio is the technology that allows dynamic spectrum access i.e., opportunistic spectrum usage. It provides promising solutions of current challenges such as spectrum scarcity, under-utilization, interference between users. Cognitive radio has potential to sense the radio spectrum and check availability of vacant bands. Cognitive radio has a capability to change the current running environment operating parameters dynamically to make a use of available vacant bands in such a manner that trumping the traditional fixed spectrum assignment policy in order to get best spectrum utilization. Cognitive Radio Networks (CRNs) assigned a licensed bands and unlicensed bands for wireless communication. In Licensed bands, CRN allots a specific License to users for communication over the allocated band. The users of licensed bands are known as primary users (PU) which have priority to access the channel anytime dynamically. The users of cognitive radio are known as secondary users (SU), they have potential to access the unused licensed channel without interrupting the PU. Primary users are legitimate users of the radio spectrum while secondary users can access the unused licensed bands intelligently and opportunistically without causing interference to legitimate users. The unused licensed spectrum band is used by secondary users are called unlicensed band (Giupponi, Galindo-Serrano, Blasco, & Haykin, 2010; Thomson, & Reed, 2009).

Working Process of CRN

CRN has four functional blocks by which CR user can access channel. These functional blocks are: Spectrum Sensing, Spectrum Decision, Spectrum Mobility/Handoff and Spectrum Sharing. Secondary users follow cognition cycle which contains all CR functional blocks in order to get better utilization of overall spectrum without causing any harmful interference (Figure 1).

1. **Spectrum Sensing:** Spectrum sensing is an active Radio Spectrum Discernment Process (RSDP). In this process, Spectrum Sensing is used to perceive unused/vacant radio spectrum bands/holes and allocate these perceiving bands/holes to Secondary users without causing any interference to other primary users. Overall Accuracy of the radio environment can be improved by using RSDP. This function assists secondary users "How to detect the unlicensed bands over the licensed bands without triggering disturbances for licensed users". Some AI techniques are used for spectrum sensing. The network performance such as accuracy, efficiency, speed, and throughput depends on spectrum sensing technique (Haykin, Thomson, & Reed, 2009).
2. **Spectrum Decision:** Based on perceiving information by RSDP, cognitive radio device (used for secondary users) take some technical decision such as on which channel to move, when to start communication, which is the nearest vacant bands/holes, and some other required parameters for communication. Cognitive radio must have objective to convey exact perceiving information in order to maintain required quality of service without triggering any disturbances at primary users ends. Besides it, CR must use a database to store sensing information on regular basis and has potential to update its operational statistics and quality of service.
3. **Spectrum Sharing:** After receiving all spectrum decision parameters, CR must be able to distribute spectrum holes per band among all secondary users intelligently and equally. The distribution of spectrum holes reasonably is called spectrum sharing. The main concern with spectrum sharing is CR has to achieve its self-promising objective in an effective way and also share the available resources in form of vacant radio spectrum holes, with secondary users. This rule decides the CR behaviour in radio environment and maintains along with traditional policy.
4. **Spectrum Mobility/Handoff:** Spectrum handoff is the process of continuing active communications over the new spectrum holes/channel. For maintaining Mobility feature in Cognitive Radio networks, they used transceiver in order to achieve flawless communication between unlicensed and licensed users. In other words, if primary user demands access to his allocated channel then cognitive radio user (or secondary user) is bound to stop its ongoing operation immediately. The cognitive user then changes the current bands/holes in radio frequency spectrum, thereby vacating allocated bands for primary users. In order to keep away interference from licensed users and to

Figure 1. Functional Blocks in Cognitive Cycle

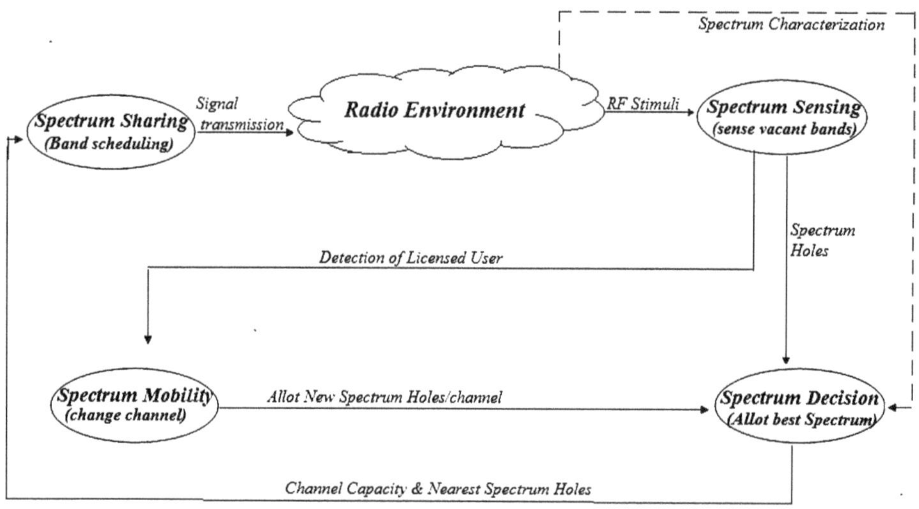

get better performance through network, CR users have to implement this function at real time. For handoff purpose, CR detects alternative vacant spectrum/holes consistently.

Benefits of CRN

Cognitive radio has some beneficial features such as:

- **Adaptive Learning Network:** CRN has the potential to adapt current operating parameters and take actions according to the working environment. Cognitive radio device has capability to learn from surrounding environment which assists the cognitive radio devices in future to take further actions, hence creating an adaptive learning environment.
- **Self-Organised Network:** CRN are self-organised networks i.e., they can access the new channel without causing interference to primary users. They have capability to switch the new channel instead of violating the spectrum policy of primary users.
- **Real Time Operation:** Cognitive Radio devices perform all their functions such as spectrum sensing, spectrum decision, spectrum sharing, and spectrum handoff at real time. They have potential to take any action on real time so that they can maintain better performance in network and utilization of radio spectrum bands.

Major Pitfalls in CRN

- Cognitive Radio networks must be very efficient at the time of Hand-off in order to avoid procrastinating in communication at primary user's ends.
- Radio device must perceive all required parameters of primary users and fade away all the usage of secondary users in order to avoid interference.
- Cognitive radio networks must have potential to differentiate between communication links of licensed users or unlicensed users.

- Primary users must always get priority to access the spectrum over secondary users without any interference.
- Decision making process is very challenging in CRN. Radio device must take a valid decision to prevent conflict between primary users and secondary users and infringing of policy.
- Security is the main concern at each functional blocks of cognitive cycle. Attackers may violate the security of cognitive radio networks, making it vulnerable to threats that may exploit the security of CRN.
- CRN must have potential to maintain traffic in the network, along with restricting Jamming signal.
- Cognitive radio devices must have memory like human brain for taking future decision intelligently.

SIGNIFICANCE OF AI IN COGNITIVE RADIO

AI is based on how human brain works intelligently towards addressing any real-world situation. Similarly, cognitive radio is a device which is simulated to human brain. This implies that cognitive radio is a device which designed on the basis of AI concepts. AI is a science and engineering concept by which we design a cognitive radio device. Cognitive radio device is inspired by human brain to take spectrum allocation decision. Cognitive radio is a combination of AI techniques and wireless communication. In cognition cycle, all functional blocks are implemented by using AI concepts and Algorithms. Cognitive radio networks work on the concept of AI. To achieve its goal, CRN must have act intelligently and efficiently. AI techniques are used to implement cognitive radio functions. The functions followed by cognitive radio are exactly same as human brain to perform tasks and to take decision regarding any issues (He, Bae, Newman, Gaeddert, Kim, Menon... & Tranter, 2010; Russell, Norvig, Canny, Malik, & Edwards, 1995; He, Bae, .Newman, Gaeddert, Kim, Menon, Morales-Tirado, Neel, Zhao, Reed, & Tranter, 2010).

- **To Sense:** To detect running status of the current working environment of cognitive radio networks.
- **To Learn:** To store the information in database (brain), after knowing the situation of current the environment.
- **To Coordinate:** To maintain the coordination with others.
- **To Store:** Keep all the knowledge and experiences of observations and to reach the objectives with the help of stored knowledge.

The cognitive radio network is a sophisticated and complicated multiuser cordless communication system which is capable of adopting desire behaviour of operating device. Cognitive Radio networks have interconnected cognitive radio devices which deploy by using the concepts of AI and work intelligently throughout communication. Cognitive radio networks provide spectrum holes intelligently but for sensing vacant spectrum band intelligently we need to use AI techniques. Cognitive radio network encapsulates the following functions:

- **SENSE:** To *perceive* the radio spectrum environment (i.e. spectrum holes) in real time via granting each user's request (i.e. received by receiver) to sense the current working environment.
- **LEARN:** To *remember* the sensed information from sensing environment and *improve* performance of each communicating radio device which is currently participating in running environment.

- **FORWARD:** To *promote* continuous communication through maintaining the coordination among multiple users.
- **RULE:** To *supervise* the coordination processes between competing devices (users) via use of efficient resource allocation algorithm.

Cognitive radio and human mind simulates with each other in many ways. To overcome the challenges of wireless networks, cognitive networks is deployed which work on the same notion of executing tasks by human brain at real time. Cognitive radio devices resolve the spectrum underutilization problem by following ways:

- By sensing radio environment to find spectrum holes in spectrum based on both time and location
- By reusing spectrum holes for unofficial users proficiently by checking all the constraints which happens in current environment

LEARNING PROBLEM IN CRN

Several learning algorithms are used by CRNs to achieve goals. Learning is the elemental notion of AI. Learning means collection of knowledge from perception. Learning is an ability to learn the system with cognition. With learning feature, a cognitive radio networks can classify, organize, analyse and generalize the information received by sensors. Learning depends on two important features of CRNs such as perception and reasoning. Without perceiving feature, cognitive radio device cannot be aware of its surrounding RF environment and also cannot be able to take decision. By learning, cognitive radio device is able to take decision which is most applicable vacant spectrum holes. Cognition, learning, and reasoning are the unique and most dependable feature of intelligent device i.e., Cognitive radio networks. Learning states that take actions based on past and present observations of surrounding environments. Learning phase in CRNs is basically used for providing best spectrum sensing functionality and helps radio devices in performing future actions. The two main learning problems are decision making problem and feature classification (Ghasemi & Sousa, 2008). Feature classification problem encounters at the time spectrum sensing and decision- making problem come into existence at time of determining the spectrum sensing policy (Figure 2).

DECISION-MAKING PROBLEM

Decision making problem occurs when we are selecting spectrum policy. The question arise "Which spectrum policy is the best policy in order to achieve efficient spectrum utilization?" In order to answer this question we have to take best decision based on encountered past and current aspects of environments. Decision making problem is the most important task in cognitive radio networks. Decision-making is selection of the best possible actions while considering all the aspects of environment. This problem is the major aspects that will be encountered very much in CR applications. Decision-making problem can be divided into two parts such as policy making and decision rules (Mitchell, 1997; Bkassiny, Li, & Jayaweera, 2013; James & Singh, 2004).

Figure 2. Typical Learning Problem in Cognitive Radio Networks

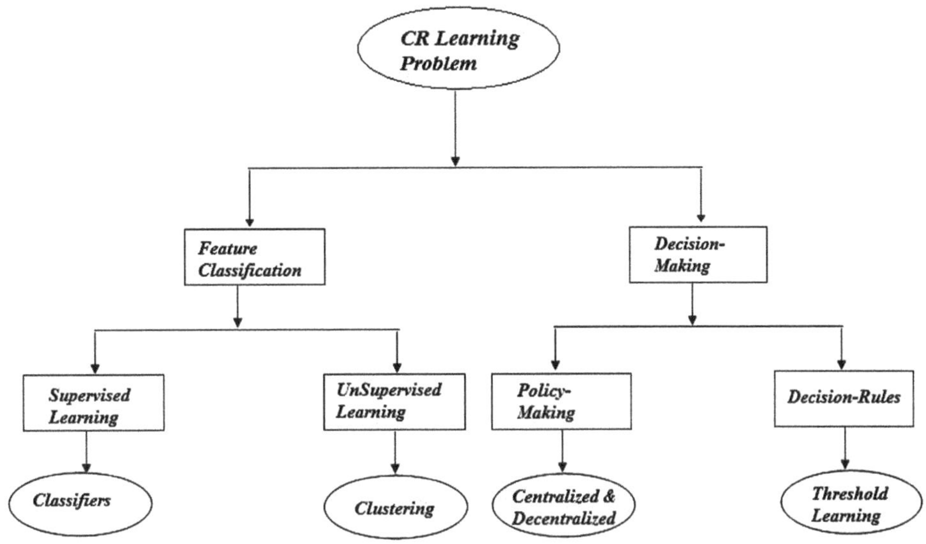

Policy- Making

Policy making states that finding optimal actions in certain time duration in order to get optimal spectrum policy. Policy making can be categorized into two groups: centralized and decentralized. In centralized, reinforcement learning technique is used to find the optimal solution. We do not have any prior knowledge to get the optimal solution. We have done some trials and get some errors. Based on trials and errors we are ready to take some actions which makes easier to get optimal spectrum policy. By Trials, an agent performs some actions to explore the current working environment but agent do not have any prior knowledge of environment. In this case, a central node is assigned to take all decisions on the basis of collected information. Centralized method increases the communication overhead and algorithms complexity. In decentralized method, similar approach is followed to find the solution but different algorithms are used to minimize communication overhead and complexity without any central node (Weingart, Sicker, & Grunwald, 2007; James & Singh, 2004).

Decision Rules

Decision rules is another form of decision making problem. It is based on hypothetical testing problem for certain observations. Decision rules works under uncertainty conditions. It works on the concept of threshold learning. Threshold learning is a technique that adapts operating parameters dynamically to satisfy the performance while learning from past experience. Decision rules used feedback signal for taking current actions. To maintain desired performance of the system, decision rules optimize the operating parameters dynamically. Decision rules should not affect the current running environment. It might introduce noise and interference level during communication (Bkassiny, Li, & Jayaweera, 2013; Cussens, 1996).

FEATURE CLASSIFICATION PROBLEM

Cognitive radio device receives so many signals from other devices working in current RF environment. The received signals may contain noise signals and redundant and unknown signals also. CR must be able to extract knowledge from received signals. The received data may be too large to be processed and may contain some other parameters which are not useful. We apply some algorithms to extract the relevant information from the learned information so that desired task can be performed by using this extracted dataset. The overall process of receiving signals and then extraction of relevant information from the received data is called feature classification problem. This problem arises at the time of spectrum sensing. This problem is also known as learning problem because at time of learning CR device have no prior knowledge of transmitted signals and they receive some unknown signals about the surroundings of RF environment. These signals may be relevant for future decision or may be noise or interference signals. At the time of spectrum sensing CR devices must be capable of extracting the relevant knowledge from the input signals. To resolve the problem of feature classification we use both supervised and unsupervised learning approach. Clustering is unsupervised approach for feature classification. We have some classifiers to classify the receiving feature named linear classifiers, quadratic classifiers, ANN, SVM. These classifiers follow the supervised learning approach to extract the relevant feature from receiving signals (Ghasemi & Sousa, 2008; Kotsiantis, 2007; James & Singh, 2004).

TYPES OF LEARNING PARADIGMS AND LEARNING ALGORITHMS

Cognitive radio networks operates with artificial intelligence as its backbone. For sensing the spectrum holes, cognitive radio devices need to be learning from the RF environment. The aim of learning is to detect the spectrum holes efficiently and intelligently. AI provides three Learning paradigms to learn about surrounding RF environment. The three types of learning paradigms are: Supervised Learning, Unsupervised Learning, and Reinforcement Learning (Figure 3).

SUPERVISED LEARNING

Supervised learning approach perceives information from the experiential instances (Figure 4). In supervised learning, the main objective is to learn target function that can be used to predict desired output. CRN operates with major objective of detecting vacant bands from RF environment. Cognitive radio devices may use supervised learning approach to learn target functions by which they can achieve their goals (Zhang & Xie, 2007; Ting-chao, 2004).

Supervised learning approach is also known as inductive learning approach or conceptual learning approach. Inductive learning refers to the procedure of learning from trained data (experiences). In supervised learning approach, we have pre-defined conceptual desired output (training data) and some inputs (testing data). The idea is to "learn" from labelled data in training data to identify the unlabelled data from test data. The basic goal of using supervised learning to design a function or procedure or program that maps unlabelled data into labelled data for providing a desired output.

Figure 3. Artificial Intelligence Learning Paradigms

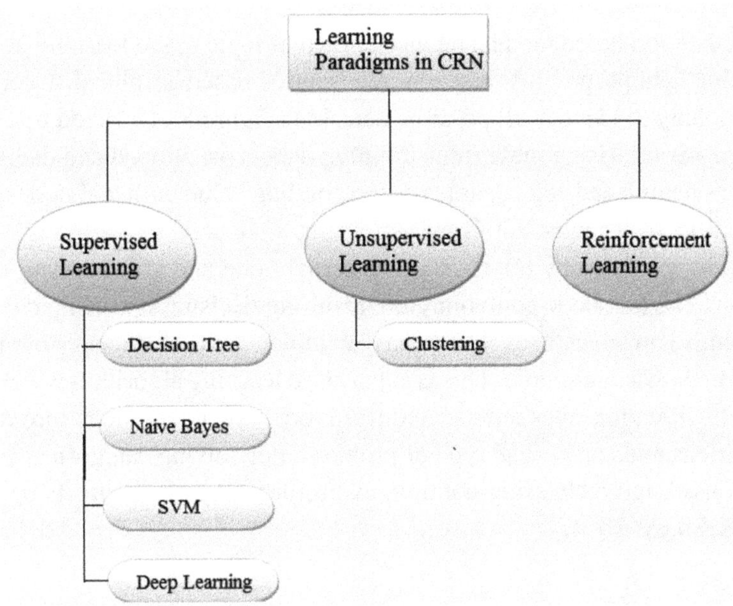

Figure 4. Supervised Learning

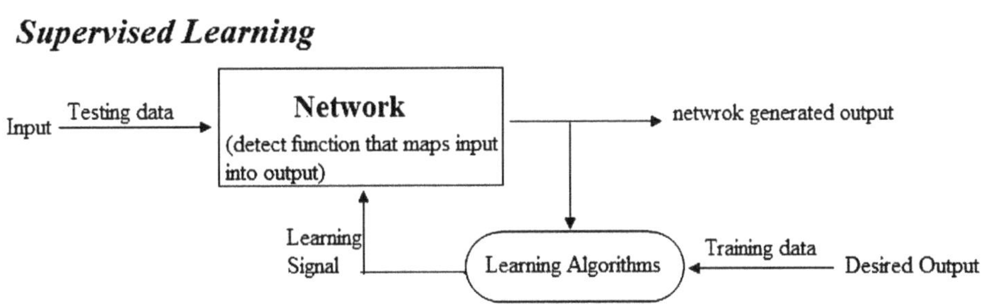

AI provides so many learning algorithms which come under supervised learning. These learning algorithms are used to reveal learning signals through which cognitive device can detect mapping functions to achieve desired goals. Some learning algorithms include logic based algorithms (e.g. decision tree), statistical learning algorithms (e.g. Naive Bayes), support vector machine, deep learning (with ANN) and some more. These algorithms used to find "learning signal" to get the target function for mapping training data into testing data. These algorithms possess their own set of rules. Cognitive radio networks use these algorithms for learning purpose. In order to achieve their goals efficiently, cognitive radio networks must be an intelligent device which sense spectrum and take decision which will not create interference between primary users and secondary users. These AI concepts tend to make CRN intelligent and aware. Following section is focused on most important supervised learning algorithms.

Decision Tree

This technique is predication based model and also known as logic based learning algorithms. Decision tree is used for training data purpose. A decision tree is a tree based graphical structure that makes use of branching methodology to identify all possible decision (Figure 5). Decision tree has nodes and two rules in the form of *yes* or *no*. Using these rules, training data is partitioned and decision tree is formed, where node represents feature and edge denotes corresponding value with leaf nodes representing sense (Karim & Rahman, 2013; Kotsiantis, 2007).

Testing data follows the paths by traversing on internal node and after visiting on leaf node it displays the final output. The leaf node containing the resulting decision is considered as "sense output". Decision tree has some main advantages such as easy to understand, produce result vigorously, and well organized structure of knowledge source. This is supervised learning algorithm which is mainly used for classification. The classification rules represented from root node to leaf node through edges. Decision tree works on continuous and categorical type of problem. For solving categorical problem, classification decision tree is used and addressing continuous problem, regression decision tree are employed extensively (James & Singh, 2004).

Naive Bayes

Naive Bayes classifier is theoretically based on statistical learning, which is considered to be the fundamental framework from statistics and functional analysis for machine learning. The major goal of statistical learning is to provide predictive function based on data and solve reasoning problem that is

Figure 5. A Sample Decision Tree

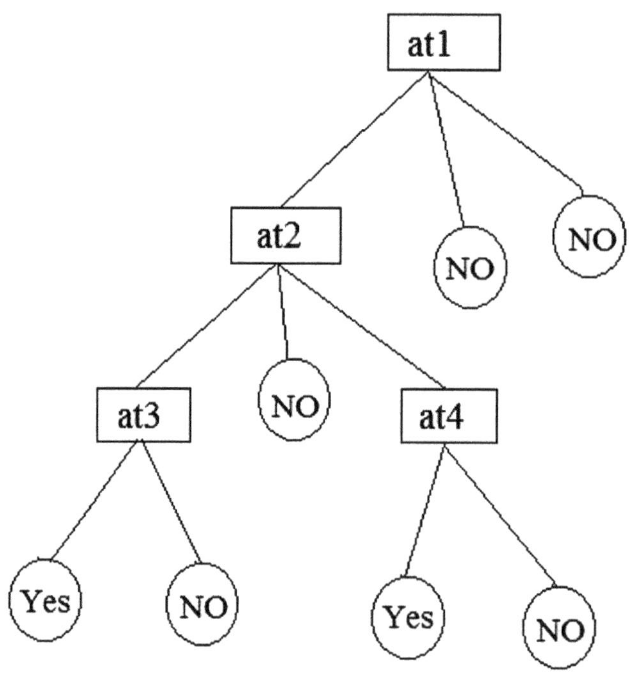

deriving knowledge. Naive Bayes classifier is based on Bayes theorem and falls under the category of probabilistic or predictive classifier. This classifier is also based on supervised learning method that deploys training data for finding output from testing data. This classifier based on Bayes theorem, maps unlabelled testing data into labelled training data. This classifier works on hypothetical situation in a predictive manner.

Cognitive radio devices may also use Naive Bayes in learning phase for spectrum sensing purpose. Cognitive radio devices is enhanced with the feature of finding spectrum holes in spectrum sensing function. In that case, pre-defined feature (training data) is used so as to find spectrum holes on basis of training information. The Bayes theorem is conditional probability that is used as function for classification. The theorem is formulated as following:

$$p\left(C_k \mid x\right) = \frac{p\left(C_k\right) p\left(x \mid C_k\right)}{p\left(x\right)} \tag{2}$$

$X = \left(x_1,..,x_n\right)$ represents some n feature and feature probabilities $p(C_k \mid x_1...x_n)$ for each k possible outcomes. All the features are independent in Naive Bayes model. Naive Bayes use very small amount of training data for estimating necessary parameters for classification (Karim & Rahman, 2013; Safavian & Landgrebe, 1991).

Support Vector Machine

Support Vector Machine (SVM) is based on supervised learning and is used for classification of data and regression. The main functionality of SVM is to design hyper plane in high dimensional space to produce optimal separation of classes for classification. SVM is often deployed to estimate separation margin in hyper plane for inseparability problem. In such cases, training data is divided into classes, which are separated by hyper plane to compute less-error classification.

Support vectors contains training elements and uses subsets of training data in decision function. It has different kernels for decision function that is basically a function which maps testing data into training data to get output. Support vector machine basically deals with pattern classification to classify different types of patterns by using decision function. Support vector machine classifier is designed on the principle of statistical theory and structural risk minimization theory. Hyper plane means finding the boundaries to differentiate between different classes. This hyper plane is also known as Margin. The hyper plane equation which is used to represent data shown below:

$$aX + bY = c \tag{1}$$

Cognitive radio devices use SVM to perform classification of unknown primary signals which can be only be classify in hyper plane (Figure 6). Cognitive radio devices have some classes of patterns which can be separated by using decision functions. SVM has different types of kernel such as linear, polynomial, and non-linear, sigmoid, RBF to classify any type of data. SVM gives the best classification performance on labelled training data and contains different patterns from unknown primary signals. The objective is to separate these patterns by constructing hyper plane in high dimensional space (Steinwart & Christmann, 2008; Mitchell, 1997; Coats & Huval, 2013).

Deep Learning

Deep learning is hierarchal learning in which we deeply learn all the parameters to sense spectrum holes from licensed bands. Deep Learning approach processing is done with multi-layer. It uses some hidden layer to process the input. One layer of output is used as another layer of input to process repeatedly. The process of deep learning of RF environment means to process all the parameters for Spectrum sensing by using multi-layer concept. Deep learning improved the performance better than any other algorithm. Deep Learning is applied on large datasets to obtain better results. Some popular techniques of deep learning are generally built from Artificial Neural Network (ANN). Deep learning is designed as a model from neural networks with multi-layer and training will be done in layer-wise manner. Deep learning approach is composed of supervised learning methods for implementing the network. A multi-layer neural network is used for solving problem with deeper insight or intelligence. ANN is an interconnection of large number of processing elements to solve the specific problem like human brain. ANN is also learned from experiences and it cannot be programmed like computer to solve the problem. In ANN, processing elements are composed in layer-wise fashion. Deep learning approach uses multi-layer of these processing elements. In figure 7, $L = \left\{L_i \mid i = 2..\left(n-1\right)\right\}$ denotes the set of hidden layers. In artificial neural network, an activation function helps for mapping input to output. In deep learning approach, cognitive radio devices sensed all relative parameters of sensing spectrum holes and then pass the information from one layer to another layer to get the most relevant parameter which provides the desired output i.e., which is the best spectrum hole in licensed bands and how many number of vacant bands are available (Zhang & Xie, 2007).

Figure 6. A simple model of SVM to classify Inseparability problem

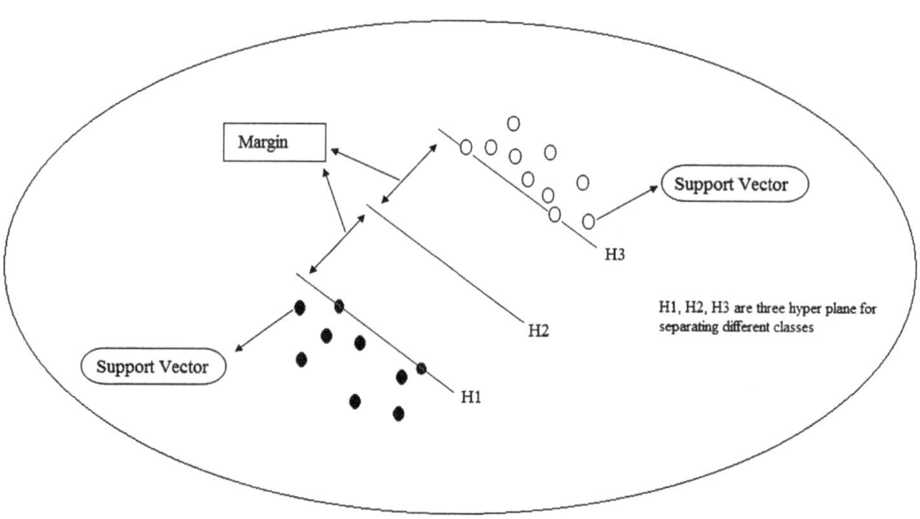

Figure 7. Deep Learning Architecture

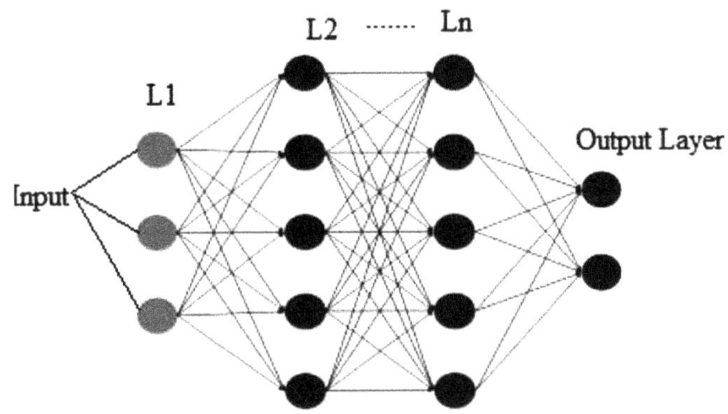

UNSUPERVISED LEARNING

Unsupervised learning approach is adapting of knowledge in absence of supervision or in case the desired response is not known. It is the process of labelling the unlabelled data without using the labelled data classes. In unsupervised learning, main objective is to learn the target function that can be used to predict success without labelled training data. In CRN, the prime objective is to detect the vacant bands from RF environment. Cognitive radio devices may also use unsupervised learning approach to learn the target function when they don't have any prior knowledge of the environment (Figure 8).

Unsupervised learning approach is also known as learning without knowledge source. In this, the network itself as much capable to detect any existing possible patterns, features, similarities in testing data. It is the process of learning from itself (without any prior information) in which algorithms are designed to organize data into groups of clusters. The input data is not labelled data and without requirement of desired output or labelled data for training purpose. Clustering being the technique of finding similarity of data into groups, is based on unsupervised learning in which grouping of similar data into clusters. It refers to problem of finding hidden structures in the labelled data. An unsupervised learning provides clustering technique for grouping of unlabelled data (Zhang & Xie, 2007; Ting-chao, 2004).

Clustering is the process of finding structure from set of unlabelled data, which helps to determine the groups of similar data from unlabelled data. Clustering makes the different groups in which data looks more similar in characteristics to each other. Specifically, it refers to the process of finding those

Figure 8. Unsupervised Learning

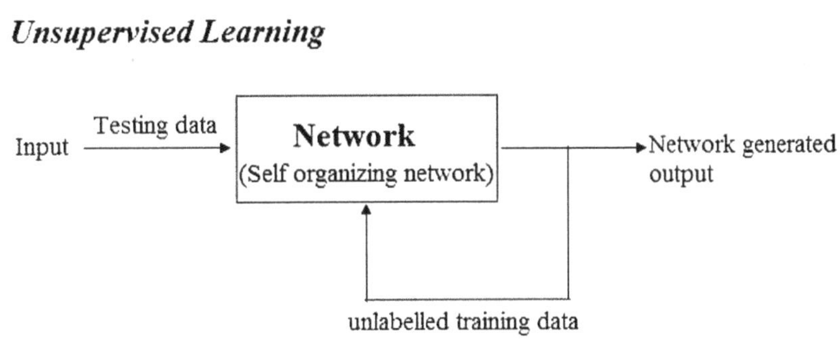

features from unlabelled data which are very similar to each other and put them in one group. Clustering can be carried out on the basis of distance, function, time, space or concept. The criteria of distance are grouping of those data which are very close to each other in distance. This type of clustering is known as distance based clustering. The criteria of concept are grouping of those data which are very close to each other by concept. This type of clustering is known as concept-based clustering. Clustering is an unsupervised approach i.e. there is no supervision for leaning about the RF environment.

Clustering has some capabilities:

- To determine input parameters that makes grouping feasible
- To maintain scalability
- To make use of unlabelled data
- To deal with huge amount of data set

Some clustering algorithms which we used to make groups or clusters:

- K-means
- Fuzzy C-means
- Hierarchical clustering
- Mixture of Gaussians

In cognitive radio networks, when radio devices have unlabelled data then they use clustering for making valuable use of that data.

REINFORCEMENT LEARNING

Reinforcement learning is the process of training the agent by rewards and punishments. It is also known as output-based learning in which the condition of success (+1) or failure (-1) indicates the reward and penalty respectively. By reinforcement learning, an agent or cognitive radio device learns by interacting with its RF environment.

The agent receives rewards to produce correct output and penalties for performing incorrect output. The agent does not learn from a human intervention but it learns by maximizing its reward points and minimizing its penalty points. The agent uses dynamic programming to make decisions that maximize its success with reduced chances of failures. Reinforcement learning is similar to supervised learning approach, with a difference that in spite of having desired output an agent has rewards and penalties to decide whether agent is performing in correct or incorrect direction. That is why it is combination of both supervised and unsupervised learning approach. Like unsupervised learning, RL agents can learn from its environment without supervision (Sutton & Barto, 1998).

Reinforcement learning allows the agent to learn its behaviour based on the feedback from its environment (Figure 9). In reinforcement learning, an agent stores the state of environment to take appropriate decision, which is why the learning approach becomes too expensive in terms of storage. Cognitive radio device may use this learning approach when dealing with RF environment. On the basis of success

Figure 9. Reinforcement Learning

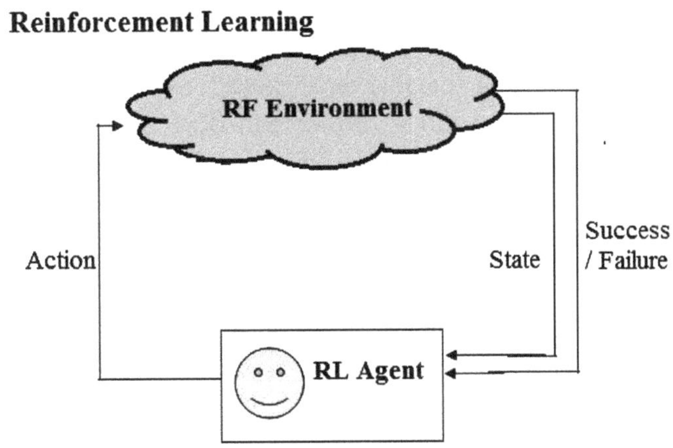

rate, devices can learn and act according to it to sense the spectrum holes without creating disturbances between primary users. Robot navigation, game playing, are the applications of reinforcement learning in which next state of agent depends on previous state and the performance of agent is judged by reward points or success rate.

DESIGNING EFFICIENT LEARNING ALGORITHMS

Designing learning algorithms require certain significant factors to be considered due to improve efficiency or performance of cognitive radio network. Apparently, learning algorithm helps radio devices to perceive information from RF environment. The process of learning information makes use of radio devices for taking future decision with respect to RF environment. Since, spectrum sensing requires associated radio devices to learn information, therefore it becomes essential to design efficient learning algorithm. Equipped with effective algorithmic design, the cognitive radio devices provide better results from even partial information perceived from RF environment. This eventually assists the devices to automatically become capable of learning unknown parameters from RF environment. Learning algorithms play an important role in CRN in improving the performance of cognitive networks, in absence of which the system throughput would decrease in spite of applying AI concepts.

Learning is necessary when the information about inputs/outputs is unknown or incomplete. Learning rules generate different activation function which helps to map inputs to outputs. The activation function must more effective and accurate. With the help of activation function we can take decision about observable environment in effective manner. Learning algorithms are very necessary part for taking future decision about RF environment by radio devices. To achieve escalation in network performance, efficiency is required in design of learning algorithms. Some factors definitely to be considered while designing learning algorithms, include learning in partially perceivable environments, multi-agent learning, and autonomous learning in unknown radio frequency environments (Bkassiny, Li, & Jayaweera, 2013; Cussens, 1996; Bkassiny, Li, & Jayaweera, 2013; James & Singh, 2004).

LEARNING IN PARTIALLY PERCEIVABLE ENVIRONMENTS

In RF environment, it will be impossible for the agent to get perfect perception of the state of current environment. Therefore, the perception remains incomplete for the agents. Unfortunately, complete observability is necessary for learning methods. Hence, radio devices make observations of the state of RF environment that may include noisy observations in sensed data. Here, the major problem posing as the ultimate challenge is incompleteness in perception of information from the environment, i.e. finding hidden state of RF environment. To overcome above problem, cognitive radio devices need learning algorithms which are as much capable to learn with incomplete information and reveal the hidden state of RF environment and provide correct output as a decision. Designing a learning model in real time which can learn from partially perceivable environments is very critical tasks for an agent.

MULTI-AGENT LEARNING

There are many radio devices which works in cognitive radio network simultaneously, which makes CRN, a multi-agent network. The entire device can perceive their information independently and take decision dependently. As we know any secondary user can communicate when vacant bands are available or finding any primary signal on that spectrum, it must put hand off its communication channel so decision must be dependable on perceiving information. It is impossible or difficult for an individual radio device to take correct decision. All the radio devices can learn RF environment at the same time. In this case, the main problem confronted is infeasibility of individual device in perception of RF environment. To overcome this problem, we need to use multi-agents learning in CRN that refers to the process of interacting RF environment by multiple interacting agents. To design a learning model for CRN, multi-agent learning needs to be ensured.

AUTONOMOUS LEARNING IN UNKNOWN RF ENVIRONMENTS

As we know cognitive radio device learn information for spectrum sensing, spectrum sharing, spectrum mobility, spectrum decision. Learning is very important process in CRN. Learning must be done automatically at every point of time so that radio devices must be capable to take correct decision about RF environment. Autonomous learning in unknown RF environment is also main factor when designing learning model for cognitive radio network. When cognitive radio devices are working in RF environment then there is need to sense the data automatically. So that devices can take decision on the basis of perceiving information. In that case autonomous learning is very important feature in unknown RF environment.

APPLICATIONS OF AI TECHNIQUES IN CRN

This section specifies certain applications of cognitive radio network, which is implemented by employing artificial intelligence techniques.

APPLICATION OF MACHINE LEARNING TO CRN

Machine learning is the ability to learn the system parameters without explicit programmed. Machine learning is the concept of artificial intelligence which provides some techniques to learn the RF environment for secondary users. Some major issues in cognitive radio can be implemented by using machine learning to avoid consequences of these issues. Some integral applications of machine learning in cognitive radio networks, include (Clancy, Hecker, Stuntebeck, & O'Shea, 2007; Rondeau, 2007; Langley & Simon, 1995; Clancy, Hecker, Stuntebeck, O'Shea, 2007):

- **Data Security:** In cognitive radio networks, all secondary users collect information about its RF environment. To provide correct information regarding the RF environment or spectrum bands, machine learning is required. CRN extends support of two types of users, including primary and secondary users. Secondary users can have access only when the licensed bands are vacant. In case, if users who is not the part of CRN, accesses information about RF environment and vacant bands could violate rules of secondary users. Attackers may not allow primary user to access the spectrum or create interference between the communications of primary users. Hence, to secure the data from the viewpoint of secondary user, machine learning technique needs to be used for storing data in real-time with inclusion of basic security techniques.
- **Personal Security:** Evidently, cognitive radio devices are associated with have some personal information that helps them to access spectrum with secure and successful handoff. Personal information include effective bands to use, handoff detection and management, etc. To secure this information from others users or attackers machine learning approach is largely used. Machine learning allows eliminating false information which is not useful or harmful in future. Radio devices must store their personal information, in a confidential manner, so that they can access RF environment as per their condition. Machine learning approaches assists in helping them to maintain their confidentiality level and access spectrum holes appropriately. Maintaining of personal information in secure manner may help CR devices to speed up the processing and improve resultant performance.
- **Fraud Detection:** Some users pretend that they are secondary users and they want to access the RF environment. This makes accessibility of information about cognitive radio networks and spectrum bands, quite vulnerable. To detect such fraud users machine learning allows to identify fraudulent tasks and associated users who are trying to take part during communication session.

APPLICATION OF ARTIFICIAL NEURAL NETWORK TO CRN

Artificial neural network is an interconnection of processing elements which executes task, similar to human brain. Cognitive radio networks perform its tasks and access information about its surrounding dynamically. To process accessing information of RF environment, application of Artificial Neural Network (ANN) is applied predominantly. Its features like improvement of efficiency, accuracy and reduced the vulnerable rate make it so usable. Cognitive radio networks need to improve its efficiency at every point. Some features of CRN can be improved by ANN, (Zhang & Xie, 2007; Ting-chao, 2004; Galindo-Serrano, Giupponi, Blasco, & Dohler, 2010) which include:

- **Primary User Signal Detection:** In cognitive radio networks, Secondary users use the vacant bands of spectrum (which is officially allotted to primary users). When primary users desire to use their licensed band, the secondary user is bound to hand off their communication onto other vacant bands. In order to avoid interference, unlicensed users need to detect all the primary user signals during communication. Initially, all the users receive signals, followed by detection of primary user signal. For this filtering, CRN use artificial neural network that classifies primary signals from all received signals.
- **Signal to Noise Ratio:** Signal-to-Noise Ratio (SNR) is another important feature of cognitive radio network. SNR must be low to achieve better efficiency in cognitive radio networks. Artificial neural networks detect weak primary signals and noise in primary signals and attempts to reduce SNR, thereby improving the signal strength to makes use of efficient networking. When primary users send signals for communication it may be weak or faded, ANN can differentiate unused bands and fade primary signals.
- **Enhance Performance:** ANN is capable to classify all feature of RF environment to enhance the performance of CRN. Cognitive radio devices find spectrum holes efficiently and detect primary users signal correctly and allow communication without creating disturbances. It will automatically enhance the performance of network. ANN can deeply learn all aspects of RF environment by using multi-layer concept to improve the accuracy and throughput of the cognitive network.

APPLICATION OF SVM TO CRN

Cognitive radio devices use SVM to perform classification of unknown primary signals which can be only be classified in hyper-plane. Cognitive radio devices have some classes of patterns which can be separated by applying decision functions. Support vector machine is a classifier which classifies those features which cannot be separable easily. Cognitive radio devices use SVM for this classification of those inseparable instances. Sometimes cognitive radio receives signals from other users which can be only filtered by using hyper plane. Filtration of those signals is necessary for taking right decision by radio devices. Some features of CRN can be improved or modified by SVM (Steinwart & Christmann, 2008). These features are:

- **Bit Error Rate:** The received bits of data stream through communication channel may be altered by noise, interference and alteration. That altered receiving bits are known as bit error. The bit error per unit time is called Bit Error Rate (BER). SVM classifies the error bit and corrects the bit from data stream by using hyper-plane. This will further assist radio devices to reduce and check error in data stream to make communication more efficient. Finding the bit error from data stream is inseparability problem, which can be solved with SVM by separating all bit error by using margin hyper-plane in high dimensional space.
- **Signal Classification:** Signal classification is the problem come from receiving some unwanted signals with primary signals. Those unwanted signals may be noise signals, some repeated signals, or other irrelevant signals, interference signals. SVM classifies all the receiving signals through hyper plane and produces the relevant signals as the output. Signal classification is very important tasks of cognitive radio. In order to avoid disturbances between primary users, suitable decisions

are required to be taken on time and based on received signals. SVM can easily separate different signals from the stream of receiving signals.

- **Eliminate Interference:** In addition to classification of received signals, SVM can also be employed to detect interference signals and eliminate them. Elimination of interference is necessary in order to avoid disturbances in the communication band of primary users. Licensed user use their spectrum whenever and wherever they want to use. SVM eliminate interference and provide good strength signals to users who are working in RF environment, simultaneously.

CONCLUSION

Artificial Intelligence (AI) techniques are widely used in Cognitive Radio Networks (CRN). In fact, cognitive radio network is designed to attain efficiency in the process of artificial intelligence. In this chapter, in-depth insight of working of CRN is elaborated for functionalities employed for artificial intelligence to support wider area of applications in real time. For implementing real time applications, cognitive radio devices need to act intelligently and take actions, decisions dynamically. Different AI techniques are used for implementation of application of CRN but it depends on requirements which techniques will be used. These techniques are also used for improvising internal features of cognitive radio networks in real time.

REFERENCES

Clancy, Hecker, Stuntebeck, & O'Shea. (2007, August). Applications of machine learning to cognitive radio networks. *IEEE Wireless Communications*, *7*, 47–52.

Bkassiny, M. (2013). *A Survey on Machine-Learning Techniques in Cognitive Radios. IEEE Communications Surveys & Tutorials, 15(3).*

Bkassiny, M., Li, Y., & Jayaweera, S. (2013). A survey on machine-learning techniques in cognitive radios. In *Communications Surveys and Tutorials* (Vol. 15). IEEE. doi:10.1109/SURV.2012.100412.00017

Clancy, C., Hecker, J., Stuntebeck, E., & O'Shea, T. (2007). Applications of machine learning to cognitive radio networks. *Wireless Communications, IEEE*, *14*(4), 47–52. doi:10.1109/MWC.2007.4300983

Coats, A., & Huval, B. (2013). Deep Learning with COTS HPS systems. *Journal of Machine Learning Research*, *28*(3), 1337–1345.

Cussens, J. (1996). Machine Learning. *IEEE Journal of Computing and Control*, *7*(4), 164–168. doi:10.1049/cce:19960402

Galindo-Serrano, A., Giupponi, L., Blasco, P., & Dohler, M. (2010). Learning from experts in cognitive radio networks: The docitive paradigm. *Proc. Fifth International Conference on Cognitive Radio Oriented Wireless Networks Communications (CROWNCOM '10)*, 1 –6. 10.4108/ICST.CROWNCOM2010.9173

Ghasemi, A., & Sousa, E. (2008, April). Spectrum sensing in cognitive radio networks: Requirements, challenges and design trade-offs. *IEEE Communications Magazine*, *46*(4), 32–39. doi:10.1109/MCOM.2008.4481338

Giupponi, L., Galindo-Serrano, A., Blasco, P., & Dohler, M. (2010, August). Docitive networks: An emerging paradigm for dynamic spectrum management. *IEEE Wireless Communications*, *17*(4), 47–54. doi:10.1109/MWC.2010.5547921

Haykin, S., Thomson, D., & Reed, J. (2009, May). Spectrum sensing for cognitive radio. *Proceedings of the IEEE*, *97*(5), 849–877. doi:10.1109/JPROC.2009.2015711

He, Bae, Newman, Gaeddert, Kim, Menon, … Tranter. (2010). A Survey of Artificial Intelligence for Cognitive Radios. *IEEE Transactions on Vehicular Technology, 59*(4).

He, A., Bae, K. K., Newman, T., Gaeddert, J., Kim, K., Menon, R., ... Tranter, W. (2010, May). A survey of artificial intelligence for cognitive radios. *IEEE Transactions on Vehicular Technology*, *59*(4), 1578–1592. doi:10.1109/TVT.2010.2043968

James, M., & Singh, S. (2004). Learning and discovery of predictive state representations in dynamical systems with reset. *Proceedings of the Twenty-First International Conference on Machine Learning (ICML)*, 417–424. 10.1145/1015330.1015359

Karim & Rahman. (2013). Decision Tree and Naïve Bayes Algorithm for Classification and Generation of Actionable Knowledge for Direct Marketing. *Journal of Software Engineering and Applications, 6*(4), 196-206.

Kotsiantis, S. B. (2007). Supervised Machine Learning: A Review of Classification Techniques. *Informatica*, *31*(3), 249–268.

Langley, P., & Simon, H. A. (1995). Applications of machine learning and rule induction. *Communications of the ACM*, *38*(11), 54–64. doi:10.1145/219717.219768

Mitchell, T. (1997). *Machine Learning: A Guide to Current Research*. McGraw Hill.

Mody, A., Sherman, M., Martinez, R., Reddy, R., & Kiernan, T. (2008). Survey of IEEE standards supporting cognitive radio and dynamic spectrum access. *IEEE Military Communications Conference (MILCOM '08)*, 1 –7. 10.1109/MILCOM.2008.4753298

Rondeau, T. W. (2007). *Application of artificial intelligence to wireless communications* (PhD thesis). Virginia Polytechnic Institute and State University.

Russell, S. J., Norvig, P., Canny, J. F., Malik, J. M., & Edwards, D. D. (1995). Artificial intelligence: A modern approach. Prentice Hall.

Safavian, S. R., & Landgrebe, D. (1991). A survey of decision tree classifier methodology. *IEEE Transactions on Systems, Man, and Cybernetics*, *21*(3), 660–674. doi:10.1109/21.97458

Steinwart, I., & Christmann, A. (2008). *Support vector machines*. Springer.

Sutton, R. S., & Barto, A. G. (1998). *Reinforcement learning: An introduction*. Cambridge Univ Press.

Weingart, T., Sicker, D. C., & Grunwald, D. (2007, August). A statistical method for reconfiguration of cognitive radios. *IEEE Wireless Communications*, *14*(4), 34–40. doi:10.1109/MWC.2007.4300981

Yu & Zhang. (2004). Study of Artificial Neural Network Model for Forecasting Urban Water Demand. *Journal of Zhejiang University, 38*(9).

Zhang, Z., & Xie, X. (2007). Intelligent cognitive radio: Research on learning and evaluation of CR based on neural network. *Proc. ITI 5th ICICT*, 33–37. 10.1109/ITICT.2007.4475612

Section 5
Practical Use of Cognitive Radio Networks

Chapter 15
Cognitive Radio Network for E-Health Systems

Shashank Gupta
Jaypee Institute of Information Technology, India

Shikha Singhal
Jaypee Institute of Information Technology, India

Adwitiya Sinha
Jaypee Institute of Information Technology, India

ABSTRACT

The world is witnessing widespread roots of ubiquitous computing across disciplines and industries. It is equipped with the ability to monitor anything from anywhere with efficient, easy, and equitable goods that offer services for everyone. This has become possible through usage of wireless technologies, which possess an extensive scope in healthcare domain. However, despite various advantages, wireless technologies are faced with distinct challenges in hospital environment. For instance, wireless devices often tend to cause electromagnetic interference to critical medical devices resulting in malfunctioning. Further, with ubiquitous computing, sensitive data about health state of patients is constantly being shared remotely from one place to another. Therefore, systems in place must address requirements of data security, and thus privacy. For this purpose, the chapter presents a collaborative study on cognitive-radio-based healthcare system, including advantages, architecture, and challenges related to implementation of cognitive radios in hospital environment.

1. INTRODUCTION

In recent times, wireless systems have gained extreme eminence in the field of healthcare and medicine. These systems aid in providing timely and flexible medical services for improving patient safety. However, due to more number of wireless devices coming up every day the wireless healthcare facilities are susceptible to problems such as spectrum scarcity. Moreover, certain existing medical devices are prone to damage from the electromagnetic interference caused by wireless transmissions from these devices.

DOI: 10.4018/978-1-5225-5354-0.ch015

To overcome these, and other similar, challenges posed by wireless communication systems this chapter suggests the use of cognitive radio networks in e-health applications. A cognitive radio system can operate on unlicensed as well as licensed bands thereby improving the utilization of radio spectrum and providing dynamic spectrum access.

This chapter aims to build an insight on the uses of cognitive radios in e-health applications. The chapter proceeds with a brief introduction on e-health in Section 3. This is followed by an overview of existing technological facilities in healthcare industry along with their shortcomings and limitations in Section 4. Section 5 talks about why there is a need for wireless devices and also mentions the challenges faced by wireless devices in a hospital environment. Further, in Section 6 the chapter details the requirements of such wireless devices and then finally moves on to explain how these requirements are well met by using cognitive radios for wireless communication in E-health applications in Section 7. Towards the end of Section 8, leveraging the advantages of cognitive radios, the chapter concludes by proposing an architecture for an E-health application. But firstly, next section presents some background literature which includes the work done by other researchers.

2. BACKGROUND STUDY

Use of cognitive radios in the healthcare industry is an intriguing topic. Many researchers have proposed various architectures for implementing a cognitive radio based e-health application. Phunchongharn, Hossain, Niyato, and Camorlinga (2010) addressed the technological requirements of a cognitive radio based wireless system in terms of electromagnetic compatibility, data security, and seamless connectivity. Further, they also put forward an architecture for implementing a cognitive radio based e-health application in which they divided the wireless devices into two categories viz. primary devices and secondary devices which use an EMI-aware handshaking protocol for accessing the wireless channel. Finally, performance evaluation of the system is carried out using simulations. Mamoon, Muzahidul-Islam, Baharun, Komaki, and Ahmed (2015) proposed an architecture for a cognitive radio based e-health application in which all the wireless devices are capable of dynamic spectrum access. They proposed a three layer model, where a cognitive base station (CBS) occupies the first layer, all the wireless devices and sink access control form the middle layer, and cognitive radio aided sensor network forms the bottom layer. Luan and Ren (2015) discuss the complete design of an e-health network including the design for the physical layer, MAC layer, and the transport layer. Some researchers have performed a detailed analysis of wearable body sensors and their wireless transmissions, and have suggested EMI-aware cognitive radio based architectures for Wide Body Area Network (WBAN) (Shen, Liu, Yu, Ma, Li, Shen, & Chen, 2013; Chávez-Santiago, Jankunas, Fomin, & Balasingham, 2014).

Various authors have also presented significant insight in the form of edited books and chapters on wireless communications, cognitive radios, and e-health systems. Some of the works are as follows. Hossain and Diamond (2008) studied the impact of fourth generation wireless access networks on e-health. They carefully analyzed the 4G wireless technology and various e-health applications and then presented a 4G heterogeneous wireless access network based architecture for implementing an intelligent software-defined radio based e-health system. Their work also analyses the task of bandwidth allocation and admission control for e-health services in a heterogeneous wireless environment. Phunchongharn, Hossain, and Camorlinga (2010) talked about the various clinical challenges in a healthcare environment and then suggested the use of wireless technologies for overcoming these challenges. Further, they mention the issues involved in using wireless technologies in a healthcare environment and suggest the use of a cognitive radio based system for the same. Ouattara, Krief, Chalouf, and Hamdi (2012) provided a method for improving spectrum sensing in cognitive radios for real-time monitoring of patients. They discuss the complete scenario addressed in patient monitoring and propose a solution by adding a prediction function to the stage of spectrum sensing. Some other works (Medeisis & Holland, 2014; Dhope, 2015) provide a detailed analysis of all the rules and regulations regarding spectrum usage and hence highlight the problems of band scarcity and spectrum utilization faced by various e-applications. They also suggest methods for implementing cognitive radio based systems for solving these problems.

After a careful analysis of advantages and drawbacks of aforementioned literature, the subsequent sections provide state of the art methods to deploy usage of wireless technologies and cognitive radios in an e-health environment.

3. SIGNIFICANCE OF E-HEALTH

E-health refers to an amalgamation of Information and Communication Technology (ICT), and healthcare and medicine. It is a recent term and was first used to convey the various promises and possibilities internet opened up for the area of healthcare and medicine. However, the true definition of e-health has a broader scope than just providing e-solutions to the arena of healthcare. As rightly suggested by Eysenbach (2001) the 'e' in e-health is not just limited to 'electronics' but also signifies *efficiency* in health care, *enhancing* quality of healthcare and medicine, *evidence* based and proven technology to be used in the healthcare industry, *empowerment* of patients, *encouragement* of a new technical relationship between healthcare professionals and patients, *education* of healthcare professionals, *enabling* information exchange between e-health devices and establishments in a standardized format, *extending* the scope of healthcare and medicine by providing online access to global services, *ethical* online professional practice, and *equitable* healthcare for people who do not have the required resources. Therefore, e-health includes much more than just electronics and healthcare. Cognitive radios provide an easy realization of these features by providing easy to use and efficient technology. However, initially, e-health was much more related to just digitalization, which definitely provided ease of use but did not address many other important issues like equitable and efficient healthcare services for all the strata of the society. The following section discusses some of the existing technologies used under the banner of e-health and then gives the modern state of the art technology.

4. CONVENTIONAL E-HEALTH APPLICATIONS

According to the data presented by Bates et al. (1998) a large number of people die every year because of either wrong prescription of medicines or misinterpreted prescriptions. Wrong prescriptions take place because the doctors are not aware of the patient's allergies since there are no records being maintained for them. Even if the records were being maintained they may not be easily accessible. Misinterpreted prescriptions can be mainly due to bad handwriting. In either case, patients suffer the adverse consequences of such clinical errors. To overcome these errors, technology's intervention in the healthcare sector becomes mandatory. For this purpose, the subsequent subsections list some of the existing e-health technologies.

4.1 Telemedicine

Telemedicine, composed of the words 'tele' and 'medicine', refers to the use of telecommunication systems for providing healthcare facilities in remote areas. Specifically, it is the use of technology to provide necessary information to patients at a distance. Telemedicine can be subdivided into two categories. Firstly, 'live telemedicine' and secondly, 'store and forward telemedicine'. In live telemedicine, a hospital professional performs an interactive examination of patients living at a distance. This is established using various audio-visual tools and it requires the presence of both the patient and the doctor at the same time. These sessions are established using high bandwidth and low latency network connections. On the other hand, store and forward telemedicine is more of an offline assessment task. In it, the images, videos and any other data regarding a patient's illness and his symptoms are stored and forwarded to the professional for an examination. The professional then stores and transmits the relevant results back to a clinic where the patient is administered. A typical telemedicine system involves both types of telemedicine. Firstly, the healthcare professional obtains relevant test results using store and forward methodology and then provides guidance on the basis of examining the test results using live telemedicine.

4.2 Computerized Physician Order Entry

Computerized Physician Order Entry (CPOE) is an electronic record of a doctor's instructions for a patient's treatment. A significant advantage of using CPOE is that it reduces the errors in treatment that take place due to bad handwriting. A typical CPOE mimics the instruction chart of a doctor and consists of instructions for carrying out treatment of a patient. These instructions are communicated to all departments. These orders also include the patient's medical history along with any new laboratory test results. It helps in providing real time monitoring of a patient from checking his allergies to prescribing doses of drugs. Despite the ease they are capable of providing in a healthcare industry, there are two drawbacks of CPOE systems. Firstly, they are very costly to install and maintain. Secondly, it takes years for a simple CPOE system to get completely trained and implemented (Kuperman & Gibson, 2003).

4.3 Electronic Health Record

Electronic Health Record (EHR) is a digital record of the personal and medical information regarding a person. Personal information includes details such as name, age, weight, etc. Medical information includes the patient's allergies, vital signs, medical history, past test results, etc. Since this data is electronically stored, therefore it is accessible everywhere. Through the use of EHR, it is easy to update the information with the current medical condition of a patient. It is also easy to track any previous illnesses or medical surgeries a person might have had and therefore the examination of long term changes in his health becomes easier. Since a single file is maintained for each patient, data extraction becomes easier.

In comparison with paper based records, EHR has an advantage of accuracy and easy forwarding. There cannot be errors due to bad handwriting. Further, if a patient were to visit multiple healthcare centers, it may not be required to get the same medical tests done again. This saves time and money and makes healthcare efficient and affordable. However, such critical data about a person also raises security concerns and therefore proper encryption methods must be used for storing and forwarding such data.

4.4 M-Health

mHealth or mobile Health is the use of mobile devices and wireless technologies for providing health-care services and collecting data about patients. The mobile devices include smartphones, tablets, smart watches, PDAs, and can be used for collecting data such as vital signs of patients, information regarding their symptoms and allergies, etc. It is similar to telemedicine except for the fact that mHealth specifically uses only mobile devices; it signifies mobile self-care. The use of mobile devices ensures dynamic monitoring of patients and their vitals. Another important application of mHealth involves the use of wireless mobile devices for educating people about the various health and hygiene issues and their preventive measures. Apart from treatment support, mHealth applications are also used for disease surveillance and epidemic outbreak monitoring.

All these e-health applications offer various advantages in a healthcare environment but are also faced with certain challenges. The next section talks about the various wireless devices, technologies, and services used in the healthcare industry and also lists some of these challenges faced by these devices.

5. WIRELESS DEVICES AND TECHNOLOGIES FOR E-HEALTH

Pervasive healthcare requires the use of wireless devices and technologies for promoting quality and affordable healthcare for everyone. People no longer need to travel long distances and spend excess money for receiving quality treatment. There exist various wireless devices and technologies which make pervasive e-health solutions possible.

With the advent of radio frequency techniques, various low-cost broadband solutions emerged over time. These wireless technologies have helped in achieving the goal of pervasive healthcare to a great extent. These include long range technologies such as Wireless Metropolitan Area Networks (WMAN), Wireless Wide Area Networks (WWAN), Wireless Local Area Networks (WLAN), and Wireless Personal Area Networks (WPAN). Some other short-range technologies include Wi-Fi, Infrared Data Association (IrDA), Radio Frequency Identification (RFID), Bluetooth, Zigbee, and Ultra wideband (UWB). Further, the Next Generation (NG) of mobile communications, which is way beyond 3G and 4G integrate the existing wireless technologies into a seamless platform and thus provide a high bandwidth and low latency ubiquitous wireless network. Such a network connection facilitates the development of portable and mobile telemedicine applications.

Further, certain wireless services help in achieving the goal of pervasive healthcare. These include Medical Implant Communications Service (MICS) and Wireless Medical Telemetry Service (WMTS). Under MICS specialized medical devices are implanted inside a person and these devices transfer important data regarding the organs of that person using very low power. Examples of these devices include implanted defibrillators and cardiac pacemakers. WMTS is used for the remote monitoring of a patient's health condition. This service allows monitoring vital signs such as body temperature, blood pressure, heart rate, etc. WMTS has two components viz. wireless wearable sensors and central monitoring systems. These wireless wearable sensors transmit critical data about the vital signs of a person to a monitoring system which can then be analyzed for medicine prescription. However, these services have been launched in the United States and are not yet accepted globally. The next section talks about the need for wireless devices in a healthcare environment.

5.1 Need for Wireless Devices and Technologies

In a healthcare environment, wireless technologies offer significant benefits which ensure reception of quality healthcare to everyone. For example, through the use of wireless technologies, patients suffering from chronic problems such as heart problems and memory loss, can be continuously monitored. There is no longer a need for a 24x7 healthcare assistant to look after such patients. Further, the time efficiency of healthcare services is improved drastically. Medical staff spends a huge amount of time manually transferring equipment and moving patients in case of emergencies. With wireless technologies at hand, more time can be devoted to quality treatment.

Further, wireless devices provide service and information accessibility. Doctors in small clinics in rural areas can seek expert guidance from more qualified professionals in hospitals in urban areas. Moreover, every information regarding a person from his allergies and blood type to its medical history can be saved electronically. A doctor no longer needs to be physically present to examine a patient and doesn't require a hard copy of its report which further takes time in printing and transferring. The doctor can simply refer to electronically saved records and even pass them around if need be, in no time. Saving medical records electronically and passing them wirelessly also reduces the casualties that occur due to human error in the healthcare industry. Therefore, if wireless technology is implemented in the healthcare industry, it will satisfy all the Es of E-health mentioned earlier. The next section talks about the challenges these wireless technologies and services face in a healthcare environment.

5.2 Challenges Faced by Wireless Devices in E-Health Applications

The wireless devices when present in a healthcare environment are faced with certain challenges. The proper functioning of these devices and determination of accurate results by them is of utmost importance. However, in the presence of other wireless devices performing communications and transmissions, these devices are prone to malfunctioning due to electromagnetic interference. Further, the devices are also prone to problems of bandwidth restriction and spectrum utilization. Some of these challenges are highlighted in the subsequent subsections.

5.2.1 Medical Band Scarcity

A considerable amount of new wireless devices and technologies have emerged over the years and so has the demand for spectrum. However, the reallocation of spectrum for these devices did not keep pace with its demand. Several applications have access to licensed bands allocated by the government but medical devices still operate in the unlicensed spectrum only. Because of this medical devices are forced to operate among other wireless devices and thus become prone to electromagnetic interference as explained in the next subsection. However, in view of this electromagnetic interference caused to medical devices, the Federal Communications Commission (FCC) in the United States defined specific bands for Wireless Medical Telemetry Service (WMTS) and Medical Device Radio communications Service (MedRadio). As discussed earlier, these services are used for the transmission of medical data regarding a person's health and the communication needs of wireless medical devices respectively. However, these are not international standards and cannot be used in other countries. Therefore, the medical devices are still faced with the problem of medical band scarcity in other countries.

5.2.2 Electromagnetic Interference

All wireless devices use radio frequency to communicate with one another and their respective base centers. This radio frequency consists of electromagnetic waves which carry the required message through the means of various forms of modulation. Now the same band of the electromagnetic spectrum is available for all wireless communications. This means that different technologies can use the radio spectrum simultaneously resulting in interference while transmitting signals and hence reception of distorted signals. For this purpose, licensing of bands was done. It ensures that certain operators operate within a fixed range of frequencies only by paying a licensing amount and hence obtaining exclusive right to transmit on the assigned channel. This licensing scheme is irrelevant for short range communications such as between a mobile device and a Wi-Fi router. Therefore, these wireless devices operate in the unlicensed band, usually the 2.4 GHz Industrial, Scientific, and Medical (ISM) band. A wireless connection can be setup anywhere using the unlicensed band for free as long as the devices using them belong to the ISM category. As a result the devices operating in the unlicensed band are bound to face interference. This electromagnetic interference is particularly harmful in a hospital like setup where various medical devices operate in this ISM band as well as a user's cellular phone.

Such electromagnetic interference (EMI) between handheld wireless devices and medical equipment can cause the latter to malfunction. Pervasive wireless communication is a fundamental aspect for achieving pervasive healthcare. However medical equipment has certain sensitive components which are not immune to EMI from other wireless devices. As a result, the efficiency of these wireless devices decreases and many times wrong results are given resulting in serious problems and sometimes death. Particularly, medical equipment is prone to electromagnetic interference because the human body and connecting wires act as antennae and these devices can measure very low-level signals (Boyle, 2006). Although many new wireless devices are being made immune to EMI still they do not provide a complete solution. With the continuous increase in the number of wireless devices and communication networks, the problem of EMI in healthcare and other sectors will continuously grow too (El Khaddar, Harroud, Boulmalf, Elkoutbi & Habbani, 2012).

5.2.3 Spectrum Utilization

As per the challenges mentioned above, spectrum utilization by medical devices becomes a challenging task. Since there is no proprietary band allocated to medical devices, therefore they have to use the unlicensed band. With the increase in the number of wireless devices, it becomes very difficult for the medical devices to do any kind of wireless transmission or communication without facing EMI. There is a need for a system which allows priority based access to the wireless channel. In a hospital environment, medical devices can be assigned higher priority and other wireless devices a lower priority. Such a system is possible with the use of cognitive radios, as has been discussed later. The next section mentions the requirements of wireless devices in e-health applications.

6. REQUIREMENTS OF WIRELESS DEVICES IN E-HEALTH APPLICATIONS

The biggest difficulty ever faced by healthcare industry was the distance between doctors/healthcare professionals and patients. With the advent of wireless technologies this problem has almost eliminated and healthcare facilities are becoming pervasive. Wireless technologies provide real time data delivery. If the doctors can be connected wirelessly to the entire hospital system, they can check test results, monitor patient's vitals and provide expert guidance without being physically present. This is even true for providing medical assistance to remote areas. However, such wireless communication systems have certain requirements when used in a healthcare environment. The following subsections list some of those.

6.1 Electromagnetic Compatibility

Medical devices suffer serious performance degradation if the EMI from other wireless devices is more than their immunity levels. This performance degradation severely impacts the QoS of the healthcare industry. Therefore, the wireless devices in a healthcare environment must be electromagnetically compatible with the medical devices. For this purpose, it is important to note a few types of electromagnetic disturbances that can cause a medical device to malfunction. These have been given by the international standard International Electrotechnical Commission (IEC) 60101-1-2.

The first type of disturbance is an electrostatic discharge. Any medical device must be able to withstand an electrostatic discharge of at least 6-8 volts. Further, these devices should operate on insulating materials such as wood, ceramic tiles, or concrete. The second type of disturbance is radiated Radio Frequency (RF) electromagnetic fields. The electric field component of an EM wave is more harmful to a medical device. Typically, a non-life supporting device should be able to withstand an electric field of 3 V/m and a life supporting device of 10 V/m. The third type of disturbance is electrical surges. IEC 60101-1-2 gives the minimum amount of electrical surge a medical device should be able to tolerate and is equal to 2 volts. The next type of disturbance is voltage dips. If the load on a power supply line increases abruptly, it causes a sudden increase in voltage being supplied to the medical devices. Certain life supporting medical devices require an uninterrupted power supply. Therefore, these devices should be able to withstand such voltage dips to some extent. Further, any form of electromagnetic emission can induce undesired high voltages in medical devices. The medical devices should be designed to withstand such emissions as well.

6.2 Coexistence of Wireless Technologies With Overlapping Frequency Bands

Several wireless technologies are used in a hospital simultaneously, ranging from WLAN and WPAN technologies for bandwidth extensive applications such as telemedicine to Bluetooth and ZigBee technologies for wearable body sensors. Many of these technologies make use of overlapping frequency bands, for example, both Bluetooth and ZigBee use the ISM band for wireless communications. Therefore, it becomes difficult to avoid EMI caused to one another and other devices operating in the same frequency band. Thus, wireless devices require efficient spectrum access techniques specifically in an industry like healthcare where quality cannot be sacrificed.

6.3 Healthcare Data Security

Healthcare data is highly secure and consists of every information regarding a person's health from his illness to minute allergies. It can be exploited by people with malicious intents. Therefore, data security is of utmost importance in a technological hospital environment. Data security doesn't refer to privacy alone but also to proper handling of data while transmission. If the data is attenuated with noise during transmission, then a patient may receive wrong the prescription from a doctor which may severely affect its health. Therefore, healthcare data should be properly encrypted before transmission from an e-health application. For example, IEEE 802.1x based authentication methods and IEEE 802.11i based key distribution mechanisms can be used in a WLAN.

6.4 Seamless Connectivity

When the medical devices are complemented with wireless features, they offer some significant applications such as remote and real time monitoring of patients through the use of telemedicine and wearable sensors respectively. However, these services require an uninterrupted network connection so that the data can be continuously transmitted and monitored even while the patient is moving. A seamless connection ensures high-quality data transmission and hence, better performance of e-health applications.

6.5 Prioritized Access to Wireless Channel

Since both the medical devices and other wireless devices operate in the same frequency bands, there are chances of electromagnetic interference. Therefore, there should be prioritized access to the wireless channel. Medical devices must have primary access and other wireless devices should have secondary access. For this purpose, this chapter proposes the use of cognitive radio technology in a healthcare environment. Cognitive radios ensure prioritized access to the wireless channel among various wireless devices in a hospital. This discussion has been detailed in the subsequent sections.

7. COGNITIVE RADIO FOR WIRELESS COMMUNICATION IN E-HEALTH SYSTEMS

Conventional spectrum management results in inefficient and low spectrum usage. Further, with more number of medical devices coming up every day, the unlicensed band is getting crowded and hence EMI prone. Cognitive radios provide a method to efficiently use the licensed band of the spectrum for wireless communication of unlicensed band devices without causing interference to the former. The fundamental aspect of cognitive radios lies in the fact that the licensed band is seldom continuously occupied. It consists of white spaces or holes. A spectrum hole is defined as a particular frequency band which is currently not being used by a licensed user for transmission. The usage of such holes by unlicensed users can overcome the challenges mentioned in previous sections and at the same time eliminate the possibility of interference to licensed users.

In a cognitive radio system, the licensed users are also termed as primary users and the unlicensed users are also termed as the secondary users. Further, a primary device is any active medical device using radio frequency for transmissions or any wireless device used for critical healthcare operations. Secondary device is any wireless device used for normal data transfer such as a tablet, PDA, etc. Before going into deeper details of the architecture of a cognitive radio system, the following section talks about the advantages of using cognitive radios.

7.1 Advantages of Cognitive Radio in E-Health Applications

A cognitive radio-based system has the ability to sense the channel for holes and then use these holes for wireless transmissions of unlicensed users. Wireless transmissions in a cognitive radio-based system take place through four steps viz. spectrum sensing, adaptive learning, spectrum decision, and finally, transmission parameter setting. These are detailed in the subsequent subsections.

7.1.1 Spectrum Sensing

A cognitive radio system has the ability to sense the radio spectrum and determine the spectrum holes and their channel capacity. In addition to the availability of spectrum space, it also senses the spectrum for measuring characteristics like power, interference and noise temperature, and various other operating restrictions. This sensing is performed not just along the dimension of space but also along the dimensions of time, phase, frequency, and code. This is done to seize every opportunity for using the unutilized portion of the licensed band along any dimension.

7.1.2 Adaptive Learning

In adaptive learning, the cognitive radio system learns about the usage of radio spectrum by primary users and secondary users respectively. This usage analysis combined with the characteristics of holes determined in the previous step helps the cognitive radio in making a decision about the requirements of secondary users for accessing the unlicensed band. These requirements include the bandwidth of transmission, required data rate, transmission mode, acceptable bit error rate, operating period and lastly data size. Using the information acquired during the adaptive learning stage, the spectrum decision stage is able to predict the future occupancy of the various frequency bands thus minimizing interference to primary users. Once the future occupancy of licensed bands and the transmission characteristics of the secondary users have been determined, a new list of future spectrum holes can also be obtained. This helps in speeding up the task of spectrum sensing.

7.1.3 Spectrum Decision

As mentioned in the previous stage, in spectrum decision, a cognitive radio system determines the secondary users' requirements and the future spectrum holes. This information helps in selecting an appropriate channel among the licensed bands for wireless transmissions of secondary users. The transmission parameters for this communication are also determined by the cognitive radio system. This is detailed in the next subsection.

7.1.4 Transmission Parameter Setting

This stage is responsible for tuning the parameters of secondary devices according to the requirements determined by spectrum decision stage. After the parameters have been tuned, the secondary devices can transmit wirelessly using the selected licensed spectrum channel. The cognitive radio keeps track of all these parameters and the spectrum holes across all dimensions at all times. This dynamic maintenance is helpful when a channel suddenly becomes unavailable. In such a case, the wireless communication is seamlessly transferred to another channel. Cognitive radio technology is faced with certain technical requirements in a healthcare environment. These are listed in the following section.

7.2 Technical Requirements of Cognitive Radio Technology

Apart from being able to perform the aforementioned significant steps, cognitive radios must fulfill certain other technical requirements as well. For instance, during spectrum sensing phase the cognitive radio controller should be able to sense the presence of any medical device in a short time, that is, the computation and processing time of the controller should be minimum. This is important because the controller can only predict the usage of the licensed band and cannot make definite decisions about its usage since the state of the spectrum can change anytime. Therefore, when the controller determines the transmission parameters for a cognitive client, it is possible that a medical device may appear in the transmission range of that device anytime. Thus, to avoid such unexpected interference, the cognitive controller should perform spectrum sensing once every sensing period. The sensing period is defined as the maximum time for which a primary user is predicted to not use the respective channel. Any unexpected appearance of a medical device should be communicated to the cognitive clients which can then switch off or hand off from the band. This increases the Quality of Service (QoS). Further, every cognitive radio client has a transmission range in which it transmits data. The client cannot transmit data beyond this range. Therefore, the cognitive radio controller should be able to sense the presence of medical devices at least within this range from a cognitive user. Now, certain medical devices do not have a fixed location. Their location depends on the area of requirement and position of the patient. This is especially the case with passive medical devices (described later). Therefore, the cognitive radio inventory system must store the real time locations of these medical devices and track and update their position at all times. This can be made possible by using some real time location tracking device such as a RFID tag associated with each device. Therefore, the controller can determine the transmission parameters for cognitive clients accordingly. The next section proposes an architecture for an e-health application.

8. ARCHITECTURE FOR E-HEALTH APPLICATIONS

An e-health application must be able to address all the challenges and requirements stated above for wireless devices and cognitive radios. It must provide data security and privacy, must be able to run on multiple hardware platforms, should support multiple data formats and communication protocols, and should also be scalable (El Khaddar, Harroud, Boulmalf, Elkoutbi, & Habbani, 2012; Al-Akkad, Pramudianto, Jahn, & Zimmermann, 2009). Keeping all these aspects in mind the following sections list the various components and network architecture for an e-health application.

8.1 Components of E-Health Medical Devices

A typical cognitive radio based system for wireless communications in a hospital environment has three components viz. the inventory system, the cognitive radio controller, and the cognitive radio client. These components have been shown in Figure 1. The inventory system is like the base station of the cognitive radio system. It stores all the information regarding all the components of the system at hand, that is, all the wireless medical devices as well as the cognitive radio controllers. Some of this information includes the location of these devices, the radio frequency immunity levels, and their transmission power. This information is used while determining the spectrum access by secondary users. Further, all this information is constantly updated.

Whenever a cognitive radio client has to do wireless transmission, it establishes a weak and non-interfering connection with the nearest cognitive radio controller. The cognitive radio controller then performs spectrum sensing and obtains the current status of all the medical devices from the inventory system. It then undergoes the stages of adaptive learning and spectrum decision to determine the transmission parameters for the cognitive radio client which initially requested communication. As per the instructions of the controller, the client tunes its parameters and performs wireless transmission in an interference free manner.

This was a general scheme for doing wireless communication in a cognitive radio enabled system. A more specific approach depends on the device at hand. The wireless devices in a cognitive radio environment can be broadly classified into three categories. Firstly, cognitive radio passive medical devices, secondly, cognitive radioactive medical devices, and lastly, cognitive radio non-medical devices. Each of these devices has different requirements and avoids electromagnetic interference differently. These requirements and specifications are given in the following subsections.

8.1.1 Cognitive Radio Passive Devices

Cognitive radio passive medical devices are those devices which do not have any wireless transmissions of their own but they have sensitive electronic components which are affected by the EMI from other wireless devices. Some of these devices include incubators, hearing aids, pacemakers, defibrillators, etc. If the operating frequency and transmission power of the nearby wireless non-passive medical devices is higher than the immunity levels of these devices, then they can cause serious EMI to these

Figure 1. Components of a cognitive radio system

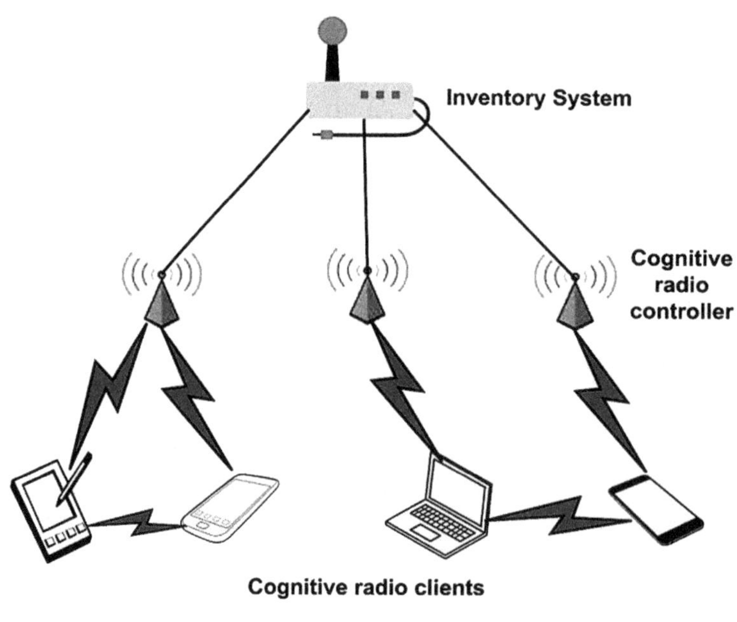

devices. Therefore cognitive radio controller needs to determine the appropriate operating frequency and transmission power for non-medical devices after sensing the presence and specifications of the passive devices so that the latter do not suffer from EMI. For this purpose, the cognitive radio controller obtains the location and immunity levels of passive devices from the cognitive radio inventory system.

8.1.2 Cognitive Radio Active Devices

Cognitive radioactive medical devices are those devices which have wireless transmissions of their own and hence are directly susceptible to EMI from other wireless devices. Some of these devices include telemetry monitors, Electrocardiograph Monitors (ECG), Electromyography Monitors (EMG), etc. If two active wireless devices use the same channel at the same time and in the same area, they can cause EMI to one another even if they do not operate in close proximity to one another. Therefore, the cognitive radio controller must sense the location of these devices and certain other parameters to determine the appropriate transmission parameters for cognitive radio clients.

The cognitive radioactive medical devices are of two types, periodically active medical devices and sporadically active medical devices. Periodically active medical devices have a regular operation pattern and therefore it is easy for a cognitive radio controller to track the activity of such devices and make spectrum usage efficient. Whereas sporadically active medical devices do not have a regular pattern of operation, and hence it is difficult for the controller to track the activity of such devices. The controller has to regularly sense the channel for their activity.

8.1.3 Cognitive Radio Non-Medical Devices

Cognitive radio non-medical devices include devices like computer monitor, television, or a radio. These devices usually exist in a homecare environment alongside other active or passive medical devices. These devices are the major sources of EMI in a hospital environment. The cognitive radio controller keeps track of these devices as cognitive radio clients and determines parametric specifications for wireless transmissions by these devices. The next section talks about some of the challenges related to the development of cognitive radio technology.

8.2 Challenges in Development of Cognitive Radio Technology

A healthcare environment may have only passive medical devices, both passive and active medical devices, or a combination of these with the non-medical devices. With the advent of pervasive healthcare, the third type of healthcare environment is more common where all type of devices exist simultaneously. When all these devices are present together, there are certain challenges which need to be addressed. Some of these prime challenges are elaborated in the subsequent section. Firstly, a medical device is not prone to EMI if it is in a shutdown state, that is, it is currently not in use and is switched off. In such a case, a cognitive radio client can operate even in close proximity to the medical device, utilize the spectrum more efficiently, and at the same time not cause harm to the medical device at hand. But for leveraging such advantage, the cognitive radio controller must be aware of the on/off state of a medical device. This works fine in case of an active device but the state of a passive device cannot be detected directly since it does not transmit any wireless signal of its own. Therefore, even though a cognitive radio-based system may be installed, the spectrum utilization may remain inefficient. Thus, there must be some effective way to detect the status of a passive medical device.

Even if there was some definite way to determine the working state of a medical device, these devices may not exhibit any regular operational patterns. Therefore, the prediction made by cognitive radio controller for the spectrum usage by an unlicensed user during the spectrum decision phase may be hindered by the unexpected appearance of a medical device. In such cases, an accurate prediction of channel usage is not possible.

Secondly, as mentioned earlier a cognitive radio medical device may occasionally change its location or may become active at a time when it was predicted to not use the licensed band. Further, as mentioned above, determining the state of passive medical devices is a difficult task. In view of these scenarios, while a cognitive radio client is wirelessly transmitting some data, a medical device may occasionally appear in its path. In such a case the client is forced to adjust its operating frequency and other parameters. This is known as spectrum handoff and can be a cause for QoS degradation. Thus, delays and losses are incurred when a cognitive radio client stops transmission and changes to a new spectrum band. Therefore, to overcome these losses the cognitive radio controller should be able to select the best band among the ones available for the clients according to their specifications.

Thirdly, in a real hospital environment, there shall exist multiple cognitive radio clients and some non-cognitive radio devices as well. In the case of multiple cognitive radio clients, the controller must ensure that the spectrum utilization remains unbiased. Further, the controller should be able to sense the presence of non-cognitive radio devices as well and provide a methodology such that their communications do not cause EMI to the cognitive radio medical devices.

Lastly, due to the unexpected appearance of medical devices in a spectrum channel, a cognitive radio client may have to stop data transmission over a particular channel. This may be a cause of hindrance in some significant task. To overcome this problem, a cognitive radio controller shall allow data transmission over multiple spectrum bands. In such a case, even if the client has to stop wireless transmission over one of the bands, it can still utilize the remaining bands. This increases the QoS but the selection of multiple bands for a client is a challenging task. Further, due to excessive channel fading and shadowing an active medical device may not be detected despite the fact that it may be transmitting data over a wireless channel. In such cases, a cognitive radio controller should be able to differentiate between a white space and a faded active medical device. The next section gives a detailed description of the network architecture and the communication protocols of a cognitive radio enabled e-health application.

8.3 Network Architecture and Communication Protocols

The main task in cognitive radio-based healthcare environment is to identify primary and secondary users. This largely depends on the area, scope and span of the application under consideration. For example, if the considered scenario is a homecare environment where there are only a few medical devices alongside other wireless telecommunication non-medical devices, then the medical devices can be treated as the primary users and the non-medical devices can be designated as the secondary users. However, if the considered scenario is a hospital environment where there are a large number of medical devices and information systems, then the wireless access of these devices needs to be prioritized among themselves without taking into account the non-medical devices. In such a hospital environment, more important life supporting devices are treated as the primary user while data transfer devices are treated as the secondary users. The architecture of this second type of system is explained in this section.

The considered system consists of two applications viz. telemedicine and a hospital information system. The telemedicine application handles real time data of remote patient monitoring and hence is sensitive to losses and delays. The hospital information system handles less significant data such as patient's personal details, facility details, etc. The considered cognitive radio system consists of a common control and data channel. Both of these channels lie in the unlicensed spectrum, and the telemedicine application is provided primary access while the information system is provided secondary access.

Each of these cognitive radio clients does wireless transmissions through the cognitive radio controller. The controller senses the environment for the presence of any other medical devices and then determines the respective transmission parameters for the requesting device. The protocol employed by the controller for this purpose is Request to Send (RTS) and Clear to Send (CTS). The controller has two radio interfaces for accessing both the channels simultaneously, whereas the clients have a single radio interface for both the channels. Since there is a common data channel for the entire system and the clients have a single radio interface for either channel, priority-based access is necessary for the data channel. As per the consideration, primary users have higher priority to access the data channel and secondary users have a lower priority. They can access the data channel only when it is not being used.

Now before making any actual transmission over the data channel, each wireless device must first setup a weak connection (an RTS message) with the cognitive radio controller which then determines the respective transmission parameters. But this weak connection should also be established using an amount of transmission power which does not cause EMI to other devices. Therefore, the cognitive radio system employs the concept of Common Control Channel Broadcasting. Common control channel broadcasting is used for broadcasting the maximum transmission power to be used by a client for transmitting RTS on the control channel. Each device has a different maximum allowable transmission power and depends on its location. Since the location of cognitive radio clients may not be static, therefore this maximum power is calculated periodically and broadcasted over the control channel regularly. While this power value is being broadcasted, no other transmission takes place in either channel to synchronize the new locations and transmission parameters of the devices with the controller.

Once the maximum transmission power has been broadcasted, a client willing to transmit data over the data channel transmits an RTS to the cognitive radio controller. The controller determines the maximum transmission power that can be used by the client on the data channel. If data transmission with the determined power is not possible without causing EMI to the neighboring devices, the request is dropped. However, if the transmission is allowable, a CTS message is communicated to the client.

The controller maintains two separate queues for primary and secondary users respectively. Once the client receives a CTS message, it changes its radio interface from the control to the data channel. If the requesting user is primary and the data channel is available, it is allowed to transmit the data. However, if the data channel is not available, the user is made to wait in the queue for primary users. If the requesting user is secondary, two cases arise depending on the state of the primary queue. Firstly, if the primary queue is empty and the data channel is free, the user is allowed to transmit the data. However, if the data channel is not free, the user waits in the queue for secondary users. Secondly, if the primary queue is not empty, the user waits for transmission in the queue for secondary users until the primary queue becomes empty. This ensures prioritized access to the data channel and thus no EMI to other devices. The communication process mentioned above has been described in the flowchart given in Figure 2.

As can be seen from the discussed architecture, a cognitive radio-based approach allows easy wireless communication in a hospital environment without causing electromagnetic interference to the neighboring medical devices. Further, cognitive radios also help in overcoming the problems of medical band

Figure 2. Flowchart of communication process employed in a cognitive radio-based system

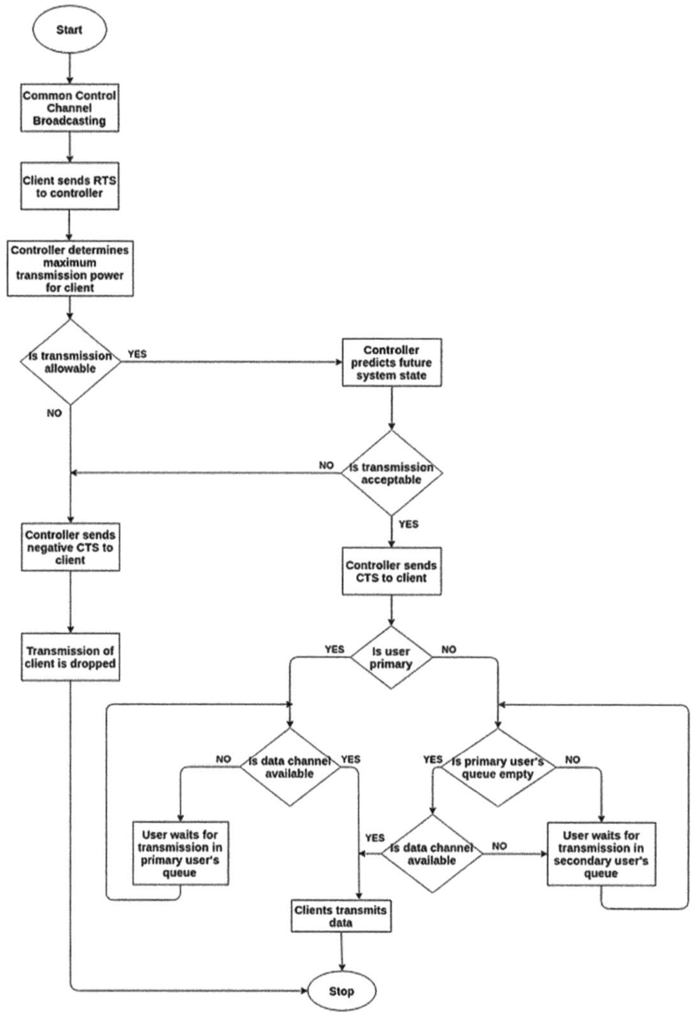

scarcity and inefficient spectrum utilization. A cognitive radio-based system also ensures a high QoS of the e-health system. The following sections mention possible future research directions and conclusion.

9. FUTURE RESEARCH DIRECTIONS

In a cognitive radio enabled hospital environment the main task of cognitive radio controller is to determine the transmission parameters for a secondary wireless user for carrying out any form of wireless communication. But sometimes there may be spectrum handoff because of the unexpected appearance of a medical device. Future research may be directed in making this spectrum handoff seamless. If the spectrum handoff is not seamless, it significantly compromises QoS. This is especially important in applications such as telemedicine which often require wireless transmission of multimedia data. Critical medical assistance is disrupted if the spectrum handoff is not seamlessly handled. Further, a single cognitive radio controller is prone to jamming attacks. It is like the brain of the cognitive radio system and therefore alternate more efficient and robust mechanisms shall be researched in the near future.

10. CONCLUSION

The chapter highlighted some of the existing wireless technologies and devices in the field of healthcare along with their requirements and the challenges faced by them. With almost every form of communication now being wireless, it has become possible to provide quality and affordable healthcare in remote areas through the use of these wireless technologies. However, use of wireless devices in a hospital environment is posed with certain challenges. This chapter discussed these challenges as well as some other technical requirements and proposed the use of cognitive radios in a healthcare environment. Further, the chapter also proposed an architecture for a cognitive radio-based e-health application.

REFERENCES

Al-Akkad, A., Pramudianto, F., Jahn, M., & Zimmermann, A. (2009, November). Middleware for building pervasive systems. IADIS AC, (1), 291-298.

Al Mamoon, I., Muzahidul-Islam, A. M., Baharun, S., Komaki, S., & Ahmed, A. (2015, March). Architecture and communication protocols for cognitive radio network enabled hospital. In *Medical Information and Communication Technology (ISMICT), 2015 9th International Symposium on* (pp. 170-174). IEEE. 10.1109/ISMICT.2015.7107522

Baker, S. D., & Hoglund, D. H. (2008). Medical-grade, mission-critical wireless networks. *IEEE Engineering in Medicine and Biology Magazine*, *27*(2), 86–95. doi:10.1109/EMB.2008.915498 PMID:18463024

Ball, M. J., & Lillis, J. (2001). E-health: Transforming the physician/patient relationship. *International Journal of Medical Informatics*, *61*(1), 1–10. doi:10.1016/S1386-5056(00)00130-1 PMID:11248599

Bates, D. W. (1998). Effect of computerized physician order entry and a team intervention on prevention of serious medication errors. *Journal of the American Medical Association*, *280*(15), 1311–1316. doi:10.1001/jama.280.15.1311 PMID:9794308

Boyle, J. (2006). Wireless technologies and patient safety in hospitals. *Telemedicine Journal and e-Health*, *12*(3), 373–382. doi:10.1089/tmj.2006.12.373 PMID:16796507

Chávez-Santiago, R., Jankunas, D., Fomin, V. V., & Balasingham, I. (2014, April). Dual-band cognitive radio for wearable sensors in hospitals. In *2014 8th International Symposium on Medical Information and Communication Technology (ISMICT),* (pp. 1-5). IEEE. 10.1109/ISMICT.2014.6825244

Dhope, T. S. (2015). *Cognitive radio networks optimization with spectrum sensing algorithms* (Vol. 44). River Publishers.

El Khaddar, M. A., Harroud, H., Boulmalf, M., Elkoutbi, M., & Habbani, A. (2012, May). Emerging wireless technologies in e-health trends, challenges, and framework design issues. In *Multimedia Computing and Systems (ICMCS), 2012 International Conference on* (pp. 440-445). IEEE. 10.1109/ICMCS.2012.6320276

Eysenbach, G. (2001). What is e-health? *Journal of Medical Internet Research, 3*(2), e20. doi:10.2196/jmir.3.2.e20 PMID:11720962

Hossain, D. N. E., & Diamond, J. (2008). *Fourth generation heterogeneous wireless access networks for eHealth services: Architecture and radio resource management. In Mobile Telemedicine: A Computing and Networking Perspective*. Auerbach Publications.

Kuperman, G. J., & Gibson, R. F. (2003). Computer physician order entry: Benefits, costs, and issues. *Annals of Internal Medicine, 139*(1), 31. doi:10.7326/0003-4819-139-1-200307010-00010 PMID:12834316

Luan, Y., & Ren, G. (2015, January). The E-health Network Design. In *2015 International Conference on Network and Information Systems for Computers (ICNISC)*, (pp. 557-559). IEEE. 10.1109/ICNISC.2015.100

Medeisis, A., & Holland, O. (2014). State-of-the-art in policy and regulation of radio spectrum. In *Cognitive Radio Policy and Regulation* (pp. 1–48). Springer International Publishing. doi:10.1007/978-3-319-04022-6_1

Ouattara, D., Krief, F., Chalouf, M. A., & Hamdi, O. (2012, November). Spectrum sensing improvement in cognitive radio networks for real-time patients monitoring. In *International Conference on Wireless Mobile Communication and Healthcare* (pp. 179-188). Springer.

Phunchongharn, P., Hossain, E., & Camorlinga, S. (2010). Cognitive radio for pervasive healthcare. *Cognitive Radio Networks, Architectures, Protocols, and Standards*, 355-388.

Phunchongharn, P., Hossain, E., Niyato, D., & Camorlinga, S. (2010). A cognitive radio system for e-health applications in a hospital environment. *IEEE Wireless Communications, 17*(1), 20–28. doi:10.1109/MWC.2010.5416346

Richard, A. (2007). *Health care unplugged: The evolving role of wireless technology*. I-health Reports.

Shen, Q., Liu, J., Yu, H., Ma, Z., Li, M., Shen, Z., & Chen, C. (2013, August). Adaptive cognitive enhanced platform for WBAN. In *2013 IEEE/CIC International Conference on Communications in China (ICCC)*, (pp. 739-744). IEEE. 10.1109/ICCChina.2013.6671208

US Food and Drug Administration. (2014). *Radio frequency wireless technology in medical devices: guidance for industry and food and drug administration staff.* Author.

Chapter 16
Cognitive Internet of Things (C-IOT)

G. Nagarajan
Sathyabama Institute of Science and Technology, Chennai, India

R. I. Minu
Jerusalem College of Engineering, Chennai, India

Jayanthiladevi A.
TechresearchNet, India

ABSTRACT

Energy conservation is one of the essential requirements for the betterment of the Earth's future. The main objective of this chapter is to provide an effective framework of smart grid (SG) using the intelligence of cognitive radio network (CRN). The SG connectivity will be enhanced by using the functionality of internet of things (IoT). Using IoT, the devices between the consumers, manufactures, and energy source provider (government) can conserve resources and economy. In this chapter, the requirement of smart grid, the need for IoT technology in SG, and the usage of CRN in SG are all elaborated. The infrastructural design pattern of cognitive radio-IoT-based smart grid is introduced in this chapter.

INTERNET OF THING

Internet of Thing (IoT) is one of the magic words used by most of the researcher around the word. It is basically interconnection of physical things with virtual object. The main motivation of this technology is to provide Smart environment to live. The fundamental infrastructure of IoT comprises of following components: IoT devices (sensors), IoT gateway, Edge processor and Massive storage as posturized in Figure 1. All these components are connected through heterogeneous communication networks (Ghassemi, Bavarian, & Lampe, 2010). The End user communication is comprised of either internet or mobile communication.

DOI: 10.4018/978-1-5225-5354-0.ch016

The notable research area in this technology are; Interfacing with the physical devices, communication and connectivity, handle the development of ubiquitous computing, unique identification, security, reliability, and interoperability.

BASIC OF COGNITIVE RADIO NETWORK

One of the challenges of IoT is communications and connectivity. This challenge can be solved by Cognitive radio network. The cognitive radio network (CRN) is an intelligent wireless design. It can monitor sense and detect the environment and reconfigure its characteristics to its requirement. The CRN in general determine the requirement parameter from the environment and adjust them for the required level satisfies the optimal communication experience of the end user. In CRN there will be two main participants as shown in Figure 2 primary user (licensed) who can transfer data whenever they required and other one is secondary user, where they can transmit data only when the primary user is silent.

Energy is one of the essential requirements of mankind which has be to preserved. Smart Grid (SG) is one of the booms to Energy industries, which has all the characteristic of conservation.

Why Cognitive Radio Networking in IOT

The important characteristic of a CRN is its ability of instantly changing the transmitter parameter based on the spectrum band availability. Thus the nodes in this network will transmit and sense the state of the channel, this process is called a dynamic spectrum accessing. If the dynamic spectrum allocation and the intelligent decision-making characteristic of a CRN is integrated with IoT framework then the performance of the IoT system would be enhanced.

Figure 1. IoT devices basic arrangement

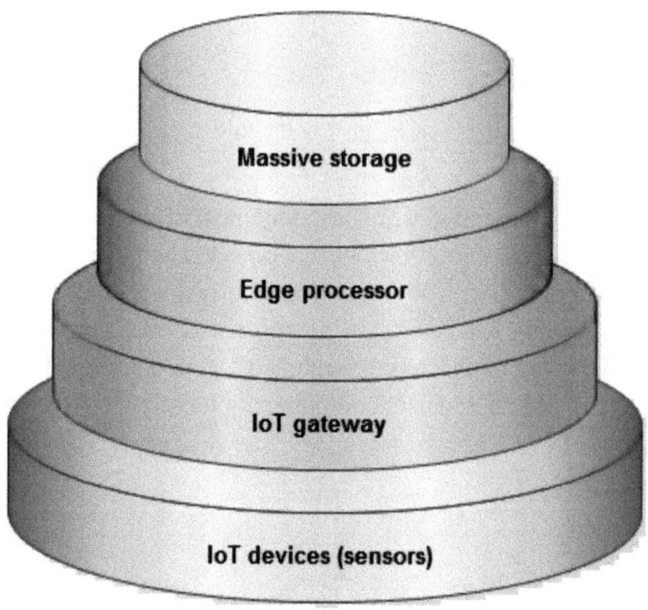

Figure 2. Cognitive Radio Network

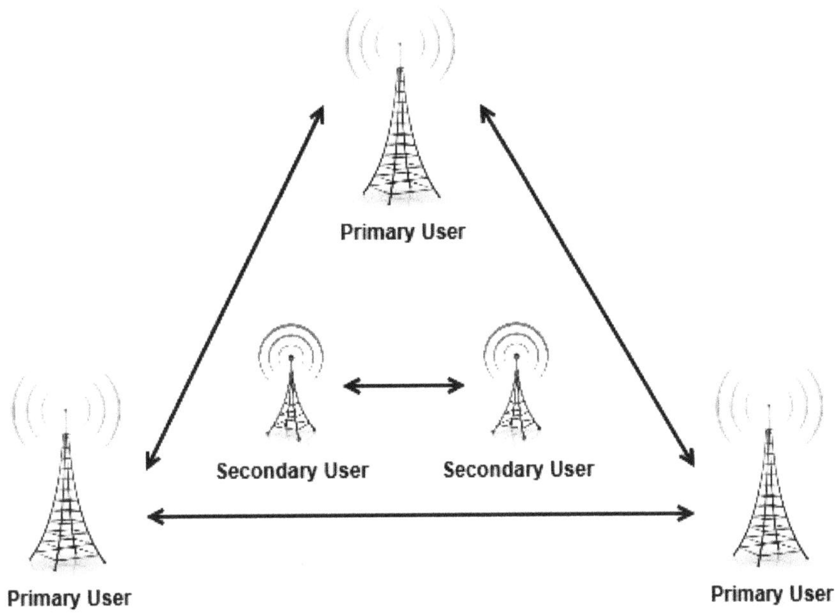

Due to the advancement of sensors and actuators the number of IoT devices will be expected to increase tremendously in the near future (Qiu, Hu, Chen, Guo, Ranganathan, Hou, & Zheng, 2011). Propositionally the number of Primary licensed network users will also be increased. Thus merging those two scenarios using CRN the IoT devices can be communication by sensing ideal channels. Most of IoT sensors were deceived to utilize the limited range, low cost communication devices such as Zigbee and Bluetooth which would not be so feasible in the near future so by using the CRN standard networking IEEE 802.2 (Wireless region access network) it would provide long range communication (Khan, Rehmani, & Reisslein, 2016). By deploying the CRN standard networking the interference can be reduced in sensor level communication. To tackle the mobility of the sensors the CRN is the boom, as it can handle seamless connectivity.

THE SMART GRID

The electric grid comprises of network of transmission lines, substations and transformers. The main purpose of an electric grid is to deliver electricity from the power plant to various industries and home. AS of June 30, 2016 (NDTV (2016)) the total power generation capacity was 3,03,118.21 MW. Indian has one of the largest operational electric grids with 329.23 GW of installed power.

To meet today's exponential demands of electricity; the digital era introduces the concept of Smart Grid. This intelligent technology allows two-way communication between the consumers and their electrical utility, also it provide full automation along the transmission lines using sensor network. Through Smart grid the utilization of electricity and the future demand can be predicted accordingly. Through this kind of predictive analysis an appropriate action can be taken to satisfy the future demand.

Benefits of Smart Grid

The future of energy supply can be secured by building the smart grid. As it reduces the losses, increases efficiency and optimized the energy demand distribution. Using the intelligent information from the Smart Grid, the energy industries can rejuvenate there proposal ideas for near future. Thus they could provide reliable and efficient contribution to the growth of economic and environmental health. The key benefits of Smart Grid are: Efficient transmission of power by an predetermine pattern, self-healing or restoration of transmission component would be quicker than anticipated, through predicate analysis the demand pattern would be determine so the power transmission would be available on demand, using smart grid it would be cost efficient and more secure. The key terminologies (Saleem, Crespi, Rehmani, & Copeland, 2017) are listed in Figure 3.

SMART GRID NETWORK INFRASTRUCTURE

Some of the remarkable technology advancement of Smart grid is advanced metering infrastructure (AMI), fault tolerance, self-healing and demand side management. For all these feature requires standard communication network between the utilities and various station. The main key terminology used in SG is two-way communication network. The networking infrastructures used by smart grid are Home area networks (HANs), Neighborhood area networks (NANs) and Wide area Network (WAN) (Saleem,

Figure 3. Benefits of Smart Grid

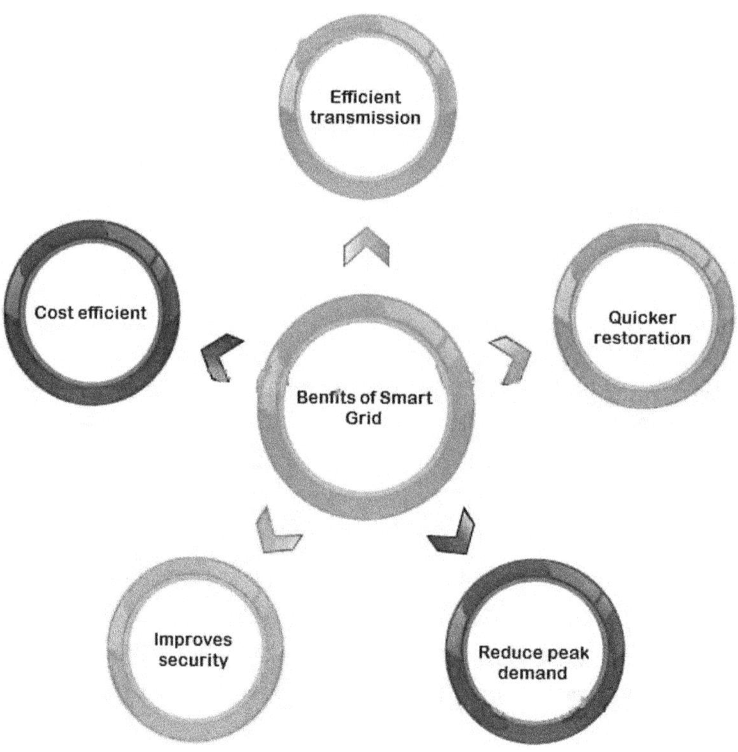

Crespi, Rehmani, & Copeland, 2017;Yu, Zhang, Gjessing, Yuen, Xie, & Guizani, 2011) as briefed in Figure 4. These three layer infrastructure provides reliable and efficient way of communication system. The HAN network design provide a infrastructure structure for connecting various user (Home) with low range devices such as radio frequency (RF), Zigbee, Bluetooth and etc., Various HAN are connected to NAN through HAN gateway. The NAN network connects the substation to HAN gateway. For a NAN it connects two different power sources: the power from the utility and the distributed renewable energy. These two power sources have two essential differences: price and availability. The WAN connects the NAN and the utility systems to transfer information.

The most challenging factors to be considered for design of communication network are handling massive stream of data form various household. Considering the population growth and number of electrical equipment in a house the number of HAN would be enormous. So we required a robust system to handle such kind of data. For the requirement of the consumer the power has to be distributed without any dispute. There would be an any predictive pattern on amount of data transfer, so we require a congestion free network traffic monitoring system. The discussed all three layer are connected with heterogeneous devices and network thus this interoperability need to be handled wisely. With this entire requirement there should not be any compromise in quality of services. The data which is been transferred contain the personal information about the consumers, these data should be preserved. In brief the challenges are shown in Figure 5.

Why Cognitive Radio Network in Smart Grid

The HAN uses a radio frequency of 2.4GHz ISM band. As this network is used by the consumer, it has to be a cost effective frequency band (Khan, Rehmani, & Reisslein, 2016). This frequency band is easily induced with external domestic noises. If the usage of Smart grid increases then these frequency range would be over crowded with interference. Thus, by using cognitive technology these kinds of interference and noises can be handled by its intelligent transmission scheme.

If the smart grid is implemented in full swing then that system would generate more than petabyte of data in the near future handling this much data using existing communication network is cumbersome.

Figure 4. Smart grid network (NAONWORKS)

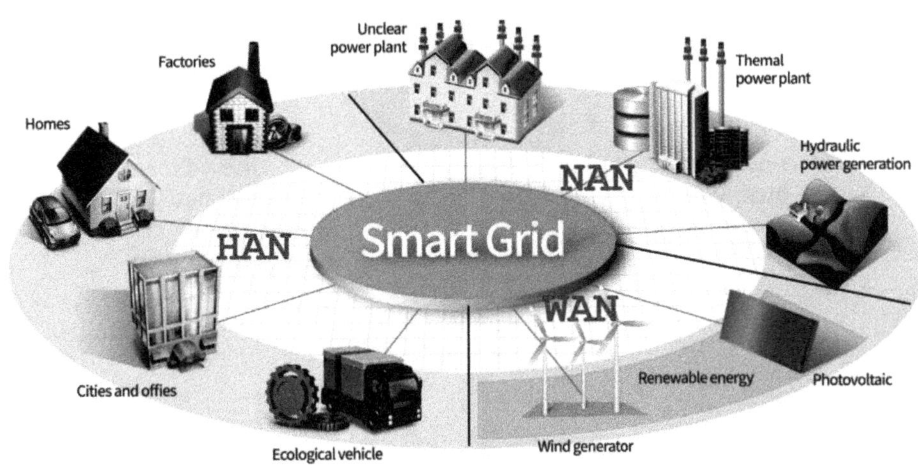

Using cognitive radio network in the smart grid will potentially improves spectrum utilization and communication capacity to support large scale data transmissions. The HAN, NAN and WAN system are all heterogeneous network (Wu, Ding, Xu, Feng, Du, Wang, & Long, 2014) with different new technologies. This technology requires smart terminals to transfer data between each terminal. The cognitive radio network functionality enables all these kinds of configuration.

The HAN requires more reliable and flexible communication technologies as it need to put up with wide range of heterogeneous devices. To communication with multiple such HAN, the NAN was is connected through a HAN gateway. This gateway should be self-configured and provided with cognitive capabilities to bend according to the surrounding environment. Using cognitive radio network the HAN gateway can communicate with external NAN through the unused licensed free spectrum.

Various NAN are communication to HAN gateway through a NAN gateway. Both these gateways are communicated through licensed spectrum. In near future due to the insufficient of bandwidth requirement both the gateway may communicate through the unlicensed access point using the advance technology of cognitive radio network. To maintain the quality of service (QoS) in unlicensed network concept like hybrid dynamic spectrum access (H-DSA) and TV white space (TVWS) can be used. Using H-DAS the throughput and capacity would improves and TVWS used to identify the interfering NAN.

Figure 5. Challenging factors in smart grid

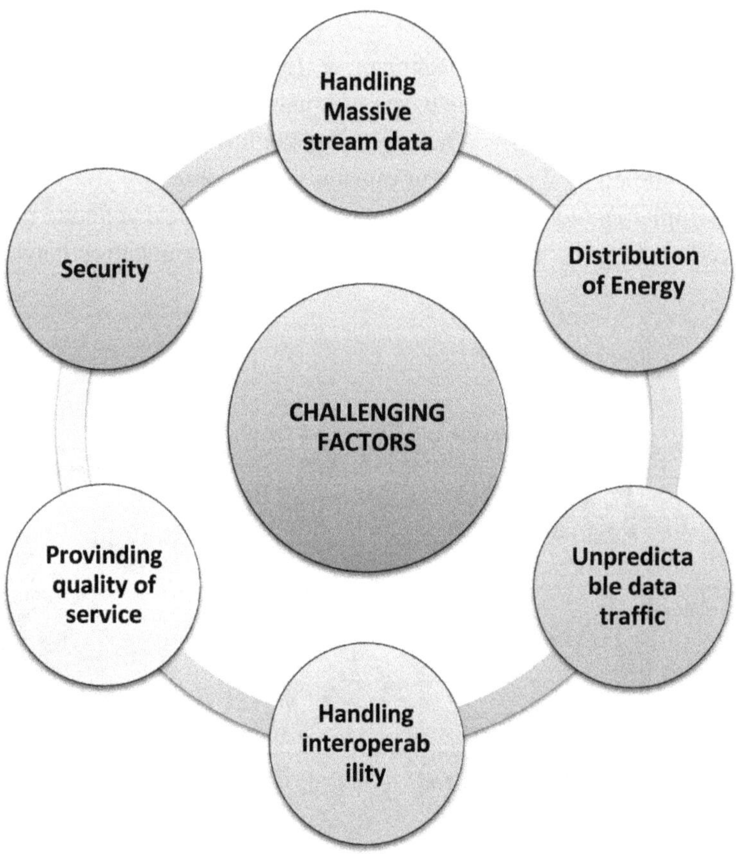

Smart Grid With Cognitive Internet of Things (C IOT)

Till this end the prologue of SG, IOT and CRN are brief and need for all this technology were specified. In this section the integration of all these technology as a whole was shown in Figure 6. In the Figure 6 the disk arrangement specifies the layer of IOT layer architecture which consists of IoT sensor devices, IoT gateway, Edge processor and Massive storage. The spherical shape represents the main building block of SG. The *household utilities* comprise of all the electrical equipment and the smart meter. The *HAN gateway* is used to connect various home networks to the NAN network which interconnect the corresponding substation with the home. The NAN in tern connects the *Power Distribution* networks which connect to various Energy source through WAN network. At last all the log data and other details are maintained in the *Cloud storage* system. The most crucial part of connectivity in smart grid aided IoT system is the interconnection of HAN gateway with the NAN and NAN gateway with the WAN. This connection could be very effective if we used CRN intelligence in spectrum analyzing and allocation.

Smart Grid Security Issues

The smart grid has various security issues (Le, Chin, & Chen, 2017) as it works over internet. Sophisticated security algorithm cannot be implementing in smart grid as the IoT device used has limited recourse. So, deployment of security algorithm with constrain is one of the open issues in smart grid technology. The data transfers through smart meter contain the personal information of the consumer, so this data has to be preserved (Shuaib et al). Some of the serious security issues are attack on smart meter information, The NAN network contains the personal log details of each HANs these data can be exploited by attackers, using effective multiplexer design the attacker can cause direct impact on communication equipment. There are fair chances of embedding wrong meter data, Unit modification and so on, which will directly impact on the financial aspect.

Figure 6. Smart Grid via Cognitive Radio IoT

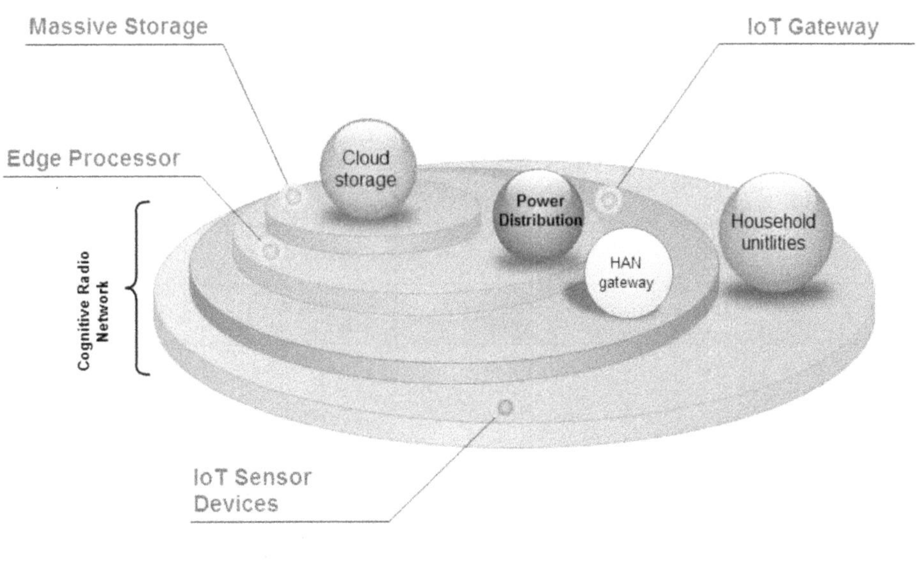

Table 1. Comparison of communication technology with respect to its features

Technology	Standard	Frequency Spectrum	Data Rate	Transmission Range	Energy Consumption	Licensed/ Non Licensed Spectrum Operation
GSM	--	900-1800 Mhz	Up to 14.4Kbps	1-10Km	Medium	Licensed
GPRS	--	900-1800 Mhz	Up to 170Kbps	1-10Km	Medium	Licensed
3G	• • IEEE 802.16e	1.92-1.98GHz,2.11-2.12GHz(Licensed)	384Kbps-2Mbps	1-10Km	Medium	Licensed
Z-Wave	IEEE 802.15.4	915 MHz	100Kbps	30m	Medium	Licensed
6LoWPAN	IEEE 802.15.4	1 GHz and 2.4GHz	250Kpbs	10-100m	Medium	Licensed
LoRaWAN	--	868/900 MHz	0.3-50Kbps	2-5Km(Urban environment); 15Km(suburban environment)	Very Low	Licensed
Mobile broadband wireless access	IEEE 802.16,	3650–3700 MHz	20 Mbps	Vehicular standard (up to 240 km/h)	Medium	Licensed
WiMAX	--	2.5GHz,3.5GHz,5.8GHz	Up to 75 Mbps	10-50Km(LOS),1-5Km(NLOS)	Medium	Both
CDMA	--	450-2100MHz	Up to 153kbps	49Km	Medium	Both
ZigBee	IEEE 802.15.4	2.4GHz,868-915MHz	250Kbps	30-50m	Low	Un Licensed
Cognitive Radio	IEEE 802.22 Standard in North America	54-862MHz	18-24Mbps	10-100Km	Low	Un Licensed
Wifi	IEEE 802.11 a/c/b/ d/g/n	2.5 and 5GHz	2-6000 Mbps	110-100K	High	Un Licensed
4G	• IEEE 802.16m	700-2500MHz	3.3Gbps for LTE-advanced	10Km	Medium	Licensed
Micro wave	--	2-40GHz	155Mbps	60Km	High	Licensed
Bluetooth	IEEE 802.15.1	2.4GHz	1Mbps	10-100m	Medium	Un Licensed
LTE	--	900MHz	300Mbps	30Km	Medium	Licensed
Satellite Communication	--	1-40GHz	1-15Mbps	28100-36000Km	High	Licensed
Powerline communication	--	1-30MHz	2-45Mbps	1-3Km	High	-
Fiber-Optical	--	3x10^{8}GHz	10Gbps to 1600Gbps	Unlimited(from 100 to 1000kmrequires 1 repeater)	Medium	Licensed
Optical communications	--	3x10^{8}GHz	Up to 100 Tbps	10-60 km	Medium	Licensed
Digital subscriber line	--	240KHz-1.5MHz	256Kbps to 10Gbps	3miles(requires 1 repeater)	Medium	Licensed

Table 2. Usage of various technologies

Technology	SG Application Areas	Application	Advantage	Limitation
GSM	Home automation	AMI, Demand response HANs, NANs, WANs.	No interference with other wireless	Low Data Rates
GPRS	Home automation	AMI, Demand response HANs, NANs, WANs.	No interference with other wireless	Low Data Rates
3G, 4G	Home automation	AMI, Demand response HANs, NANs, WANs.	No interference with other wireless	Costly spectrum Fees
Z-Wave	Home automation	HAN	No interference with other wireless technologies; reliable; low latency; scalable	Not suitable for NAN and WAN; only one propriety company for making chips; short range
6LoWPAN	Smart metering; home automation	HAN	Robust; low-power; support large mesh network topology; can be applied across various communication platforms	Require extensive training and knowledge; short range; low data rate
LoRaWAN	Management of operation and equipment; online monitoring of power transmission lines and tower	NAN,WAN	Low-power; long range; no interference with different data rates; enhance gateways capacity by creating virtual channels; low-cost secure bi-directional communication	
Cellular communications	Monitoring and management of DERs; SCADA	HAN; NAN; WAN	Wide area coverage; improved QoS	Network congestion due to high density; critical for emergency applications; non-guaranteed service in unfavorable conditions (e.g., wind storms)
Wireless mesh	Monitoring and controlling of DERs; automation and protection of substation automation	HAN; NAN	Low cost; self-healing; self-organization; high scalability; high data rate	Prone to interference and fading; network coverage problem in rural areas due to low density of smart meters; prone to loop problem due to inclusion of multiple relay nodes
Mobile broadband wireless access	Broadband communication for electric vehicles; SCADA system; wireless backhaul for SG monitoring	NAN; WAN	Low latency; high mobility; high bandwidth	Communication infrastructure not readily available; high cost; moderate data rate
WiMAX	Real time pricing; automatic meter reading; outage detection and restoration	AMI, Demand response NANs, WANs.	Long range; high data rate	Trade-off between performance and distance; costly RF hardware; low frequencies are already licensed and require leasing; high frequencies do not penetrate obstacles, Not widespread

continued on following page

Table 2. Continued

Technology	SG Application Areas	Application	Advantage	Limitation
ZigBee	Energy monitoring; smart lightning; home automation; automatic meter reading	AMI, Demand response HANs	16 channels each with 5 MHz of bandwidth in 2.4 GHz spectrum; low power usage; low complexity; low deployment cost	Low data rates; limited energy of battery; low processing capabilities; short range
Cognitive Radio(IEEE 802.22 Standard in North America)	Smart metering, SCADA	NANs, WANs AMI, Demand response HANs	High data rate, long rang, Immunity to interference, Extensive data coverage, Adaptive power level.	Passive operation(depending on PU)
Micro wave	Transfer trip between DER and distributed substation	AMI,Demand response HANs	Long distance coverage; high data rate; high bandwidth	Prone to multipath interference and precipitation, Low Data Rates, short range
Bluetooth	Home automation	AMI,Demand response HANs	Low power consumption	Low data rate; short range; weak security; prone to interference from IEEE 802.11 WLANs
Power line communication	Low voltage distribution; automatic meter reading	AMI	Cost effective; low installation cost; wide availability; utility's own ownership and control; dedicated network	Noisy and harsh medium; sensitive to disturbances; signal quality gets affected by the type and number of devices, wiring distance between nodes and network topology; cost of ownership; complexity of management, Harsh noisy channels
Fiber-Optical	Physical network infrastructure	HANs, NANs	Long distance communication; ultra-high bandwidth; robustness against radio and electromagnetic interference; high reliability	High investment cost
Optical communications	Physical network infrastructure	WAN; NAN	Long distance communication; ultra-high bandwidth; robustness against radio and electromagnetic interference; high reliability	High deployment cost of fiber installation; high cost of terminal equipment; difficult to upgrade; not suitable for metering applications
Digital subscriber line	Smart metering	HANs, NANs, WANs.	High speed; low latency; low installation cost; high data rate, high capacity; long range; wide availability	Quality is distance dependent; high installation cost in low density (rural) areas; unreliable, Low Data Rates

SOLUTION AND RECOMMENDATION

For interconnecting all this technology there are different types of communication technologies (Yang, Ping, Aijaz, & Aghvami, 2016) available. Mostly effective technology is listed in the Table 1. According to the design pattern the researcher can use any these technology in there C-IoT based SG design. Table 2 briefs the application usage of all those communication technologies with respect to SG. The Zigbee,

Bluetooth and 6LoWPAN are low range communication device. These devices would be cost effective but period of life spam is very short. The Table 3 gives the overall list of acronyms.

Some of the notable wireless IoT Communication technologies used are: Z-Waves, it is a low powered radio frequency short range communication technology. Mainly used for home automation. 6LowPAN, it is an IP (internet Protocol) based communication technology. This network used IPv6 addressing features. Mainly used in smart meter design. LoRaWAN, it is a battery operated low power wide area network communication technology. It is a long communication range devices. Zigbee, a well-known technology for its cost efficient and low range.

Some of the wireless IoT communication technologies listed are: Bluetooth, it is an IEEE standard communication technology works on short range. Cellular communication, all generation such as 2G, 3G and 4G communication protocol all come under this technology. Wireless mesh, connection of wireless nodes mostly used in rural areas. WiMAX, it is an IEEE standard communication protocol used for microwave communication frequency range.

CONCLUSION AND FUTURE SCOPE

The research directions of IoT based smart grid systems are classified with respect to IoT smart grid and with respect to cognitive radio smart grid. In general, the IOT device has short life span and they have inadequate memory capacity, when these devices are deployed in smart grid house hold utilities these will lead to serious issues. Thus, design of reliable and efficient IoT smart devices is one of the

Table 3. List of acronyms and corresponding definitions

Acronyms Definitions	Acronyms Definitions
AMI	Advanced Metering Infrastructure
GPS	Global Positioning System
GSM	Global System for Mobile
HAN	Home Area Network
IoT	Internet of Things
LTE	Long Term Evolution
LoRaWAN	Low Power Wide Area Network
NAN	Neighborhood Area Network
SG	Smart Grid
WiFi	Wireless Fidelity
WiMAX	Worldwide inter-operability for Microwave Access
WMAN	Wireless Metropolitan Area Network
WAN	Wide Area Network
6LoWPAN	IPv6 over Low-power Wireless Personal Area Networks

open issues. As in each layer different set of communication network and protocol are used there is no any standard procedure to frame a communication network. If smart grid is implemented worldwide it requires a standard protocol. As the data gather by smart grid is a high velocity stream data it required a robust networking router to avoid congestion and loss of data. For effective data transmission through CRN a predication-based spectrum sensing and analyzing algorithm are required. The CRN requires a scalable and energy efficient gateway design to meet the required load.

REFERENCES

Ghassemi, A., Bavarian, S., & Lampe, L. (2010, October). Cognitive radio for smart grid communications. In *Smart Grid Communications (SmartGridComm), 2010 First IEEE International Conference on* (pp. 297-302). IEEE.

Khan, A. A., Rehmani, M. H., & Rachedi, A. (2017). Cognitive-radio-based Internet of Things: Applications, architectures, Spectrum related functionalities, and future research directions. *IEEE Wireless Communications*, 3.

Khan, A. A., Rehmani, M. H., & Reisslein, M. (2016). Cognitive radio for smart grids: Survey of architectures, spectrum sensing mechanisms, and networking protocols. *IEEE Communications Surveys and Tutorials*, *18*(1), 860–898. doi:10.1109/COMST.2015.2481722

Le, T. N., Chin, W. L., & Chen, H. H. (2017). Standardization and Security for Smart Grid Communications Based on Cognitive Radio Technologies—A Comprehensive Survey. *IEEE Communications Surveys and Tutorials*, *19*(1), 423–445. doi:10.1109/COMST.2016.2613892

Naonworks. (n.d.). *Smart Grid*. Retrieved from http://www.naonworks.com/old/inc_html/sub2_3.html

NDTV. (2016). *India's Total Power Generation Capacity Crosses 300 GW Mark*. Retrieved from http://www.ndtv.com/india-news/indias-total-power-generation-capacity-crosses-300-gw-mark-

Qiu, R. C., Hu, Z., Chen, Z., Guo, N., Ranganathan, R., Hou, S., & Zheng, G. (2011). Cognitive radio network for the smart grid: Experimental system architecture, control algorithms, security, and microgrid testbed. *IEEE Transactions on Smart Grid*, *2*(4), 724–740. doi:10.1109/TSG.2011.2160101

Saleem, Y., Crespi, N., Rehmani, M. H., & Copeland, R. (2017). *Internet of Things-aided Smart Grid: Technologies, Architectures, Applications, Prototypes, and Future Research Directions*. arXiv preprint arXiv:1704.08977

Shuaib, K., Barka, E., Al Hussien, N., Abdel-Hafez, M., & Alahmad, M. (2016). Cognitive radio for smart grid with security considerations. *Computers*, *5*(2), 7. doi:10.3390/computers5020007

Wu, Q., Ding, G., Xu, Y., Feng, S., Du, Z., Wang, J., & Long, K. (2014). Cognitive internet of things: A new paradigm beyond connection. *IEEE Internet of Things Journal*, *1*(2), 129–143. doi:10.1109/JIOT.2014.2311513

Yang, Z., Ping, S., Aijaz, A., & Aghvami, A. H. (2016). A Global Optimization-Based Routing Protocol for Cognitive-Radio-Enabled Smart Grid AMI Networks. *IEEE Systems Journal*.

Yu, R., Zhang, Y., Gjessing, S., Yuen, C., Xie, S., & Guizani, M. (2011). Cognitive radio based hierarchical communications infrastructure for smart grid. *IEEE Network*, *25*(5), 6–14. doi:10.1109/MNET.2011.6033030

Chapter 17
Internet of Things Using Software-Defined Network and Cognitive Radio Network

Karm Veer Arya
ABV-Indian Institute of Information Technology and Management, India

Rajasi Gore
MNNIT Allahabad, India

ABSTRACT

Internet of things (IoT) is a new area of advanced networking where things or devices share information among each other though collected sensor data and perform tasks together. This chapter is a detailed description of IoT and its architecture, challenges, security, applications, and future scope. Since the devices are heterogeneous in nature, middleware are used for giving them a platform to communicate in homogeneous manner. Service requirement of IoT middleware is discussed further in the chapter. Furthermore, recent technologies integrated with IoT like software-defined networks and cognitive radios are a part of this chapter.

INTRODUCTION TO IOT

Internet of Things or IoT is a complex network wherein machines and objects are connected through the internet. IDC has predicted that IoT will be expanded to 212 billion devices globally till 2020. The technical report on Cluster of European research projects on the Internet of Things define the 'Things' active entities in social processes, business, and information systems which have the interaction and communication among themselves. They also exchange data and information with the environment by autonomous reaction to the events happening in physical world with or without human intervention.

Wireless communication has come up so far that enabled the joining of things and making communication possible among them through Internet as discussed in paper of Gil, R. (2014). Cellular communication that is happening use licensed frequency bands provided by Federal communication commission (FCC) – US and Conference of postal and telecommunication administration (CEPT)- Europe. Radio

DOI: 10.4018/978-1-5225-5354-0.ch017

waves propagate better especially inside buildings and hence, is being used for communication widely. Unlicensed bands are provided to ISM (Industrial, Scientific and Management applications) that vary from country to country. 433 MHz, 868 MHz, 915 MHz and 2.4 GHz are the unlicensed bands used popularly for wireless communication. 2.4 GHz is widely used in all regions due to its high throughput and serve large networks.

Devices connected with internet have to use TCP/IP protocol stack to be able to exchange data but if devices are connected in local network can use non-IP. Local network would then be connected to internet through Gateways. The use of IP provides flexibility and devices can add or remove applications without changing anything in home network. Local network is formed only for one specific application. Let us take the example of fire alarm system which uses the wireless sensors to form a local network with one function that sends the activation message to temperature controller. However, if the message has to be relayed through controller then the phone would be connected to TCP/IP.

The design complexity and huge size of TCP/IP stack poses the disadvantage as bigger data packets lead to more power consumption in data transmission. These reasons promote the design of proprietary protocols for communication by local networks. The wide range development of wireless network processors and wireless microcontrollers (WMCU's) has reduced the complexity of TCP/IP to make it more attractive. So there are high chances that proprietary protocols would be replaced by TCP/IP in future.

Proprietary protocol use lower ISM frequency bands and are referred to as Sub-1 GHz solutions. Proprietary NAN is created to relay meter reading. The security system, industrial control and monitoring etc. are the other popular applications developed on propriety protocols. To connect to the IOT, Sub-1 GHz systems need the application layer internet gateway. Internet of Things (IoT) may also be implemented on wired networks but drawbacks like cable cost and space complexity makes wireless communication a better option.

Popular wireless technologies that can be used in IOT are:

- **Bluetooth:** It can act as gateway in IoT application.
- **Zigbee:** It is based on IEEE 802.15.4 link layer standard, low power and low cost technology. It requires application level gateway to connect to IoT.
- **Wi-Fi:** This has been a major barrier for IoT due to its complexity and high power consumption.

Therefore, the type of wireless technology to be used will be decided according to IoT application.

ELEMENTS OF IOT

- **RFID:** This is a technology which enables making of microchips which support wireless data communication. As an electronic barcode is on things to identify it uniquely, this helps in identification automatically. Two types of RFID: In *Passive RFID* power does not come from battery but it comes from interrogation signals when reader tells the id to RFID. In *Active RFID* Power is generated by its battery and starts the communication automatically.
- **Wireless Sensor Network (WSN):** It is an efficient network comprising of low power, low cost sensor nodes communicating wirelessly. The components of WSN are:
- Hardware: Sensor node which consist of sensor, preprocessing unit and a battery for power.

- **WSN Communication Stack:** Communication between nodes is very much required for exchange of data. The WSN stack helps in multi hop communication. It should also work as a gateway to communicate with outside world.
- **WSN Middleware:** It is tool which integrates applications to an infrastructure. On a network, it acts like a glue and bridge the gap between heterogeneous resources in dynamic way.
- **Secure Data Aggregation:** It is a method of collecting data from each sensor node such that there is no redundancy and we also get required information.
- **Addressing Scheme:** The sensor nodes that are being connected to WSN has a unique identifier each. The few critical features of unique addressing are uniqueness, reliability, persistence and scalability.
- **Visualization:** As everything is moving towards being smarter, attractive and easy visualization is to be created for the better understanding of a layman on IoT. It is a critical element as it allows interaction of user with the environment. Data can also be converted to knowledge which can be used for decision making. This encompasses both event detection and visualization of the associated raw and modeled data, with information represented according to the needs of the end-user.

ACTORS AND GENERAL ARCHITECTURE

Different types of actors performing in IoT as discussed by (Gubbi, Buyya, Marusic, & Palaniswami, 2013) are:

- **Owner:** This type of actor has all the rights to deliver services and beneficiaries either temporarily or permanently.
- **User:** These actors use the services.
- **Inventor:** These actors provide services.
- **Store Manager:** These actors have the right to remove rights before service. It also builds an IoT referential that is identifier for each of the things in the network.

Each object is connected with different actors and these actors are interested in different kind of information. For example, let us place an anemometer connected to a windmill. As the windmill will generate energy, anemometer will tell the value. If it does not, the owner would be thinking of the two possibilities, either no wind or anemometer has broken down. The user is only concerned with the information. Furthermore, providing details of all implemented anemometers at a given area is a work of store manager. It creates a referential- metadata of each object having identification, actors linked to it and access rights to each of these objects.

The main actors used in IoT systems are shown in Figure 1. On one side the object user that actually uses the object and object owner that owns and maintains the object. On the other side is the inventor, which specifies, designs and market the objects. At the immediate level, is the store manager responsible for delivering, marketing services related to IoT. It relies on visualized network and IT infrastructure.

Figure 1. User Centric Architecture of IoT

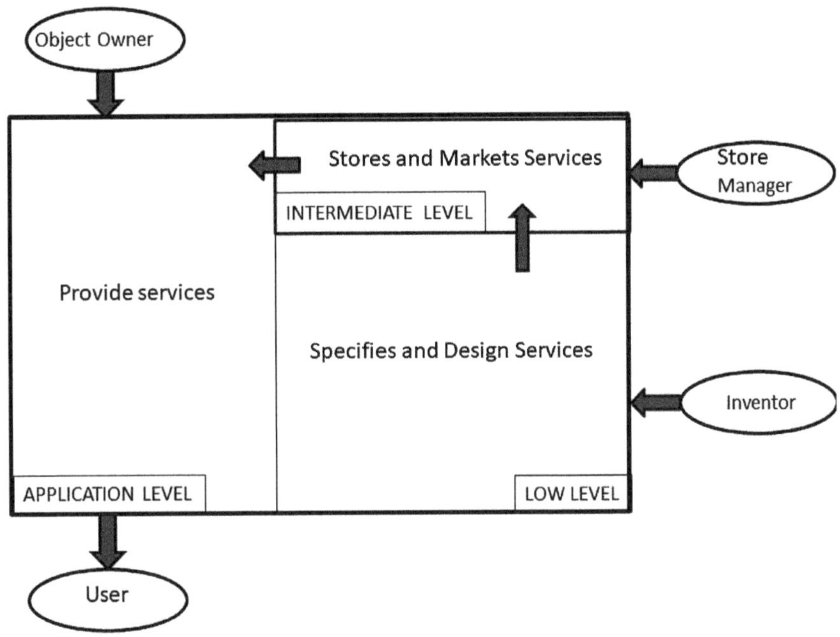

FEATURES OF IOT

The desired characteristics of IoT are listed below as in paper (Gubbi, Buyya, Marusic, & Palaniswami, 2013) which are necessary for any IoT applications:

- **Automation:** This means collection, processing of data, inferences to context, collaborations with IoT and decision making should be done automatically by IoT applications and not manually.
- **Intelligence:** The system should be intelligent means it should be able to adapt any kind of situation efficiently. This feature should be embedded in the IoT application through training and learning.
- **Dynamicity:** IoT ecosystem is mobile that means the system can take any place. So IoT ecosystem should adapt and merge efficiently. It should dynamically recognize the objects and adapt in the ecosystem according to the environment. And so, this characteristic should be there in any IoT application.
- **Zero Configurations:** This feature says that no device has to be configured before being integrated to the IoT system. Plug and Play feature should be available for easy integration.

MIDDLEWARE

Middleware is a software that acts as a bridge between operating system or database and application on a network. It serves to "glue" together separate often complex and already existing programs. In more general term, it is a software that facilitates exchange of data between application programs within the same environment over different network. It is often complemented to remove integration complexities providing seamless connectivity.

Based on previously described characteristics, a set of requirements for a middleware to support the IoT is outlined in paper by Razzaque (2016). As follows, these requirements are grouped into two sets: Service requirement are the services a middleware should provide and Architecture requirement is the system architecture middleware should support.

Service Requirements

These are the services middleware has to provide to IoT infrastructure. There are mainly two types of services- functional and non-functional.

Functional Services

- **Resource Discovery:** Heterogeneous hardware devices (e.g., RFID tags, sensors, sensor mote, and smartphones), devices' power and memory, analogue to digital converter devices (A/D), the communications module available on those devices, and infrastructural or network level information (e.g., network topology and protocols), and the services provided by these devices are resources in IoT. As the IoT environment is dynamic, resource assumption and knowledge of resource availability is invalid. Any hardware device can join the ecosystem and so resource discovery should be automated. Importantly, every device must announce its presence and resource it provides. Human intervention in resource discovery is infeasible. This is different model from centralized system, where resource publication, discovery and management are dealt by dedicated servers. There should be efficient distribution of discovery load on servers in IoT and it should scale well that means discover large amount of resources.
- **Resource Management:** Application should be provided with services that monitors resource usage, allocate resources in a fair manner and resolve resource conflicts. In IoT architecture, the middleware should have spontaneous resource composition and re-composition to satisfy application.
- **Data Management:** Data are key in IoT applications. In the IoT, data refer mainly to sensed data or any network infrastructure information of interest to applications. An IoT middleware needs to provide data management services to applications, including data acquisition, data processing (including preprocessing), and data storage. Preprocessing may include data filtering, data compression, and data aggregation.
- **Event Management:** There is massive number of events generated in IoT applications, which should be managed as an integral part of an IoT middleware. Event management transforms simple observed events into meaningful events. It should provide real-time analysis of high-velocity data so that downstream applications are driven by accurate, real-time information, and intelligence.
- **Code Management:** Deploying code in an IoT environment is tough, and it needs middleware support. Code allocation and code migration services are required in particular. Code allocation selects the set of devices or sensor nodes to be used to accomplish a user or application level task. Code migration transfers one node/device's code to another one, potentially reprogramming nodes in the network. By code migration service code gets portable.

Non-Functional Services

- **Scalability:** An IoT middleware need to be scalable to accommodate increasing number of devices in the IoT network and increased services. IPV6 is a very scalable solution to deal with the addressing of such huge number of things for unique identification.
- **Real Time or Timeliness:** The services of middleware should be real time when correctness of operator depends not only on logical correctness but also on time in which it is performed. As IoT deals with many real time applications, on time delivery of services is critical. Delayed information can be useless.
- **Reliability:** In the presence of failure, middleware should remain operational. Every component need to be reliable to achieve overall reliability.
- **Availability:** In IoT, middleware must be available every time especially for mission critical applications. Recovery time after failure should be minimum. Reliability and availability requirement should work together to ensure fault tolerance.
- **Security and Privacy:** Information can be disclosed due to context awareness in middleware. So security need to be considered in both functional and non-functional features. It should also preserve owner's privacy.
- **Ease-of-Deployment:** Since an IoT middleware is deployed manually, deployment should not require expert knowledge or support. Complicated installation and setup procedures must be avoided.
- **Popularity:** An IoT middleware should be continuously upgraded so that future enhancement in on IoT provides easy testing and deployment.

Architectural Requirements

The architectural requirements are designed to support application developers:

- **Programming Abstraction:** Middleware should provide an API for application developers for ease of development. For the application or service developer, high-level programming interfaces need to isolate the development of the applications or services from the operations provided by the underlying, heterogeneous IoT infrastructures. The level of abstraction, the programming paradigm, and the interface type all need to be considered when defining an API.

The level of abstraction refers to view of the system by developer like individual node/device level, system level. The programming paradigm deals with type of model for developing or programming the applications or services. The interface type defines the style of the programming interface. For instance, descriptive interfaces offer SQL-like languages for data query, XML-based specification files for context configuration.

- **Interoperable:** Heterogeneous devices are being connected to the IoT and their number is increasing on fast pace so A middleware should work with heterogeneous devices/technologies/applications, without additional effort manually. Heterogeneous components must be able to exchange data and services. Interoperability in a middleware can be viewed from network, syntactic, and semantic perspectives, each of which must be catered for in an IoT. A network should exchange information across different networks, potentially using different communication technologies. Syntactic interoperation should allow for heterogeneous formatting and encoding structures of any exchanged information or service. Semantic interoperability refers to the meaning of information or a service, and should allow for interchange between heterogeneous set of devices and services in IoT.

- **Service-Based:** A middleware must provide services and it should be flexible so that more number of services can be added anytime. A service-based middleware provides abstractions for the complex underlying hardware through a set of services like data management, reliability, and security and more needed by applications. It also provides easy environment for application development.
- **Adaptive:** A middleware needs to be adaptive so that it can evolve to fit itself into changes in its environment or circumstances. In the IoT, there is frequent change in the network and its environment and in addition to this application-level demands or context are also likely to change frequently. To ensure user satisfaction and effectiveness of the IoT, a middleware needs to dynamically adapt or adjust itself to fit all such variations.
- **Context-Aware:** The IoT's middleware architecture needs to be aware of the context of users, devices and the environment and use these for effective and essential services provided to users.
- **Autonomous:** It means self-governed. Devices/technologies/applications should be enabled to interact and communicate among themselves without direct human intervention. Use of intelligence including autonomous agents, embedded intelligence, predictive, and proactive approaches in middleware can fulfil this requirement.
- **Distributed:** A large-scale IoT is likely to be geographically distributed, and so a centralized view or middleware implementation will not be sufficient to support many distributed services or applications. A middleware implementation needs to support functions that are distributed across the physical infrastructure of the IoT. Event-based middleware like EMMA and GREEN perform well in mobile and reactive applications, but have limited interoperability, adaptability, and context-awareness. Finally, application-specific middleware like MUSIC and SOCRADES are optimized for an application or a group of applications, and may not be suitable and effective for other applications. Although the existing middleware solutions address many requirements associated with middleware in IoTs, some requirements and related research issues remain relatively unexplored. There is significant scope for future work in this area.

COMMUNICATION: PROTOCOL STACK APPROACH

The IoT is seen as *a global network of connected devices having identities and virtual personalities operating in smart spaces and using intelligent interfaces to communicate within social, environmental, and user contexts*. When two machines or devices are communicating, exchanging information over same network, it is called machine to machine (M2M) communication. When devices are also communicating with business server applications then these are used for IoT. Researchers have even come up with a standard protocol stack for M2M communication which will be using number of technologies. Generally, M2M networks can be divided into two broad domains:

1. **Capillary M2M:** This M2M communication typically is conducted between machines in short range through wireless technologies or in long range through gateway. It is a low power and lossy

network (LLN) because it has high packet loss ratio, requires high reliability and uses low powered technologies.

2. **Cellular M2M Networks:** It is a communication in a cellular network devices consisting of a cell which provides power and a SIM card for unique identification. Their communication with each other is similar to communication between user devices with short range and less or no mobility.

Figure 2 shows the standardized protocol stack for capillary M2M communications discussed by (Adnan, & Aghvami, 2015). The physical layer follows IEEE 802.15.4-2006 which is a low powered radio technology and works efficiently on low energy. At MAC layer IEEE 802.15.4e, a new developed protocol is proposed which uses time synchronization channel hopping for tackling the problems of fading and interferences and also uses rigid slot in scheduling for achieving a greater energy efficiency. At network layer, 6lowPAN protocol is proposed for connection between M2M devices as it provides IPV6 packets to move over low power link layer by special encapsulation and header compression. At transport layer, UDP and TCP are used. At last, at application level, CoAP is adopted as it makes the IoT system interoperable with HTTP so as to get easy integration of new devices and applications in the system.

Cognitive M2M Communications

M2M communications employing cognitive radio technology is supposed to rule the IoT in future but it would be a challenge for spectrum scarcity.

Dynamic spectrum access in which spectrum usage is according to best utilization and alternate spectrum can also be used dynamically.

- It addresses problem of heterogeneity and efficiency.
- It shows up new application areas in M2M communications.

Figure 2. Protocol Stack for M2M communication

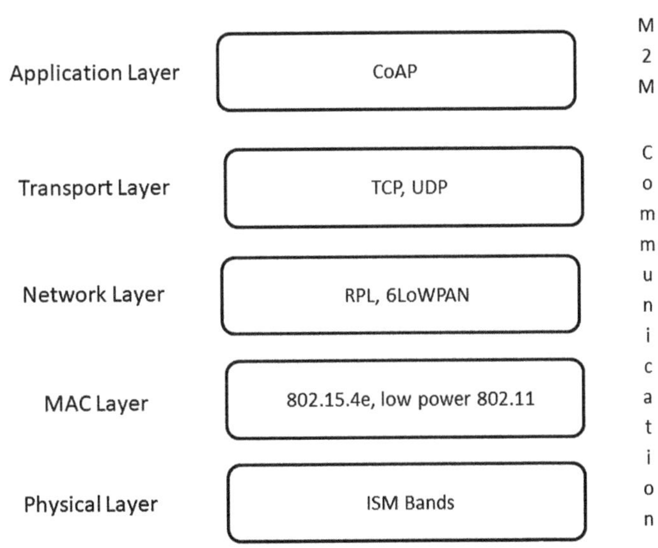

CHALLENGES IN IOT DEPLOYMENT

- **Heterogeneity:** The IoT consist of many heterogeneous things having different functionality, resolution, operating system, speed etc. connected through internet. Degree of complexity gets huge when we have to connect many such devices to form a complex network.
- **Scalability:** IoT network is getting bigger and more complex as many devices are being invented and are being attached to network. This also increases amount of data, handling it is a big challenge.
- **Interoperability:** Devices in IoT network are treated as actors and which comprise of human and non-human objects and each actor would be playing role according to the situation. Seamless interaction is challenging between these devices.
- **Security and Privacy:** As we saw that many actors are participating and huge data is being shared, privacy of each actor and security of each data has to be provided. So authentication and information security is a challenge in such a complex network.

DIAT: DISTRIBUTED ARCHITECTURE

DIAT by (Sarkar, 2012), is an architecture which deals with interoperability- different nodes of different configuration communicating with each other. Communication within similar nodes is easy as it follows same standard, same protocols but in IoT where heterogeneous devices have to be integrated with the help of internet and then made to communicate with each node is very complex and challenging.

So the basic idea is to make heterogeneous devices to look homogeneous. With the help of DIAT it is possible, by extracting all the services from each object and then grouping them together, coordinating services according to defined goal. The service request need to be properly studied for creation and management of these services. Thus, DIAT architecture as shown in Figure 3 is divided into three layers according to the functions performed by them:

1. Virtual object layer (VOL),
2. Composite virtual object layer (CVOL) and,
3. Service layer (SL).

Virtual Object Layer (VOL)

This layer converts real world objects into virtual objects (VO) by encapsulating its features and functionalities. This is how heterogeneity of objects made homogeneous through virtual representation. Accessing real object through virtual objects that means mapping of virtual object to its real object is done by a universally defined method. This layer deals with the problem of heterogeneity and interoperability.

Composite Virtual Object Layer (CVOL)

This layer manages all the virtual objects created by VOL. and according to goal to be accomplished, different objects are grouped together. CVOL takes care of creation of group of virtual objects and co-

Figure 3. Layered architecture of DIAT

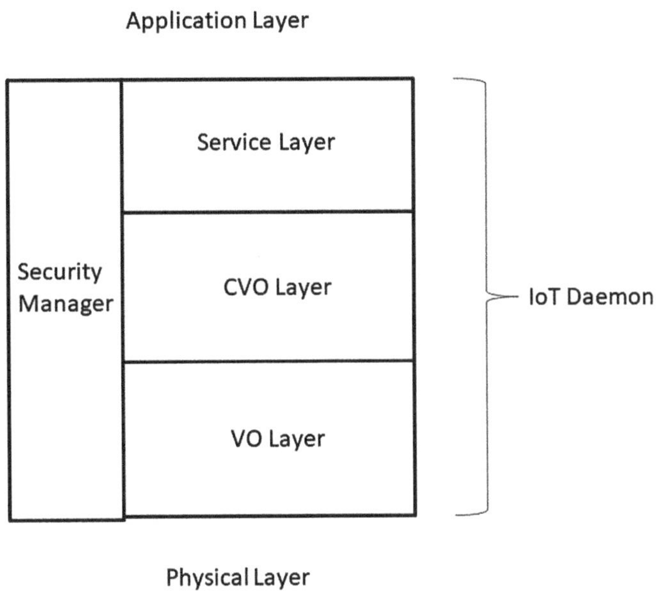

ordinates them according to their joint motive. For grouping virtual objects, it needs to identify them hence, CVOL also conduct discovery mechanism.

Service Layer (SL)

This layer works upon services of all the objects- it creates them according to the service request and manages them according to the task or goal to be completed. The task is executed by VO at CVOL. If VO is not able to provide service, other VO can also be dynamically chosen, this is important feature known as approximate service.

Security Management (SM)

This module provides security to each layer discussed above. It provides authentication as well as secured data communication. It also maintains privacy of owners. It has list of security policies that has to be followed by the components of this architecture.

IoT Daemon

Integration of all the functions each layer performs forms an IoT daemon. It also follows security policies and cognitive functionalities.

DATA AGGREGATION

It is a process in which messages from different sources are aggregated or put together in a single datagram. This reduces energy consumption because it reduces transmission power of each message together to transmission power of single datagram. Two approaches of data aggregation were defined that are Application dependent data aggregation (ADDA) and application independent data aggregation (AIDA). In ADDA, as name says, it needs application specific information and then create efficient aggregation network structure accordingly whereas in AIDA no such information is required. It just embeds the functions in the underlying network and does not provide optimality.

A new approach was needed for IoT, which could provide efficient data aggregation with capabilities to create custom communication structure. So Hitch Hicker developed a model combining aggregation with RPC (Remote procedure call) which has its own benefits like request reply protocol which is bidirectional, low overhead and same semantics as of a local procedure call. Even bindings can be dynamically modified where remote configuration is supported.

Following are Hitch Hiker characteristics:

To provide aggregation, it appends low priority data in overlay and the overlay is created using unused payload of high priority datagram.

Multi Hop overlay data aggregation is created using metadata.

Route discovery in multi hop overlay network uses central Meta manager for binding data aggregation overlay.

Hitch Hiker 2.0 by (Ramachandran, 2016) expands the previous version of Hitch Hiker with Ad-hoc Hitch Hiker uses route discovery algorithm for finding route and not by the help of central Meta manager.

SECURITY IN IOT

As we already know that to support interoperability, devices connected in the IOT network has to follow a same protocol suite. As TLS is used in transport layer of computer networks and has been a high success in providing security. So (Loong, Kumar, & Tschofenig, 2014) thought of using DTLS to provide security to devices of IoT. The only drawback in DTLS is that it is not suitable when there are many constraints over devices or network but research community is still trying hard to create a single security suite based on DTLS. CoAP would be secured by DTLS as HTTP is secured by TLS, through handshaking phase for authentication and key exchange phase for secured end to end communication. The different security mods of IoT device are shown below in Figure 4.

- **Datagram Transport Layer Security**: DTLS performs authentication, key exchange and encapsulating data for secured transmission with agreed algorithm and keying material used. CoAP is already provided with keying material when it is manufactured or when it is being configured.

IoT devices can be in four security modes based on its configuration:

1. **NoSec:** No security is provided to the devices that means no DTLS is required but still IP layer security IPSec has to be implemented. Coap sends message over UDP (connectionless and unreliable)

Figure 4. Security modes of IoT device

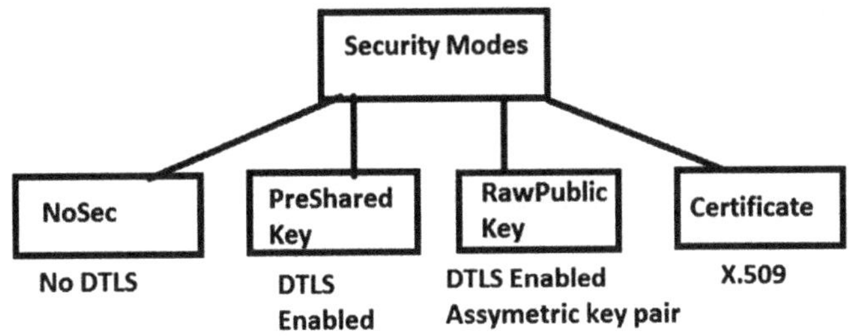

2. **PreSharedKey:** DTLS is used for securing transport layer and in it, for authentication, PSK (pre-SharedKey) authentication is applied. Keys has list of nodes that would be using it and these nodes would be having list of keys that can be used. Communication of devices with other devices would have unique for each.
3. **RawPublicKey:** Security is provided using DTLS and asymmetric key pair is assigned to each device. Public key is not certified but private key is. Each device also has list of all devices it is communicating with.
4. **Certificate:** DTLS, Asymmetric key pair-same as RawPublicKey. The only difference is x.509 certificate let the device get unique public key. Devices also has trust anchors list which can validate other certificates coming from "to be communicating" devices.

FUTURE CHALLENGES IN IOT AND SOFTWARE DEFINED NETWORK

In future a lot more devices would be invented for sure and these would be connected with the help of internet to form IoT application with the help of software defined network (SDN). The layerd architecture of SDN is depicted in Figure 5. For such a huge number of devices, huge bandwidth need to be reserved for communication between them. As those devices can be geographically implemented anywhere they would be in wireless communication and no standard wireless protocol could be followed by devices. Furthermore, overloaded network could make the network slow and inefficient in the future where speed and ease would be most important.

In this integration of SDN with IoT architecture as shown in the Figure 6, the main role is of *IoT Controller*. More the workload, more complex would be the functionality of controller. It is high level controller connected with low level SDN controller and the integration is through IoT agent. After getting the model of underlying network and services/ responses, IoT controller takes the decision accordingly.

Proposed Approach

IoT controller looks like a single functional block but it has various functional modules and even can be extended to IoT structure. (Araniti, Cosmas, Iera, Molinaro, Morabito, & Orsino, 2014) proposes a new function that can be added over IoT to the IoT overlay and no new connection has to be established with SDN controller. Neither would it affect any element of the network.

Figure 5. SDN Architecture

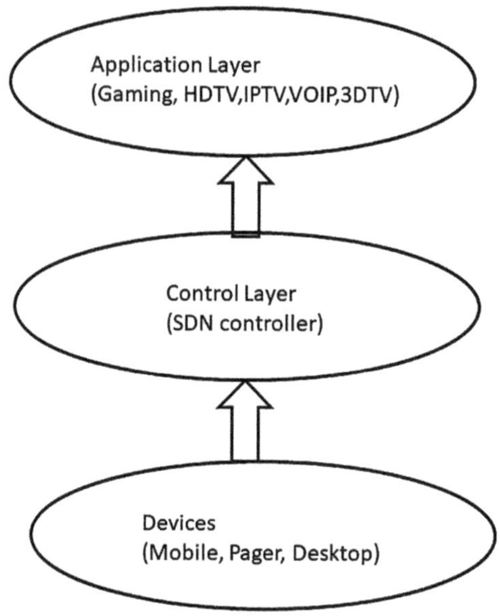

Communication will start only when requester would request to send data to another object and it should know IPV6 address of the receiver. For such communication, a path is formed by SDN virtually through routing table. IoT controller would find out route using routing algorithm and create forwarding rules according to the request type and then provide the rules to SDN controller. Requests are sent through IoT agents to the IoT controller providing necessary details for communication before actual communication starts.

By communication interest, IoT controller locates the receiver as it has all the details of each object in the system and then find route and form rules.

Figure 6. Integration of IoT with SDN

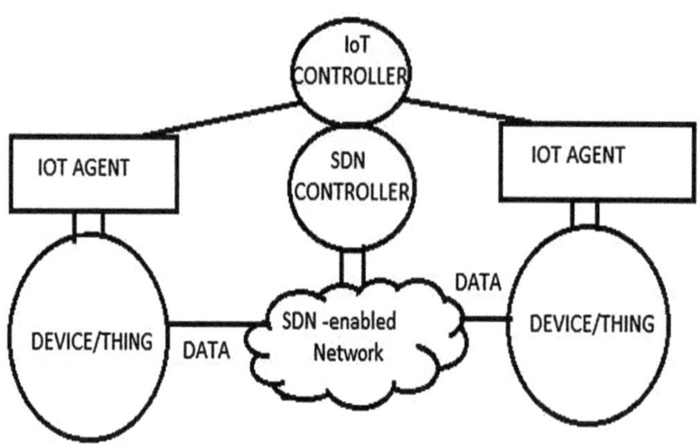

EXAMPLE

The setup of communication using SDN is described through following example:

1. The view of a global communication is composed of an object A, identified by IdA, that wants to communicate with an object B, identified by IdB, and NetA and NetB that represent the networks to which A and B are correspondingly connected.
2. When idA sends a message to IdB, NetA receives it, sends it to the SDN controller that in turn sends it to the IoT controller.
3. The IoT controller finds IdB in the IoT network graph and figures out that it is in a different island so the path between the two objects crosses the Internet.
4. The IoT controller uses or reserves new IPv6 addresses for IdA and IdB, and saves mappings among them (IdA = IPv6_A, IdB = IPv6_B).
5. Now, the IoT controller builds the forwarding rules and sends them to the SDN controller.
6. The SDN controller sets those rules to the underlying forwarding elements so that the packages coming from A will be forwarded to the gateway of NetA that in turn translates or embeds those packets into IPv6 packets, using the mapping information provided by the IoT controller, before sending them to the destination network through the Internet.
7. The gateway of NetB will extract the packets, apply other transformations if specified, and forward them to B. The inverse process occurs when B sends a packet to A.

COGNITIVE RADIO NETWORK AND IOT

Internet of Things (IoT) is based on wireless communications between nodes. For this communication, Fixed Spectrum Access (FSA) policy was being used traditionally. In this policy, the available spectrums of certain bandwidth are assigned to one or more licensed users to fully exploit it. Due to increase in rate of communication, the available spectrums in most of the countries have been fully allocated which has causes problem of spectrum scarcity. Recent research has also discovered that the assigned spectrums are having low utilization. To deal with these problems of wireless communication spectrum challenge, Dynamic Spectrum Access discussed by (Akildiz, Lo, & Balakrishan, 2011), came into action. In this approach, primary users (PUs) are licensed users of a spectrum and Secondary users (SUs) are allocated spectrum of PU temporarily when PU is not using his spectrum. In this way, both the problems are solved and spectrum utilization efficiency is significantly improved. SU need to sense the network and status of PUs to find out the unused spectrum. These SUs are therefore called as Cognitive Radios (CR) (Figure 7). Due to different cognitive abilities, there are two cognitive spectrum access models as discussed in paper by (Shah, Zhang, & Maple, 2013):

- **The Opportunistic Spectrum Access Model:** The CR user detects spectrum holes, that is the spectrum portions being allocated to some PU but are unused for a time period. After detection of holes, the CR user reconfigures itself to operate in the detected hole. During the operation, CR user needs to monitor the spectrum and vacate as soon as the PU is active. This method was first proposed by MITOLA under the title "spectrum pooling". Later, this model was introduced by

Defense Advanced Research Projects Agency (DARPA) in its Next-Generation Communications (XG) Program. It is also referred as "spectrum overlay".

- **The Concurrent Spectrum Access Model:** Also known as "spectrum sharing" or "spectrum underlay", this model has three components, A CR user, A CR transmitter and PU. A CR user coexists with an active PU in a licensed band till CR transmitter refrains its transmitting power. The interference that is caused by CR transmitter to the primary Rx should be below a tolerable threshold for CR user to operate in licensed spectrum. This model requires the CR Transmitter to predict the interference power level that is received at a particular location.

CRN contains more than one CR node and each CR node may have same or different cognitive capabilities. A new paradigm of IOT has included sensing as a service when IOT has been used for smart cities. This model has four main layers. First layer is sensor owners (SSO) layer, the second layer is sensor publishers (SPs) layer, the third layer is extended service providers (ESPs) layer, and finally, the fourth layer is sensor data consumer layer. As sensing is fundamental service for information transfer in IoT, According to by (Khan, Rehmani, & Rachedi, 2016), CRN can benefit to great extend with its nodes of cognitive capabilities. IoT and CRN are both evolving technologies and will be great research aspects in future.

APPLICATION OF IOT

IoT or internet of things is collection and connection of different types of devices through internet. It is "Coming Up" research work with basics being revised. Making its standard architecture is also very complex task as we do not know what type of devices, link, and topology in the network. Different applications would follow different architecture of IoT according to the type of devices it is connected to and type of service it is providing. Designing of IoT system is itself very complex when we have to integrate a large number of heterogeneous devices and make it communicable according to service of application.

Figure 7. Internet of things with CRN paradigm

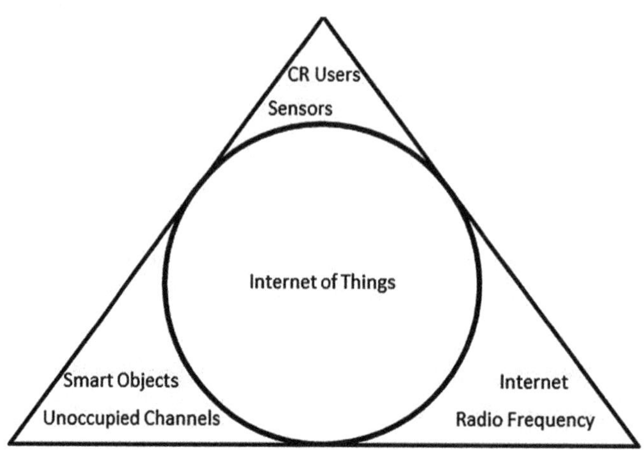

One of the applications by a paper of (Zanella, 2014) is based on Urban Iot or Smart City. Smart city vision would help in administration of city and would be able to implement advance technologies on whole for the citizens. Other application is satellite based IoT, term coined by (Sanctis, 2016). This IoT takes information from sensors, which are geographically distributed, through actuators. By this we can get information from the places where human cannot try to reach. For this IOT, the following has to be considered. Firstly, sensors and satellite to communicate that is they should be interoperable. Secondly, satellite should support IPV6 addressing. Lastly, a good radio management algorithm has to be implemented on satellite based IoT.

CONCLUSION

This chapter provides an insight into Internet of Things(IoT) discussing its architecture, middleware services, applications, security and future work. The technologies used by internet of things like software defined network and cognitive radios are also discussed. Challenges provided by IoT are being worked upon. The near future is expected to have research on middleware construction, security improvement, increasing number of technologies being integrated with IoT and application of Iot on a large scale. IoT is a vast and trending area of research and its research state is changing on a fast pace. This chapter is to provide basic vision of IoT that would be helpful for the future research purpose.

REFERENCES

Adnan, A., & Aghvami, A. H. (2015). Cognitive machine-to-machine communications for Internet-of-Things: A protocol stack perspective. *IEEE Internet of Things Journal*, *2*(2), 103–112. doi:10.1109/JIOT.2015.2390775

Akildiz, I. F., Lo, B. F., & Balakrishan, R. (2011). Cooperative spectrum sensing in cognitive radio networks: A survey. *Physical Communication*, *1*(4), 40–62. doi:10.1016/j.phycom.2010.12.003

Araniti, G., Cosmas, J., Iera, A., Molinaro, A., Morabito, R., & Orsino, A. (2014). OpenFlow Over Wireless Networks: Performance Analysis. In *Proceedings of IEEE International Symposium on Broadband Multimedia Systems and Broadcasting (BMSB)*. Beijing, China: IEEE.

Gil, R. (2014). *Wireless connectivity for the Internet of Things*. Europe 433: 868MHz.

Gubbi, J., Buyya, R., Marusic, S., & Palaniswami, M. (2013). Internet of Things (IoT): A vision, architectural elements, and future directions. *Future Generation Computer Systems*, *29*(7), 1645–1660. doi:10.1016/j.future.2013.01.010

Khan, A. A., Rehmani, M. H., & Rachedi, A. (2016). When Cognitive Radio meets the Internet of Things. In *Proceedings of IEEE International Wireless Communications and Mobile Computing Conference (IWCMC)*, (pp. 469-474), Paphos, Cyprus: IEEE.

Loong, K. S., Kumar, S. S., & Tschofenig, H. (2014). Securing the internet of things: A standardization perspective. *IEEE Internet of Things Journal*, *1*(3), 265–275. doi:10.1109/JIOT.2014.2323395

Ramachandran, G. S., Proença, J., Daniels, W., Pickavet, M., Staessens, D., Huygens, C., ... Hughes, D. (2016). Hitch Hiker 2.0: A binding model with flexible data aggregation for the Internet-of-Things. *Journal of Internet Services and Applications*, *7*(1), 2–15. doi:10.118613174-016-0047-7

Razzaque, M. A., Milojevic-Jevric, M., Palade, A., & Clarke, S. (2016). Middleware for internet of things: A survey. *IEEE Internet of Things Journal*, *3*(1), 70–95. doi:10.1109/JIOT.2015.2498900

Sanctis, M. D. (2016). Satellite communications supporting internet of remote things. *IEEE Internet of Things Journal*, *3*(1), 113–123. doi:10.1109/JIOT.2015.2487046

Sarkar, C. (2012). *DIAT: A scalable distributed architecture for IoT*. Academic Press.

Shah, M. A., Zhang, S., & Maple, C. (2013). Cognitive radio networks for Internet of Things: Applications, challenges and future. In *Proceedings of IEEE 19th International Conference Automation and Computing (ICAC)*, (pp. 1-6). London: IEEE.

Zanella, A. (2014). Internet of things for smart cities. *IEEE Internet of Things Journal, 1*(1), 22-32.

Compilation of References

Adnan, A., & Aghvami, A. H. (2015). Cognitive machine-to-machine communications for Internet-of-Things: A protocol stack perspective. *IEEE Internet of Things Journal*, *2*(2), 103–112. doi:10.1109/JIOT.2015.2390775

Adoum, B. A., & Jeoti, V. (2010, May). Cyclostationary feature based multiresolution spectrum sensing approach for DVB-T and wireless microphone signals. In *Computer and Communication Engineering (ICCCE), 2010 International Conference on* (pp. 1-6). IEEE. 10.1109/ICCCE.2010.5556803

Akildiz, I. F., Lo, B. F., & Balakrishan, R. (2011). Cooperative spectrum sensing in cognitive radio networks: A survey. *Physical Communication*, *1*(4), 40–62. doi:10.1016/j.phycom.2010.12.003

Akyildiz, Lo, & Ravikumar. (2011). Cooperative spectrum sensing in cognitive radio networks: A survey. *Elsevier Journal on Physical Communication*, 40-62.

Akyildiz, I. F., Lee, W. Y., Vuran, M. C., & Mohanty, S. (2006). NeXt generation/dynamic spectrum access/cognitive radio wireless networks: A survey. *Computer Networks*, *50*(13), 2127–2159. doi:10.1016/j.comnet.2006.05.001

Al Mamoon, I., Muzahidul-Islam, A. M., Baharun, S., Komaki, S., & Ahmed, A. (2015, March). Architecture and communication protocols for cognitive radio network enabled hospital. In *Medical Information and Communication Technology (ISMICT), 2015 9th International Symposium on* (pp. 170-174). IEEE. 10.1109/ISMICT.2015.7107522

Al-Akkad, A., Pramudianto, F., Jahn, M., & Zimmermann, A. (2009, November). Middleware for building pervasive systems. IADIS AC, (1), 291-298.

Alexander, M. (2010). *Cognitive Radio Communications and Networks - Principles and Practice*. Elsevier Inc.

Ali, A., & Hamouda, W. (2017). Advances on Spectrum Sensing for Cognitive Radio Networks: Theory and Applications. *IEEE Communications Surveys and Tutorials*, *19*(2), 1277–1304. doi:10.1109/COMST.2016.2631080

Aparna, P. S., & Jayasheela, M. (2012). Cyclostationary feature detection in cognitive radio using different modulation schemes. *International Journal of Computers and Applications*, *47*(21).

Araniti, G., Cosmas, J., Iera, A., Molinaro, A., Morabito, R., & Orsino, A. (2014). OpenFlow Over Wireless Networks: Performance Analysis. In *Proceedings of IEEE International Symposium on Broadband Multimedia Systems and Broadcasting (BMSB)*. Beijing, China: IEEE.

Arikan, E. (2009). Channel Polarization: A Method for Constructing Capacity-Achieving Codes for Symmetric Binary-Input Memoryless Channels. *IEEE Transactions on Information Theory*, *55*(7), 3051–3073. doi:10.1109/TIT.2009.2021379

Atapattu, S., Tellambura, C., & Jiang, H. (2014). *Energy detection for spectrum sensing in cognitive radio*. New York, NY: Springer. doi:10.1007/978-1-4939-0494-5_2

Aulakh, I. K., & Vig, R. (2014). Optimization of SU's probability of false alarm for dynamic spectrum access in cognitive radio. In *International Conference on Computing for Sustainable Global Development* (pp. 710–715). New Delhi, India: Academic Press. 10.1109/IndiaCom.2014.6828053

Aziz, B., Traore, S., Nafkha, A., & Le Guennec, D. (2014). Spectrum sensing for cognitive radio using multicoset sampling. *Proc. IEEE Global Commun. Conf. (GLOBECOM)*, 816–821. 10.1109/GLOCOM.2014.7036909

Babich, F., Noschese, M., & Vatta, F. (2016). Analysis and Design of Rate Compatible LDPC Codes. In *IEEE 27th Annual International Symposium on Personal, Indoor, and Mobile Radio Communications* (pp. 1-6). Valencia, Spain: IEEE. 10.1109/PIMRC.2016.7794684

Badoi, C. I., Prasad, N., Croitoru, V., & Prasad, R. (2011). 5G based on cognitive radio. *Wireless Personal Communications*, *57*(3), 441–464. doi:10.100711277-010-0082-9

Bagwari, A., & Singh, B. (2012, November). Comparative performance evaluation of spectrum sensing techniques for cognitive radio networks. In *Computational Intelligence and Communication Networks (CICN), 2012 Fourth International Conference on* (pp. 98-105). IEEE. 10.1109/CICN.2012.66

Bagwari, A., Kanti, J., & Tomar, G. S. (2016). *New Cooperative Spectrum Detection Technique in Cognitive Radio Networks*. Academic Press.

Bagwari, A., & Tomar, G. S. (2013). Cooperative spectrum sensing with adaptive double-threshold based energy detector in cognitive radio networks. *Wireless Personal Communications*, *73*(3), 1005–1019. doi:10.100711277-013-1244-3

Bagwari, A., Tomar, G. S., & Verma, S. (2013). Cooperative spectrum sensing based on two-stage detectors with multiple energy detectors and adaptive double threshold in cognitive radio networks. *Canadian Journal of Electrical and Computer Engineering*, *36*(4), 172–180. doi:10.1109/CJECE.2014.2303519

Baker, S. D., & Hoglund, D. H. (2008). Medical-grade, mission-critical wireless networks. *IEEE Engineering in Medicine and Biology Magazine*, *27*(2), 86–95. doi:10.1109/EMB.2008.915498 PMID:18463024

Ball, M. J., & Lillis, J. (2001). E-health: Transforming the physician/patient relationship. *International Journal of Medical Informatics*, *61*(1), 1–10. doi:10.1016/S1386-5056(00)00130-1 PMID:11248599

Baraniuk, R. G. (2007). Compressive sensing. *IEEE Signal Processing Magazine*, *24*(4), 118–121. doi:10.1109/MSP.2007.4286571

Bates, D. W. (1998). Effect of computerized physician order entry and a team intervention on prevention of serious medication errors. *Journal of the American Medical Association*, *280*(15), 1311–1316. doi:10.1001/jama.280.15.1311 PMID:9794308

Bavarian, S., & Cavers, J. K. (2007). Reduced Complexity Belief Propagation Algorithm Based on Iterative Groupwise Multiuser Detection. In *Canadian Conference on Electrical and Computer Engineering* (pp. 466-468). Vancouver, BC, Canada: Academic Press. 10.1109/CCECE.2007.123

Bayhan. (2013). *Overview of Cognitive Radio Basics and Spectrum Sensing*. CN-S2013.

Benkhouya, R., Chana, I., & Hadi, Y. (2016). Correction of sensing imperfections using channel coding in cognitive radio networks while keeping sensing time low. *International Conference on Advanced Communication Systems and Information Security* (pp. 1-5). Marrakech, Morocco: Academic Press. 10.1109/ACOSIS.2016.7843938

Benkhouya, R., Chana, I., & Hadi, Y. (2017). Investigation on BER sensitivity to design SNR of Bhattacharrya bounds based construction. *Journal of Theoretical & Applied Information Technology*, *95*(7), 1398–1404.

Berlekamp, E. R. (1968). *Algebraic coding theory*. Dunfermline, UK: McGraw-Hill Education.

Bhalla, M. R., & Bhalla, A. V. (2010). Generations of mobile wireless technology: A survey. *International Journal of Computers and Applications*, *5*(4).

Bhatti, F. A., Rowe, G. B., & Sowerby, K. W. (2012). Spectrum Sensing using Principal Component Analysis. In *Proceedings of IEEE Wireless Communications and Networking Conferences* (pp. 725-730). Paris, France: IEEE.

Bhowmick, Abhijit, Binod, Sanjay, & Sumit. (2016). Performance of Cognitive Radio Network with Novel Hybrid Spectrum Access Schemes. *Wireless Personal Communications*, *91*(2), 541–560. doi:10.100711277-016-3476-5

Bhowmick, Abhijit, Nallagonda, Srinivas, Sanjay, & Sumit. (2014). Spectrum sensing with censoring of double threshold based cognitive radios in rayleigh fading. In *Communications (NCC), 2014 Twentieth National Conference on* (pp. 1-6). IEEE.

Bhowmick, Abhijit, Nallagonda, Srinivas, Sanjay, & Sumit. (2015). Cooperative Spectrum Sensing with Double Threshold and Censoring in Rayleigh Faded Cognitive Radio Network. *Wireless Personal Communications*, *84*(1), 251–271. doi:10.100711277-015-2606-9

Bhowmick, A., & Sanjay, D. R. (2016). Sensing throughput trade-off for an energy efficient cognitive radio network under faded sensing and reporting channel. *International Journal of Communication Systems*, *29*(7), 1208–1218. doi:10.1002/dac.3087

Bhuvaneswari, P. T. V. (2016). Spectrum Sensing in Cognitive Radio Networks: A Survey. *Introduction to Cognitive Radio Networks and Applications, 83*.

Biglieri, E., Goldsmith, A. J., Greenstein, L. J., & Mandayam, N. B. (2012). *Principles of Cognitive Radio*. Cambridge University Press. doi:10.1017/CBO9781139236850

Bishnu, A., & Bhatia, V. (2015). On performance analysis of IEEE 802.22 (PHY) for COST-207 channel models. In *Proceedings of IEEE Conference on Standards for Communications and Networking* (pp. 229-234). Tokyo, Japan: IEEE. 10.1109/CSCN.2015.7390449

Bkassiny & Jayaweera. (2014). Robust, Non-Gaussian Wideband Spectrum Sensing in Cognitive Radios. *IEEE Transactions on Wireless Communications, 13*(11), 6410-6421.

Bkassiny, M. (2013). *A Survey on Machine-Learning Techniques in Cognitive Radios. IEEE Communications Surveys & Tutorials, 15(3).*

Bkassiny, M., Li, Y., & Jayaweera, S. (2013). A survey on machine-learning techniques in cognitive radios. In *Communications Surveys and Tutorials* (Vol. 15). IEEE. doi:10.1109/SURV.2012.100412.00017

Blair-Stahn, N. (2000). *Random Permutation Matrices*. Academic Press.

Bountouris. (2013). *Spectrum Sensing Techniques for Cognitive Radios* (Thesis). Technical University of Crete, Greece, Department of Electronic and Computer Engineering.

Boyle, J. (2006). Wireless technologies and patient safety in hospitals. *Telemedicine Journal and e-Health*, *12*(3), 373–382. doi:10.1089/tmj.2006.12.373 PMID:16796507

Brodersen, R. W., Wolisz, A., Cabric, D., Mishra, S. M., & Willkomm, D. (2004). *Corvus: a cognitive radio approach for usage of virtual unlicensed spectrum*. Berkeley Wireless Research Center (BWRC) White paper, 1-21.

Bushell-Embling, D. (2016). *Huawei achieves 27Gbps 5G speeds with Polar Code*. Retrieved July 26, 2017, from https://www.telecomasia.net

Cabric, D., & Brodersen, R. W. (2005, September). Physical layer design issues unique to cognitive radio systems. In *Personal, Indoor and Mobile Radio Communications, 2005. PIMRC 2005. IEEE 16th International Symposium on* (Vol. 2, pp. 759-763). IEEE. 10.1109/PIMRC.2005.1651545

Cabric, D., Mishra, S. M., & Brodersen, R. W. (2004, November). Implementation issues in spectrum sensing for cognitive radios. In *Signals, systems and computers, 2004. Conference record of the thirty-eighth Asilomar conference on* (Vol. 1, pp. 772-776). Ieee. 10.1109/ACSSC.2004.1399240

Cabric, D., Tkachenko, A., & Brodersen, R. W. (2006, October). Spectrum sensing measurements of pilot, energy, and collaborative detection. In Military communications conference, 2006. MILCOM 2006. IEEE (pp. 1-7). IEEE. doi:10.1109/MILCOM.2006.301994

Candes, E., & Romberg, J. (2005). *l1-magic: Recovery of sparse signals via convex programming*. Retrieved from www.acm.caltech.edu/l1magic/downloads/l1magic.pdf

Carothers, N. L. (2000). *Real analysis*. Cambridge University Press. doi:10.1017/CBO9780511814228

Cavalcanti, D., & Ghosh, M. (2008, May). Cognitive radio networks: Enabling new wireless broadband opportunities. In *Cognitive Radio Oriented Wireless Networks and Communications, 2008. CrownCom 2008. 3rd International Conference on* (pp. 1-6). IEEE.

Chávez-Santiago, R., Jankunas, D., Fomin, V. V., & Balasingham, I. (2014, April). Dual-band cognitive radio for wearable sensors in hospitals. In *2014 8th International Symposium on Medical Information and Communication Technology (ISMICT),* (pp. 1-5). IEEE. 10.1109/ISMICT.2014.6825244

Chen, X., Jing, T., Huo, Y., Li, W., & Cheng, T. (2012). Achievable Transmission Capacity of Cognitive radio Network with cooperative relaying. *Proc. IEEE CROWNCOM*, 1-6.

Chen, H.-S., Gao, W., & Daut, D. G. (2009). Spectrum Sensing for OFDM Systems Employing Pilot Tones. *IEEE Transactions on Wireless Communications*, *8*(12), 5862–5870. doi:10.1109/TWC.2009.12.080777

Chen, Y., & Tang, Z. (2012). Effect of spectrum sensing errors on the performance of OFDM-based cognitive radio transmission. *IEEE Transactions on Wireless Communications*, *11*(6), 2342–2350. doi:10.1109/TWC.2012.040412.111744

Chen, Z., Luan, T., & Zhang, X.-D. (2012). Sensing orthogonal frequency division multiplexing systems for cognitive radio with cyclic prefix and pilot tones. *IET Communications*, *6*(1), 97–106. doi:10.1049/iet-com.2010.1038

Chunhua, S., & Wei, Z., & Ben.(2007). Cooperative spectrum sensing for cognitive radios under bandwidth constraints. In *Wireless Communications and Networking Conference, 2007.WCNC 2007. IEEE* (pp. 1-5). IEEE.

Clancy, Hecker, Stuntebeck, & O'Shea. (2007, August). Applications of machine learning to cognitive radio networks. *IEEE Wireless Communications*, *7*, 47–52.

Clancy, C., Hecker, J., Stuntebeck, E., & O'Shea, T. (2007). Applications of machine learning to cognitive radio networks. *Wireless Communications, IEEE*, *14*(4), 47–52. doi:10.1109/MWC.2007.4300983

Coats, A., & Huval, B. (2013). Deep Learning with COTS HPS systems. *Journal of Machine Learning Research*, *28*(3), 1337–1345.

Cordeiro, C., Challapali, K., Birru, D., & Shankar, S. (2005, November). IEEE 802.22: the first worldwide wireless standard based on cognitive radios. In *New Frontiers in Dynamic Spectrum Access Networks, 2005. DySPAN 2005. 2005 First IEEE International Symposium on* (pp. 328-337). IEEE.

Cordeiro, C., Challapali, K., & Ghosh, M. (2006). Cognitive PHY and MAC Layers for Dynamic Spectrum Access and Sharing of TV Bands. *Proceedings of First International Workshop on Technology and Policy for Accessing Spectrum*. 10.1145/1234388.1234391

Corderio, C., Challapali, K., Birru, D., & Shankar, S. (2006). IEEE 802.22: An introduction to the first wireless standard based on cognitive radios. *Journal of Communication*, *1*(1), 38–47.

Cussens, J. (1996). Machine Learning. *IEEE Journal of Computing and Control*, *7*(4), 164–168. doi:10.1049/cce:19960402

Dhar Roy, S., Bhowmick, A., Kundu, S., & Chandra, A. (2015). Double threshold-based cooperative spectrum sensing for a cognitive radio network with improved energy detectors. *IET Communications*, *9*(18), 2216–2226. doi:10.1049/iet-com.2014.1098

Dhope, T. S. (2015). *Cognitive radio networks optimization with spectrum sensing algorithms* (Vol. 44). River Publishers.

Di Benedetto, M., Hua, Y., Kaiser, T., & Wang, X. (2008). Cognitive Radio Technology. *IEEE Signal Processing Magazine*, *25*(6), 10–198. doi:10.1109/MSP.2008.930499

Digham, F. F., Alouini, M. S., & Simon, M. K. (2007). On the energy detection of unknown signals over fading channels. IEEE Transactions on Communications, 55(1), 21-24.

Ding, D., Wang, J., Wu, Q., Zhang, L., Song, F., Zou, Y., ... Feng, S. (2014). Robust Spectrum Sensing With Crowd Sensors. *IEEE Transactions on Communications*, *62*(9), 3129–3143. doi:10.1109/TCOMM.2014.2346775

Divakaran, K., Manikandan, N., & Hari, S. (2011). Wavelet based spectrum sensing techniques for cognitive radio-A survey. *International Journal of Computer Science & Information Technology*, *3*(2), 123–137. doi:10.5121/ijcsit.2011.3209

Dodson, N., Bradford, G. J., & Laneman, J. N. (2010, November). A high performance rf transceiver implementation. In *Proceedings of the SDR 10 Technical Conference and Product Exposition* (pp. 652-658). Academic Press.

Donoho, D. L. (2006). Compressed sensing. *IEEE Transactions on Information Theory*, *52*(4), 1289–1306. doi:10.1109/TIT.2006.871582

Doyle, L. (2009). *Essentials of cognitive radio*. Cambridge University Press. doi:10.1017/CBO9780511576577

Ejaz, W. (2013). I3S: Intelligent spectrum sensing scheme for cognitive radio networks. *EURASIP Journal on Wireless Communications and Networking*, (1), 1–10. doi:10.1186/1687-1499-2013-26

Ejaz, W., Hasan, N. U., & Kim, H. S. (2012). SNR-based adaptive spectrum sensing for cognitive radio networks. *International Journal of Innovative Computing, Information, & Control*, *8*(9), 6095–6105.

El Khaddar, M. A., Harroud, H., Boulmalf, M., Elkoutbi, M., & Habbani, A. (2012, May). Emerging wireless technologies in e-health trends, challenges, and framework design issues. In *Multimedia Computing and Systems (ICMCS), 2012 International Conference on* (pp. 440-445). IEEE. 10.1109/ICMCS.2012.6320276

El-Khamy, S. E., El-Mahallawy, M. S., & Youssef, E. N. S. (2013, January). Improved wideband spectrum sensing techniques using wavelet-based edge detection for cognitive radio. In *Computing, Networking and Communications (ICNC), 2013 International Conference on* (pp. 418-423). IEEE. 10.1109/ICCNC.2013.6504120

El-Khamy, S. E., Abdel-Malek, M. B., & Kamel, S. H. (2014, April). An improved reconstruction technique for wavelet-based compressive spectrum sensing using genetic algorithm. In *Radio Science Conference (NRSC), 2014 31st National* (pp. 99-106). IEEE. 10.1109/NRSC.2014.6835066

El-Khamy, S. E., Abd-el-Malek, M. B., & Kamel, S. H. (2017). A stationary wavelet transform approach to compressed spectrum sensing in cognitive radio. *International Journal of Communication Systems*, *30*(7), e3140. doi:10.1002/dac.3140

El-Khamy, S., Abdel-Malek, M., & Kamel, S. (2013, April). C9. A New Wideband Spectrum Sensing Technique based on Compressive Sensing, Wavelet Edge Detection and Parallel Processing. In *Radio Science Conference (NRSC), 2013 30th National* (pp. 202-210). IEEE.

Eysenbach, G. (2001). What is e-health? *Journal of Medical Internet Research, 3*(2), e20. doi:10.2196/jmir.3.2.e20 PMID:11720962

Fanan, A. M., Riley, N. G., Mehdawi, M., Ammar, M., & Zolfaghari, M. (2014). Survey: A Comparison of Spectrum Sensing Techniques in Cognitive Radio. *Int'l Conference Image Processing, Computers and Industrial Engineering (ICICIE'2014).*

Farhang-Boroujeny, B. (2008, May). Filter Bank Spectrum Sensing for Cognitive Radios. *IEEE Transactions on Signal Processing*, *56*(5), 1801–1811. doi:10.1109/TSP.2007.911490

FCC. (2002). *Et docket no 02-135 spectrum policy task force report*. FCC.

FCC. (2003). *Docket No 03-222 Notice of proposed rule making and order*. FCC.

Font-Segura, J., & Wang, X. (2010). GLRT-Based Spectrum Sensing for Cognitive Radio with Prior Information. *IEEE Transactions on Communications*, *58*(7), 2137–2146. doi:10.1109/TCOMM.2010.07.090556

Force, S. (2002). Spectrum policy task force report. Federal Communications Commission ET Docket 02, vol. 135.

Freewimaxinfo.com. (2017). *Advantages of cognitive radio networks*. Available at: http://freewimaxinfo.com/advantages-of-cognitive-radio-networks.html

Fullmer, C. L., & Garcia-Luna-Aceves, J. (1997). Solutions to hidden terminal problems in wireless networks. *Proceedings of the ACM SIGCOMM97 Conf. on Applications, Technologies, Architectures, and Protocols for Computer Communication, 97*, 39-49. 10.1145/263105.263137

Galindo-Serrano, A., Giupponi, L., Blasco, P., & Dohler, M. (2010). Learning from experts in cognitive radio networks: The docitive paradigm. *Proc. Fifth International Conference on Cognitive Radio Oriented Wireless Networks Communications (CROWNCOM '10)*, 1 –6. 10.4108/ICST.CROWNCOM2010.9173

Gallager, R. G. (1963). *Low-density parity-check codes*. MIT Press.

Ganesan, G., & Li, Y. (2005, December). Agility improvement through cooperative diversity in cognitive radio. In *Global Telecommunications Conference, 2005. GLOBECOM'05. IEEE* (Vol. 5, pp. 5-pp). IEEE. 10.1109/GLOCOM.2005.1578213

Ganesan, G., & Li, Y. (2005). Cooperative spectrum sensing in cognitive radio networks. In *Proceedings of IEEE Symposium on New Frontiers Dynamic Spectrum Access Networks* (pp. 137-143). Baltimore, MD: IEEE. 10.1109/DYSPAN.2005.1542628

Garhwal, A., & Bhattacharya, P. P. (2011). A Survey on Spectrum Sensing Techniques in Cognitive Radio. *International Journal of Computer Science & Communication Networks, 1*(2), 196-206.

Garhwal, A., & Bhattacharya, P. P. (2012). *A Survey on Dynamic Spectrum Access Techniques for Cognitive Radio.* Academic Press.

Ghasemi, A., & Sousa, E. S. (2005). Collaborative spectrum sensing for opportunistic access in fading environments. *Proc. of 1st IEEE Symp. New Frontiers in Dynamic Spectrum Access Networks*, 131-136.

Ghasemi, A., & Sousa, E. S. (2005). Collaborative Spectrum Sensing for Opportunistic Access in Fading Environments. In *Proceedings of IEEE Symposium on New Frontiers Dynamic Spectrum Access Networks* (pp. 131-136). Baltimore, MD: IEEE. 10.1109/DYSPAN.2005.1542627

Ghasemi, A., & Sousa, E. S. (2008). Spectrum sensing in cognitive radio networks: Requirements, challenges and design trade-offs. *IEEE Communications Magazine*, *46*(4), 32–39. doi:10.1109/MCOM.2008.4481338

Ghassemi, A., Bavarian, S., & Lampe, L. (2010, October). Cognitive radio for smart grid communications. In *Smart Grid Communications (SmartGridComm), 2010 First IEEE International Conference on* (pp. 297-302). IEEE.

Gil, R. (2014). *Wireless connectivity for the Internet of Things.* Europe 433: 868MHz.

Giorgetti, A., Chiani, M., Dardari, D., Piesiewicz, R., & Bruck, G. H. (2008, September). The cognitive radio paradigm for ultra-wideband systems: The *European project EUWB. In Ultra-Wideband, 2008. ICUWB 2008. IEEE International Conference on* (Vol. 2, pp. 169-172). IEEE.

Giupponi, L., Galindo-Serrano, A., Blasco, P., & Dohler, M. (2010, August). Docitive networks: An emerging paradigm for dynamic spectrum management. *IEEE Wireless Communications*, *17*(4), 47–54. doi:10.1109/MWC.2010.5547921

Giweli, N., Shahrestani, S. A., & Cheung, H. N. (2016). Selection of spectrum sensing method to enhance Qos in cognitive radio networks. *International Journal of Wireless and Mobile Networks*, *8*(1), 39–50. doi:10.5121/ijwmn.2016.8104

Gorodnitsky, I. F., & Rao, B. D. (1997). Sparse signal reconstruction from limited data using FOCUSS: A re-weighted minimum norm algorithm. *IEEE Transactions on Signal Processing*, *45*(3), 600–616. doi:10.1109/78.558475

Gradshteyn, I. S., & Ryzhik, I. M. (2007). *Table of integrals, series and products.* Elsevier.

Gray, R. M. (2006). Toeplitz and circulant matrices: A review. *Foundations and Trends® in Communications and Information Theory*, *2*(3), 155-239.

Gubbi, J., Buyya, R., Marusic, S., & Palaniswami, M. (2013). Internet of Things (IoT): A vision, architectural elements, and future directions. *Future Generation Computer Systems*, *29*(7), 1645–1660. doi:10.1016/j.future.2013.01.010

Guo, C., Zhang, T., Zeng, Z., & Feng, C. (2006). Investigation on Spectrum Sharing Technology Based on Cognitive Radio. In *Proceedings of IEEE International Conference on Communications and Networking in China* (pp. 1-5). Beijing, China: IEEE. 10.1109/CHINACOM.2006.344901

Haddadi, S., Saeedi, H., & Navaie, K. (2013). Channel coding adoption versus increasing sensing time in secondary service to manage the effect of imperfect spectrum sensing in cognitive radio networks. In *Iran Workshop on Communication and Information Theory* (pp. 1–5). Tehran, Iran: Academic Press. 10.1109/IWCIT.2013.6555757

Haggstrom, P. (2016). *The good, the bad and the ugly of kernels: why the Dirichlet kernel is not a good kernel.* Academic Press.

Han, C., Harrold, T., Armour, S., Krikidis, I., Videv, S., Grant, P. M., ... Le, T. A. (2011). Green radio: Radio techniques to enable energy-efficient wireless networks. *IEEE Communications Magazine*, *49*(6), 46–54. doi:10.1109/MCOM.2011.5783984

Han, N., Shon, S., Chung, J. H., & Kim, J. M. (2006). Spectral Correlation Based Signal Detection Method for Spectrum Sensing in IEEE 802.22 WRAN Systems. In *Proceedings of IEEE International Conference on Advanced Communication Technology* (vol. 3). Gangwon-Do, South Korea: IEEE.

Hattab & Ibnkahla. (2014). Multiband Spectrum Access: Great Promises for Future Cognitive Radio Networks. *Proceedings of the IEEE*, 102(3), 282-306.

Haykin, S. (2005). Cognitive radio: Brain-empowered wireless communications. *IEEE Journal on Selected Areas in Communications*, *23*(2), 201–220. doi:10.1109/JSAC.2004.839380

Haykin, S., & Setoodeh, P. (2015). Cognitive Radio Networks: The Spectrum Supply Chain Paradigm. *IEEE Transactions on Cognitive Communications and Networking*, *1*(1), 3–28. doi:10.1109/TCCN.2015.2488627

Haykin, S., Thomson, D., & Reed, J. (2009, May). Spectrum sensing for cognitive radio. *Proceedings of the IEEE*, *97*(5), 849–877. doi:10.1109/JPROC.2009.2015711

He, Bae, Newman, Gaeddert, Kim, Menon, … Tranter. (2010). A Survey of Artificial Intelligence for Cognitive Radios. *IEEE Transactions on Vehicular Technology, 59*(4).

He, Y., Ratnarajah, T., Xue, J., Yousif, E. H., & Sellathurai, M. (2014, May). Optimal decision threshold for eigenvalue-based spectrum sensing techniques. In *Acoustics, Speech and Signal Processing (ICASSP), 2014 IEEE International Conference on* (pp. 7734-7738). IEEE. 10.1109/ICASSP.2014.6855105

He, A., Bae, K. K., Newman, T., Gaeddert, J., Kim, K., Menon, R., ... Tranter, W. (2010, May). A survey of artificial intelligence for cognitive radios. *IEEE Transactions on Vehicular Technology*, *59*(4), 1578–1592. doi:10.1109/TVT.2010.2043968

Holland, O., Bogucka, H., & Medeisis, A. (2015). *Opportunistic spectrum sharing and white space access: The practical reality*. John Wiley & Sons. doi:10.1002/9781119057246

Horev, I., Yger, F., & Sugiyama, M. (2017). Geometry-aware principal component analysis for symmetric positive definite matrices. *Springer Machine Learning*, *106*(4), 93–522.

Hossain, E., Niyato, D., & Han, Z. (2009). *Dynamic spectrum access and management in cognitive radio networks.* Cambridge University Press. Retrieved from http://freewimaxinfo.com/advantages-of-cognitive-radio-networks.html

Hossain, E., Niyato, D., & Han, Z. (2009). *Dynamic spectrum access and management in cognitive radio networks.* Cambridge University Press. Retrieved from http://www.radio-electronics.com/info/rf-technology-design/cognitive-radio-cr/cooperative-spectrum-sensing.php'

Hossain, D. N. E., & Diamond, J. (2008). *Fourth generation heterogeneous wireless access networks for eHealth services: Architecture and radio resource management. In Mobile Telemedicine: A Computing and Networking Perspective*. Auerbach Publications.

Hou, S., & Qiu, R. C. (2014). Kernel Feature Template Matching for Spectrum Sensing. *IEEE Transactions on Vehicular Technology*, *63*(5), 2258–2271. doi:10.1109/TVT.2013.2290866

Höyhtyä, Hekkala, Katz, & Mämmelä. (2007). Spectrum Awareness: Techniques and Challenges for Active Spectrum Sensing. *Cognitive Wireless Networks,* 353-372.

Huang, X., Han, T., & Ansari, N. (2015). On green-energy-powered cognitive radio networks. *IEEE Communications Surveys and Tutorials*, *17*(2), 827–842. doi:10.1109/COMST.2014.2387697

Iacobucc. (2013). *Reconfigurable Radio Systems: Network Architectures and Standards*. Wiley Publications.

Ibnkahla, M. (2014). *Cooperative Cognitive Radio Networks: The Complete Spectrum Cycle*. CRC Press.

Jafarpour, S., Schapire, R. E., & Cevher, V. (2011, May). Compressive sensing meets game theory. In *Acoustics, Speech and Signal Processing (ICASSP), 2011 IEEE International Conference on* (pp. 3660-3663). IEEE. 10.1109/ICASSP.2011.5947144

Jaiswal, M., Sharma, A. K., & Singh, V. (2013). A Survey on Spectrum Sensing Techniques for Cognitive Radio. *Conference on Advances in Communication and Control Systems*, 647-660.

James, M., & Singh, S. (2004). Learning and discovery of predictive state representations in dynamical systems with reset. *Proceedings of the Twenty-First International Conference on Machine Learning (ICML)*, 417–424. 10.1145/1015330.1015359

Javed, M., & Siddiqui, A. T. (2017). Transformation of Mobile Communication Network from 1G to 4G and 5G. *International Journal of Advanced Research in Computer Science*, *8*(3).

Jayaweera. (2014). Robust, Non-Gaussian Wideband Spectrum Sensing in Cognitive Radios. *IEEE Transactions on Wireless Communications, 13*(11), 6410-6421.

Jolliffe, I. T. (1986). *Principal Component Analysis*. Springer-Verlag. doi:10.1007/978-1-4757-1904-8

Jondral, F. K. (2005). Software-defined radio: Basics and evolution to cognitive radio. *EURASIP Journal on Wireless Communications and Networking*, (3): 275–283.

Joshi, G. P., Nam, S. Y., & Kim, S. W. (2013). Cognitive radio wireless sensor networks: Applications, challenges and research trends. *Sensors (Basel)*, *13*(9), 11196–11228. doi:10.3390130911196 PMID:23974152

K, M., & S, R. (2014). *Improved Sensing Accuracy using Enhanced Energy Detection Algorithm with Secondary User Cooperation in Cognitive Radios*. Retrieved from http://www.ijcnis.org/index.php/ijcnis/article/view/533

Kaabouch, N. (Ed.). (2014). *Handbook of Research on Software-Defined and Cognitive Radio Technologies for Dynamic Spectrum Management*. IGI Global.

Kamel, S. H., El-Khamy, S. E., & Abdel-Malek, M. B. (2014, August). An improved compressed wideband spectrum sensing technique based on stationary wavelet transform in Cognitive Radio systems. In *General Assembly and Scientific Symposium (URSI GASS), 2014 XXXIth URSI* (pp. 1-4). IEEE. 10.1109/URSIGASS.2014.6929237

Kang, X., Garg, H., Liang, Y.-C., & Zhang, R. (2010). Optimal power allocation for OFDM-based cognitive radio with new primary transmission protection criteria. *IEEE Transactions on Wireless Communications*, *9*(6), 2066–2075. doi:10.1109/TWC.2010.06.090912

Kanojia, D., Motwani, M., Khan, M. J., & Yadav, M. (2012). Beyond 3G: 4G Mobile Communication. *International Journal of Innovations in Engineering and Technology*, *1*(2).

Kanti, J., & Tomar, G. S. (2016). Various Sensing Techniques in Cognitive Radio Networks: A Review. *Spectrum, 9*(1).

Kanti, J., Bagwari, A., & Tomar, G.S. (2015). Novel Spectrum detector for IEEE 802.22 Wireless Regional Area Network. *International Journal of Smart Device and Appliance, 3*(2), 9-25.

Karim & Rahman. (2013). Decision Tree and Naïve Bayes Algorithm for Classification and Generation of Actionable Knowledge for Direct Marketing. *Journal of Software Engineering and Applications, 6*(4), 196-206.

Khan, A. A., Rehmani, M. H., & Rachedi, A. (2017). Cognitive-radio-based Internet of Things: Applications, architectures, Spectrum related functionalities, and future research directions. *IEEE Wireless Communications*, 3.

Khan, F., & Nakagawa, K. (2013, June). Comparative study of spectrum sensing techniques in cognitive radio networks. In *Computer and Information Technology (WCCIT), 2013 World Congress on* (pp. 1-8). IEEE. 10.1109/WCCIT.2013.6618728

Khan, A. A., Rehmani, M. H., & Rachedi, A. (2016). When Cognitive Radio meets the Internet of Things. In *Proceedings of IEEE International Wireless Communications and Mobile Computing Conference (IWCMC)*, (pp. 469-474), Paphos, Cyprus: IEEE.

Khan, A. A., Rehmani, M. H., & Reisslein, M. (2016). Cognitive radio for smart grids: Survey of architectures, spectrum sensing mechanisms, and networking protocols. *IEEE Communications Surveys and Tutorials*, *18*(1), 860–898. doi:10.1109/COMST.2015.2481722

Khan, M. H., & Barman, P. C. (2015). 5G-Future Generation Technologies Of Wireless Communication Revolution 2020. *American Journal of Engineering Research*, *4*, 206–215.

Kim, H., & Shin, K. G. (2010). In-Band Spectrum Sensing in IEEE 802.22 WRANs for Incumbent Protection. *IEEE Transactions on Mobile Computing*, *9*(12), 1767–1779. doi:10.1109/TMC.2010.169

Kim, K., Akbar, I. A., Bae, K. K., Um, J.-S., Spooner, C.-M., & Reed, J. H. (2007). Cyclostationary Approaches to Signal Detection and Classification in Cognitive Radio. In *Proceedings of IEEE Symposium on New Frontiers Dynamic Spectrum Access Networks* (pp. 212-215). Dublin, Ireland: IEEE. 10.1109/DYSPAN.2007.35

Koenig, W., Nolte, K., Gebert, J., Demestichas, P., Stavroulaki, V., & Saatsakis, A. (2009, September). Introducing cognitive systems in the B3G world: The E 3 approach to the" invisible" network of the future. In *Personal, Indoor and Mobile Radio Communications, 2009 IEEE 20th International Symposium on* (pp. 1-5). IEEE.

Kokare & Kamble. (2014). Spectrum Sensing Techniques in Cognitive Radio Cycle. *International Journal of Engineering Trends and Technology, 9*(1).

Kotsiantis, S. B. (2007). Supervised Machine Learning: A Review of Classification Techniques. *Informatica*, *31*(3), 249–268.

Kumar, R. (2014). Analysis of Spectrum Sensing Techniques in Cognitive Radio. *International Journal of Information and Computation Technology, 4*(4), 437-444.

Kumari. (2016). Spectrum Sensing Techniques for Cognitive Radio Networks: A Review. *International Journal of Innovative Research in Computer and Communication Engineering, 4*(3).

Kuperman, G. J., & Gibson, R. F. (2003). Computer physician order entry: Benefits, costs, and issues. *Annals of Internal Medicine*, *139*(1), 31. doi:10.7326/0003-4819-139-1-200307010-00010 PMID:12834316

Langley, P., & Simon, H. A. (1995). Applications of machine learning and rule induction. *Communications of the ACM*, *38*(11), 54–64. doi:10.1145/219717.219768

Lee, J.-H., Baek, J.-H., & Hwang, S.-H. (2008). Collaborative Spectrum Sensing using Energy Detector in Multiple Antenna System. *10th International Conference on Advanced Communication Technology ICACT, 1*, 427 - 430.

Lee, D. J., & Jang, M. S. (2009). Optimal spectrum sensing time considering spectrum handoff due to false alarm in cognitive radio networks. *IEEE Communications Letters*, *13*(12), 899–901. doi:10.1109/LCOMM.2009.12.091448

Lee, W. Y., & Akyildiz, I. F. (2008). Optimal spectrum sensing framework for cognitive radio networks. *IEEE Transactions on Wireless Communications*, *7*(10).

Lee, X. R., Chen, C. L., Chang, H. C., & Lee, C. Y. (2014). A 7.92 Gb/s 437.2 mW Stochastic LDPC Decoder Chip for IEEE 802.15.3c Applications. In *IEEE Transactions on Circuits and Systems I* (Vol. 62, pp. 507–5016). IEEE Circuits and Systems Society.

Lei, Z., & Chin, F. P. S. (2010). Sensing OFDM Systems Under Frequency-Selective Fading Channels. *IEEE Transactions on Vehicular Technology*, *59*(4), 1960–1968. doi:10.1109/TVT.2010.2042139

Le, T. N., Chin, W. L., & Chen, H. H. (2017). Standardization and Security for Smart Grid Communications Based on Cognitive Radio Technologies—A Comprehensive Survey. *IEEE Communications Surveys and Tutorials*, *19*(1), 423–445. doi:10.1109/COMST.2016.2613892

Liang, Y. C., Hoang, A. T., & Chen, H. H. (2008). Cognitive radio on TV bands: A new approach to provide wireless connectivity for rural areas. *IEEE Wireless Communications*, *15*(3).

Liang, Y. C., Zeng, Y., Peh, E. C., & Hoang, A. T. (2008). Sensing-throughput tradeoff for cognitive radio networks. *IEEE Transactions on Wireless Communications*, *7*(4), 1326–1337. doi:10.1109/TWC.2008.060869

Li, J., & Zhang, D. (2008). Reduced-complexity belief propagation decoding for low-density parity-check codes. *Electronics Letters*.

Lin, F., Qiu, R. C., Hu, Z., Hou, S., Browning, J. P., & Wicks, M. C. (2012). Generalized FMD Detection for Spectrum Sensing under Low Signal-to-Noise Ratio. *IEEE Communications Letters*, *16*(5), 604–607. doi:10.1109/LCOMM.2012.030512.112164

Ling-ling, Z., Jian-guo, H., & Cheng-kai, T. (2011). Novel Energy Detection Scheme in Cognitive Radio. *IEEE conference on Signal Processing, Communications and Computing*, 1-4.

Lin, J., Xiong, C., & Yan, Z. (2015). Reduced Complexity Belief Propagation Decoders for Polar Codes. In *IEEE Workshop on Signal Processing Systems* (pp. 1-6). Hangzhou, China: IEEE. 10.1109/SiPS.2015.7344984

Lin, Y. E., Liu, K. H., & Hsieh, H. Y. (2013). On Using Interference-Aware Spectrum Sensing for Dynamic Spectrum Access in Cognitive Radio Networks. *IEEE Transactions on Mobile Computing*, *12*(3), 461–474. doi:10.1109/TMC.2012.16

Liu, S. Q., Hu, B. J., & Wang, X. Y. (2012). Hierarchical cooperative spectrum sensing based on double thresholds energy detection. *IEEE Communications Letters*, *16*(7), 1096–1099. doi:10.1109/LCOMM.2012.050112.120765

Liu, X., Jia, M., Gu, X., & Tan, X. (2013). Optimal Periodic Cooperative Spectrum Sensing Based on Weight Fusion in Cognitive Radio Networks. *Sensors (Basel)*, *13*(4), 5251–5272. doi:10.3390130405251 PMID:23604027

Liu, X., & Sai Shankar, N. (2006). Sensing-based opportunistic channel access. *Mobile Networks and Applications*, *11*(4), 577–591. doi:10.100711036-006-7323-x

Loong, K. S., Kumar, S. S., & Tschofenig, H. (2014). Securing the internet of things: A standardization perspective. *IEEE Internet of Things Journal*, *1*(3), 265–275. doi:10.1109/JIOT.2014.2323395

Loyka, S. L. (2001). *Channel Capacity of MIMO Architecture Using the Exponential Correlation Matrix. IEEE Communications Letters*, *5*(9), 369–371.

Lpez-Bentez, M., & Casadevall, F. (2012). Improved energy detection spectrum sensing for cognitive radio. *IET Communications, 6*(8), 785.

Luan, Y., & Ren, G. (2015, January). The E-health Network Design. In *2015 International Conference on Network and Information Systems for Computers (ICNISC)*, (pp. 557-559). IEEE. 10.1109/ICNISC.2015.100

Lu, L., Zhou, X., Onunkwo, U., & Li, G. Y. (2012). Ten years of research in spectrum sensing and sharing in cognitive radio. *EURASIP Journal on Wireless Communications and Networking*, *2012*(1), 28. doi:10.1186/1687-1499-2012-28

Lunden, J., Koivunen, V., Huttunen, A., & Poor, H. V. (2007). Spectrum Sensing in Cognitive Radios Based on Multiple Cyclic Frequencies. In *Proceedings of IEEE International Conference on Cognitive Radio Oriented Wireless Network and Communications* (pp. 37-43). Orlando, FL: IEEE. 10.1109/CROWNCOM.2007.4549769

Ma, J., Li, G. Y., & Juang, B. H. (2009). Signal processing in cognitive radio. *Proceedings of the IEEE*, *97*(5), 805–823. doi:10.1109/JPROC.2009.2015707

Maleki, S., Pandharipande, A., & Leus, G. (2010, March). Two-stage spectrum sensing for cognitive radios. In *Acoustics Speech and Signal Processing (ICASSP), 2010 IEEE International Conference on* (pp. 2946-2949). IEEE. 10.1109/ICASSP.2010.5496149

Marcus, M. J. (2005). Unlicensed cognitive sharing of TV spectrum: The controversy at the federal communications commission. *IEEE Communications Magazine*, *43*(5), 24–25. doi:10.1109/MCOM.2005.1453413

Marques, P., Rodriguez, J., Delaere, S., Delahaye, P., Lecroart, B., Gundlach, M., . . . Noguet, D. (2013, June). Shared use of radio spectrum in the eu: From research projects to standards. In *Communications Workshops (ICC), 2013 IEEE International Conference on* (pp. 1249-1254). IEEE.

Medeisis, A., & Holland, O. (2014). State-of-the-art in policy and regulation of radio spectrum. In *Cognitive Radio Policy and Regulation* (pp. 1–48). Springer International Publishing. doi:10.1007/978-3-319-04022-6_1

Mehta, Kumar, & Saini. (2013). Comparison of Spectrum Sensing Techniques in Cognitive Radio Networks. *International Journal on Electronics & Communication Technology, 4*(3), 33-37.

Mehta, T., Kumar, N., & Saini, S. S. (2013). Comparison of Spectrum Sensing Techniques in Cognitive Radio Networks. *IJECT, 4*.

Mehta, T., Kumar, N., & Saini, S. S. (2013). Comparison of Spectrum Sensing Techniques in Cognitive Radio Networks. *International Journal of Electronics & Communication Technology, 4*(3), 33-37.

Merry, R. J. E., & Steinbuch, M. (2005). *Wavelet theory and applications.* literature study, Eindhoven University of Technology, Department of Mechanical Engineering, Control Systems Technology Group.

Migliore, M. D. (2014). A simple introduction to compressed sensing/sparse recovery with applications in antenna measurements. *IEEE Antennas & Propagation Magazine*, *56*(2), 14–26. doi:10.1109/MAP.2014.6837061

Mishra, S. M., Sahai, A., & Brodersen, R. W. (2006, June). Cooperative sensing among cognitive radios. In *Communications, 2006. ICC'06. IEEE International conference on* (Vol. 4, pp. 1658-1663). IEEE. 10.1109/ICC.2006.254957

Mishra, A. K., & Johnson, D. L. (2014). *White Space Communications: Advances, Developments and Engineering Challenges*. Springer.

Mitchell, T. (1997). *Machine Learning: A Guide to Current Research.* McGraw Hill.

Mitola Iii, J. (2001). Cognitive radio for flexible mobile multimedia communications. *Mobile Networks and Applications*, *6*(5), 435–441. doi:10.1023/A:1011426600077

Mitola, J. (2000). *Cognitive Radio: An Integrated Agent Architecture for Software Defined Radio* (PhD). KTH. Retrieved from http://www.citeulike.org/group/17835/article/9342875

Mitola, J. (2000). *Cognitive radio---an integrated agent architecture for software defined radio.* Academic Press.

Mitola, J., & Maguire, G. Q. (1999). Cognitive radio: making software radios more personal. *IEEE Personal Communications, 6*(4), 13-18.

Mody, A., Sherman, M., Martinez, R., Reddy, R., & Kiernan, T. (2008). Survey of IEEE standards supporting cognitive radio and dynamic spectrum access. *IEEE Military Communications Conference (MILCOM '08)*, 1 –7. 10.1109/MILCOM.2008.4753298

Mounika, Chandra, & Kumar. (2013). Spectrum Sensing Techniques and Issues in Cognitive Radio. *International Journal of Engineering Trends and Technology, 4*(4).

Musavian, L., & Aïssa, S. (2007). Ergodic and outage capacities of spectrum sharing systems in fading channels. *Proc. IEEE GLOBECOM*, 3327–3331.

Musavian, L., & Aissa, S. (2009). Fundamental capacity limits of cognitive radio in fading environments with imperfect channel information. *IEEE Transactions on Communications*, *57*(11), 3472–3480. doi:10.1109/TCOMM.2009.11.070410

Nallagonda, Sanjay, Sumit, Ferrari, & Raheli. (2013a). Cooperative spectrum sensing with censoring of cognitive radios in Rayleigh fading under majority logic fusion. *Proc. IEEE NCC*, 1-5. 10.1109/NCC.2013.6487926

Nallagonda, Sanjay, Sumit, Ferrari, & Raheli. (2013b). Performance of MRC fusion based cooperative spectrum sensing with censoring of cognitive radios in Rayleigh fading channel. *Proc. IEEE IWCMC*, 30- 35.

Nallagonda, S., Roy, S. D., & Kundu, S. (2013). Performance evaluation of cooperative spectrum sensing scheme with censoring of cognitive radios in Rayleigh fading channel. *Wireless Personal Communications*, *70*(4), 1409–1424. doi:10.100711277-012-0756-6

Nallagonda, Srinivas, & Sanjay, Sumit, Ferrari, & Raheli. (2017). Censoring-based Cooperative Spectrum Sensing with Improved Energy Detectors and Multiple Antennas in Fading Channels. *IEEE Transactions on Aerospace and Electronic Systems*.

Naonworks. (n.d.). *Smart Grid*. Retrieved from http://www.naonworks.com/old/inc_html/sub2_3.html

Narendar, M., Vinod, A. P., Madhukumar, A. S., & Krishna, A. K. (2010, April). An algorithm for spectrum sensing in cognitive radio using tree-structured filter bank. In *Telecommunications (ICT), 2010 IEEE 17th International Conference on* (pp. 418-424). IEEE. 10.1109/ICTEL.2010.5478782

Nason, G. P., & Silverman, B. W. (1995). The stationary wavelet transform and some statistical applications. *Lecture Notes in Statistics*, 281-281.

NDTV. (2016). *India's Total Power Generation Capacity Crosses 300 GW Mark*. Retrieved from http://www.ndtv.com/india-news/indias-total-power-generation-capacity-crosses-300-gw-mark-

Needell, D. (2009, November). Noisy signal recovery via iterative reweighted l1-minimization. In *Signals, Systems and Computers, 2009 Conference Record of the Forty-Third Asilomar Conference on* (pp. 113-117). IEEE. 10.1109/ACSSC.2009.5470154

Neel, J. O. D. (2006). *Analysis and design of cognitive radio networks and distributed radio resource management algorithms* (Doctoral dissertation).

Networking, C. V. (2017). *Global mobile data traffic forecast update, 2016–2021*. Cisco white paper.

Ouattara, D., Krief, F., Chalouf, M. A., & Hamdi, O. (2012, November). Spectrum sensing improvement in cognitive radio networks for real-time patients monitoring. In *International Conference on Wireless Mobile Communication and Healthcare* (pp. 179-188). Springer.

Pandharipande, A., & Linnartz, J. P. (2007, June). Performance analysis of primary user detection in a multiple antenna cognitive radio. In *Communications, 2007. ICC'07. IEEE International Conference on* (pp. 6482-6486). IEEE. 10.1109/ICC.2007.1072

Pandi, N., & Kumar, A. (2017). *A Review on Cognitive Radio for Next Generation Cellular Network and its Challenges*. Academic Press.

Peha, J. A. (2005). Approaches to spectrum sharing. *IEEE Communications Magazine*, *43*(2), 10–12. doi:10.1109/MCOM.2005.1391490

Peh, E., & Liang, Y.-C. (2007). Optimization for Cooperative Sensing in Cognitive Radio Networks. In *Proceedings of IEEE Wireless Communications and Networking Conference* (pp. 27-32). Paris, France: IEEE. 10.1109/WCNC.2007.11

Pereira, V., & Sousa, T. (2004). *Evolution of Mobile Communications: from 1G to 4G*. Department of Informatics Engineering of the University of Coimbra.

Peterson, W. W., & Weldon, E. G. (1972). *Error-correcting codes*. MIT Press.

Phunchongharn, P., Hossain, E., & Camorlinga, S. (2010). Cognitive radio for pervasive healthcare. *Cognitive Radio Networks, Architectures, Protocols, and Standards*, 355-388.

Phunchongharn, P., Hossain, E., Niyato, D., & Camorlinga, S. (2010). A cognitive radio system for e-health applications in a hospital environment. *IEEE Wireless Communications*, *17*(1), 20–28. doi:10.1109/MWC.2010.5416346

Pillay, N., & Xu, H. J. (2013). Eigen value-based spectrum sensing using the exact distribution of the maximum eigenvalue of a Wishart matrix. *IET Signal Processing*, *7*(9), 833–842. doi:10.1049/iet-spr.2012.0320

Polo, Y. L. (2011). *Compressive wideband spectrum sensing for cognitive radio applications*. TU Delft, Electrical Engineering, Mathematics and Computer Science, Telecommunications.

Poole, I. (2017). *IEEE 802.22 Physical Layer / Air Interface*. Available at:http://www.radio-electronics.com/info/wireless/ieee-802-22/physical-layer-air-interface.php

Poole, I. (2017). *IEEE 802.22 rf-technology-design/cognitive-radio-cr/cooperative-spectrum-sensing*. Available at: http://www.radio-electronics.com/info/rf-technology-design/cognitive-radio-cr/cooperative-spectrum-sensing.php

Prasad, B. Sanjay, & Sumit. (2014). Outage and SEP of secondary user with imperfect channel estimation and primary user interference. *Electronics, Computing and Communication Technologies (IEEE CONECCT), 2014 IEEE International Conference on*, 1-6.

Prasad, B., Bhowmick, A., Dhar Roy, S., & Kundu, S. (2016). Performance of cognitive relay network with novel hybrid spectrum access schemes with imperfect CSI. *International Journal of Communication Systems*, *29*(11), 1761–1776. doi:10.1002/dac.3142

Prinima & Jyoti. (2016). Evolution of Mobile Communication Network: from 1G to 5G. *International Journal of Innovative Research in Computer and Communication Engineering, 4*(4).

Qin, Z., Gao, Y., & Parini, C. G. (2016). Data-assisted low complexity compressive spectrum sensing on real-time signals under sub-nyquist rate. *Proc. IEEE Trans. Wireless Commun., 15*(2), 1174–1185. 10.1109/TWC.2015.2485992

Qin, M., Guo, J., Bhatia, A., Fàbregas, A. G. I., & Siegel, P. H. (2017). Polar code constructions based on LLR evolution. *IEEE Communications Letters*, *21*(6), 1221–1224. doi:10.1109/LCOMM.2017.2656126

Qiu, R. C., Hu, Z., Chen, Z., Guo, N., Ranganathan, R., Hou, S., & Zheng, G. (2011). Cognitive radio network for the smart grid: Experimental system architecture, control algorithms, security, and microgrid testbed. *IEEE Transactions on Smart Grid*, *2*(4), 724–740. doi:10.1109/TSG.2011.2160101

Quan, Z. (2009, March). Optimal Multiband Joint Detection for Spectrum Sensing in Cognitive Radio Networks. *IEEE Transactions on Signal Processing*, *57*(3), 1128–1140. doi:10.1109/TSP.2008.2008540

Quan, Z., Shellhammer, S. T., Zhang, W., & Sayed, A. H. (2009). Spectrum Sensing by Cognitive Radios at Very Low SNR. In *Proceedings of IEEE Global Communications* (pp. 1-6). Honolulu, HI: IEEE. 10.1109/GLOCOM.2009.5426262

Ramachandran, G. S., Proença, J., Daniels, W., Pickavet, M., Staessens, D., Huygens, C., ... Hughes, D. (2016). Hitch Hiker 2.0: A binding model with flexible data aggregation for the Internet-of-Things. *Journal of Internet Services and Applications*, *7*(1), 2–15. doi:10.118613174-016-0047-7

Rashidi, M., Haghighi, K., Owrang, A., & Viberg, M. (2011). A wideband spectrum sensing method for cognitive radio using sub-Nyquist sampling. In *Digital Signal Processing Workshop and IEEE Signal Processing Education Workshop* (pp. 30–35). Sedona, AZ: IEEE. doi:10.1109/DSP-SPE.2011.5739182

Razzaque, M. A., Milojevic-Jevric, M., Palade, A., & Clarke, S. (2016). Middleware for internet of things: A survey. *IEEE Internet of Things Journal*, *3*(1), 70–95. doi:10.1109/JIOT.2015.2498900

Richard, A. (2007). *Health care unplugged: The evolving role of wireless technology*. I-health Reports.

Rondeau, T. W. (2007). *Application of artificial intelligence to wireless communications* (PhD thesis). Virginia Polytechnic Institute and State University.

Russell, S. J., Norvig, P., Canny, J. F., Malik, J. M., & Edwards, D. D. (1995). Artificial intelligence: A modern approach. Prentice Hall.

Safavian, S. R., & Landgrebe, D. (1991). A survey of decision tree classifier methodology. *IEEE Transactions on Systems, Man, and Cybernetics*, *21*(3), 660–674. doi:10.1109/21.97458

Sahai, A., & Danijela, C. (2005). Spectrum sensing: fundamental limits and practical challenges. *In Proc. International Symposium on New Frontiers in Dynamic Spectrum Access Networks.2005. DySPAN 2005 first IEEE international symposium on* (vol. 41). IEEE.

Sahai, A., & Cabric, D. (2005). Spectrum sensing: fundamental limits and practical challenges. In *Proceedings of IEEE Symposium on New Frontiers Dynamic Spectrum Access Networks*. Baltimore, MD: IEEE.

Sahai, Hoven, Mishra, & Tandra. (2006). Fundamental tradeoffs in robust spectrum sensing for opportunistic frequency reuse. *IEEE Journal on Selected Areas in Communications*.

Saleem, Y., Crespi, N., Rehmani, M. H., & Copeland, R. (2017). *Internet of Things-aided Smart Grid: Technologies, Architectures, Applications, Prototypes, and Future Research Directions*. arXiv preprint arXiv:1704.08977

Sanctis, M. D. (2016). Satellite communications supporting internet of remote things. *IEEE Internet of Things Journal*, *3*(1), 113–123. doi:10.1109/JIOT.2015.2487046

Sarkar, C. (2012). *DIAT: A scalable distributed architecture for IoT*. Academic Press.

Sengupta, S., Chatterjee, M., & Kwiat, K. (2009). *Interference aware spectrum allocation in IEEE 802.22 wireless mesh networks*. Air Force Research Lab.

Shah, M. A., Zhang, S., & Maple, C. (2013). Cognitive radio networks for Internet of Things: Applications, challenges and future. In *Proceedings of IEEE 19th International Conference Automation and Computing (ICAC)*, (pp. 1-6). London: IEEE.

Shannon, C. E. (1949). Communication in the presence of noise. *Proceedings of the IRE*, *37*(1), 10–21. doi:10.1109/JRPROC.1949.232969

Sharma, S. K., Chatzinotas, S., & Ottersten, B. (2013). Eigenvalue-Based Sensing and SNR Estimation for Cognitive Radio in Presence of Noise Correlation. *IEEE Transactions on Vehicular Technology*, *62*(8), 3671–3683. doi:10.1109/TVT.2013.2260834

Sharma, S. K., Chatzinotas, S., & Ottersten, B. (2014). Maximum Eigenvalue Detection for Spectrum Sensing Under Correlated Noise. In *Proceedings of IEEE International Conference on Acoustic, Speech, and Signal Processing* (pp. 7268-7272). Florence, Italy: IEEE. 10.1109/ICASSP.2014.6855011

Sharswat, V., Godara, P., Bishnu, A., & Bhatia, V. (in press). Measurement and Identification of White Space using Low Cost SDR for Rural Wireless Broadband. In *IEEE International Conference on Wireless Communications, Vehicular Technology, Information Theory, Aerospace and Electronic System*. Hyderabad, India: IEEE.

Shellhammer, S. J. (2008). Spectrum sensing in IEEE 802.22. *IAPR Wksp. Cognitive Info. Processing*, 9-10.

Shellhammer, S., Tawil, V., Chouinard, G., Muterspaugh, M., & Ghosh, M., (2006). *Spectrum Sensing Simulation Model*. IEEE 802.22-06/0028r10, 2006.

Shen, Q., Liu, J., Yu, H., Ma, Z., Li, M., Shen, Z., & Chen, C. (2013, August). Adaptive cognitive enhanced platform for WBAN. In *2013 IEEE/CIC International Conference on Communications in China (ICCC)*, (pp. 739-744). IEEE. 10.1109/ICCChina.2013.6671208

Shojaeifard, A., Zarringhalam, F., & Shikh-Bahaei, M. (2011). Joint Physical Layer and Data Link Layer Optimization of CDMA-Based Networks. *IEEE Transactions on Wireless Communications*, *10*(10), 3278–3287. doi:10.1109/TWC.2011.080311.101301

Shuaib, K., Barka, E., Al Hussien, N., Abdel-Hafez, M., & Alahmad, M. (2016). Cognitive radio for smart grid with security considerations. *Computers*, *5*(2), 7. doi:10.3390/computers5020007

Sobron, I., Diniz, P. S. R., Martins, W. A., & Velez, M. (2015). Energy Detection Technique for Adaptive Spectrum Sensing. *IEEE Transactions on Communications*, *63*(3), 617–627. doi:10.1109/TCOMM.2015.2394436

Song, C., Alemseged, Y. D., Tran, H. N., Villardi, G., Sun, C., Filin, S., & Harada, H. (2010). *Adaptive Two Thresholds Based Energy Detection For Cooperative Spectrum Sensing*. Paper presented at the 2010 7th IEEE Consumer Communications and Networking Conference.

Sridhar, B., & Srinivasulu, D. T. (2013). A Novel High Resolution Spectrum Sensing Algorithm for Cognitive Radio Applications. *IOSR Journal of Electronics and Communication Engineering*, *8*(4), 30-38. Retrieved from http://www.radio-electronics.com/info/rf-technology-design/cognitive-radio-cr/cooperative-spectrum-sensing.php

Steinwart, I., & Christmann, A. (2008). *Support vector machines*. Springer.

Stevenson C. R., Cordeiro, C., Sofer, E., & Chouinard, G., (2006). *Functional Requirements for the 802.22 WRAN Standard*. IEEE 802.22-05/0007r47, Jan. 2006.

Stevenson, C. R., Chouinard, G., Lei, Z., Hu, W., Shellhammer, S. J., & Caldwell, W. (2009). IEEE 802.22: The first cognitive radio wireless regional area network standard. *IEEE Communications Magazine*, *47*(1), 130–138. doi:10.1109/MCOM.2009.4752688

Stolte, N. (2002). *Rekursive codes mit der Plotkin-Konstruktion und ihre Decodierung* (Unpublished doctoral dissertation). Technische Universität Darmstadt, Germany.

Stotas, S., & Nallanathan, A. (2011, October). Enhancing the capacity of spectrum sharing cognitive radio networks. *IEEE Transactions on Vehicular Technology*, *60*(8), 3768–3779. doi:10.1109/TVT.2011.2165306

Subhedar & Birajdar. (2011). Spectrum sensing techniques in cognitive radio networks: A survey. *International Journal of Next-Generation Networks, 3*(2).

Subhedar & Birajdar. (2011). Spectrum Sensing Techniques in Cognitive Radio Networks: A Survey. *International Journal of Next-Generation Networks*, *3*(2).

Subhedar, M., & Birajdar, G. (2011, June). Spectrum sensing techniques in cognitive radio networks: A survey. *International Journal of Next-Generation Networks*, *3*(2), 37–51. doi:10.5121/ijngn.2011.3203

Sun, H., Nallanathan, A., Wang, C.-X., & Chen, Y. (2013). Wideband spectrum sensing for cognitive radio networks: A survey. IEEE Wireless Communication, 20(2), 74–81.

Sun, C., Zhang, W., & Letaief, K. B. (2007). Cluster-Based Cooperative Spectrum Sensing in Cognitive Radio Systems. In *Proceedings of IEEE International Conference on Communications* (pp. 2511-2515). Glasgow, UK: IEEE. 10.1109/ICC.2007.415

Sun, H., Laurenson, D. I., & Wang, C. X. (2010). Computationally tractable model of energy detection performance over slow fading channels. *IEEE Communications Letters*, *14*(10), 924–926. doi:10.1109/LCOMM.2010.090710.100934

Sun, H., Nallanathan, A., Wang, C.-X., & Chen, Y. (2013, April). Wideband spectrum sensing for cognitive radio networks: A survey. *IEEE Wireless Communications*, *20*(2), 74–81. doi:10.1109/MWC.2013.6507397

Suraweera, H. A., Smith, P. J., & Shafi, M. (2010). Capacity limits and performance analysis of cognitive radio with imperfect channel knowledge. *IEEE Transactions on Vehicular Technology*, *59*(4), 1811–1822. doi:10.1109/TVT.2010.2043454

Sutton, P. D., Lotze, J., Nolan, K. E., & Doyle, L. E. (2007). Cyclostationary signature detection in multipath Rayleigh fading environments. *Proceedings of IEEE International conference on Cognitive Radio Oriented Wireless Networks and Communication*. 10.1109/CROWNCOM.2007.4549833

Sutton, R. S., & Barto, A. G. (1998). *Reinforcement learning: An introduction*. Cambridge Univ Press.

Taherpour, A., Nasiri-Kenari, M., & Gazor, S. (2010). Multiple antenna spectrum sensing in cognitive radios. *IEEE Transactions on Wireless Communications*, *9*(2), 814–823. doi:10.1109/TWC.2009.02.090385

Tandra, R., & Sahai, A. (2005, June). Fundamental limits on detection in low SNR under noise uncertainty. In *Wireless Networks, Communications and Mobile Computing, 2005 International Conference on* (Vol. 1, pp. 464-469). IEEE. 10.1109/WIRLES.2005.1549453

Tandra, R., & Sahai, A. (2008). SNR walls for signal detection. *IEEE Journal of Selected Topics in Signal Processing*, *2*(1), 4–17. doi:10.1109/JSTSP.2007.914879

Tang, H. (2005, November). Some physical layer issues of wide-band cognitive radio systems. In *New frontiers in dynamic spectrum access networks, 2005. DySPAN 2005. 2005 first IEEE international symposium on* (pp. 151-159). IEEE.

Tawil, V. (2006). *51 Captured DTV Signal*. Available: http://grouper.ieee.org/groups/802/22/Meetingsdocuments/2006May

Tian, Z., & Giannakis, G. (2007). Compressive Sensing for Wideband Cognitive Radios. *Proc. IEEE Int'l. Conf. Acoustics, Speech, and Sig.*, 1357–60.

Tian, Z., & Giannakis, G. B. (2006, June). A wavelet approach to wideband spectrum sensing for cognitive radios. In *Cognitive Radio Oriented Wireless Networks and Communications, 2006. 1st International Conference on* (pp. 1-5). IEEE. 10.1109/CROWNCOM.2006.363459

Tian, Z., & Giannakis, G. B. (2007, April). Compressed sensing for wideband cognitive radios. In *Acoustics, Speech and Signal Processing, 2007. ICASSP 2007. IEEE International Conference on* (Vol. 4, pp. IV-1357). IEEE. 10.1109/ICASSP.2007.367330

Tian, Z., Leus, G., & Lottici, V. (2010). Compressed Sensing Techniques for Dynamic Resource Allocation in Wideband Cognitive Networks. *Proc. IEEE International Workshop on Signal Processing Advances in Wireless Communications (SPAWC)*. 10.1109/SPAWC.2010.5670984

Tomar, G., Bagwari, A., & Kanti, J. (Eds.). (2016). *Introduction to Cognitive Radio Networks and Applications*. CRC Press. doi:10.1201/9781315367545

Treichler, Davenport, Laska, & Baraniuk. (2011). Dynamic range and compressive sensing acquisition receivers. *Proceedings Defense Applications Signal Processing (DASP)*, 1–7.

Tropp, J. A., & Gilbert, A. C. (2007). Signal recovery from random measurements via orthogonal matching pursuit. *IEEE Transactions on Information Theory*, *53*(12), 4655–4666. doi:10.1109/TIT.2007.909108

Tsutsu, H., & Morikawa, Y. (2012). An lp norm minimization using auxiliary function for compressed sensing. *Proc. IMECS*.

US Food and Drug Administration. (2014). *Radiofrequency wireless technology in medical devices: guidance for industry and food and drug administration staff*. Author.

Visotsky, E., Kuffner, S., & Peterson, R. (2005). On Collaborative Detection of TV Transmissions in Support of Dynamic Spectrum Sharing. In *Proceedings of IEEE Symposium on New Frontiers Dynamic Spectrum Access Networks* (pp. 338-345). Baltimore, MD: IEEE. 10.1109/DYSPAN.2005.1542650

Vladeanu, C., Nastase, C.-V., & Martian, A. (2016). Energy Detection Algorithm for Spectrum Sensing Using Three Consecutive Sensing Events. *IEEE Wireless Communications Letters*, *5*(3), 284–287. doi:10.1109/LWC.2016.2543723

Wang, Ghosh & Challapali. (2011). Emerging Cognitive Radio Applications: A Survey. *IEEE Communications Magazine*, 74-81.

Wang, W. (2009). Spectrum Sensing for Cognitive Radio. *Proceedings of the IEEE Computer Society*, 110-112.

Wang, B., & Liu, K. J. R. (2011). Advances in Cognitive Radio Networks: A Survey. *IEEE Journal of Selected Topics in Signal Processing*, *5*(1), 5–23. doi:10.1109/JSTSP.2010.2093210

Wang, H., Yang, E. H., Zhao, Z., & Zhang, W. (2009). Spectrum sensing in cognitive radio using goodness of fit testing. *IEEE Transactions on Wireless Communications*, *8*(11), 5427–5430. doi:10.1109/TWC.2009.081586

Wang, J., & Zhang, Q. T. (2009, April). A multitaper spectrum based detector for cognitive radio. In *Wireless Communications and Networking Conference, 2009. WCNC 2009. IEEE* (pp. 1-5). IEEE. 10.1109/WCNC.2009.4917513

Wei, L., & Tirkkonen, O. (2011). Analysis of Scaled Largest Eigenvalue Based Detection for Spectrum Sensing. In *Proceedings of IEEE International Conference on Communications* (pp. 1-5). Kyoto, Japan: IEEE. 10.1109/icc.2011.5962520

Weingart, T., Sicker, D. C., & Grunwald, D. (2007, August). A statistical method for reconfiguration of cognitive radios. *IEEE Wireless Communications*, *14*(4), 34–40. doi:10.1109/MWC.2007.4300981

Wells, N. (2011). *DVB-T2 in relation to the DVB-x2 Family of Standards*. Paper presented at the ATSC Symposium on Next Generation Broadcast Television, Rancho Las Palmas, Rancho Mirage, CA.

Wikimedia Foundation Inc. (2008). *Software Defined Radio*. Wikipedia.

Wu, Q., Ding, G., Xu, Y., Feng, S., Du, Z., Wang, J., & Long, K. (2014). Cognitive internet of things: A new paradigm beyond connection. *IEEE Internet of Things Journal*, *1*(2), 129–143. doi:10.1109/JIOT.2014.2311513

Y¨ucek, T., & Arslan, H¨. (2009). *A Survey of Spectrum Sensing Algorithms for Cognitive Radio Applications. IEEE Communications Surveys & Tutorials, 11(1).*

Yang, C. Y., & Ku, M. K. (2008). LDPC coded OFDM modulation for high spectral efficiency transmission. *4th European Conference on Circuits and Systems for Communications*. 10.1109/ECCSC.2008.4611693

Yang, Z., Ping, S., Aijaz, A., & Aghvami, A. H. (2016). A Global Optimization-Based Routing Protocol for Cognitive-Radio-Enabled Smart Grid AMI Networks. *IEEE Systems Journal*.

Yu & Zhang. (2004). Study of Artificial Neural Network Model for Forecasting Urban Water Demand. *Journal of Zhejiang University, 38*(9).

Yucek & Arslan. (2009). A Survey of Spectrum Sensing Algorithms for Cognitive Radio Applications. *Proceedings of the IEEE Communications Surveys & Tutorials, 11*(1), 116-130.

Yucek, T., & Arslan, H. (2009). A Survey of Spectrum Sensing Algorithms for Cognitive Radio Applications. *IEEE Communications Surveys and Tutorials*, *11*(1), 116–130. doi:10.1109/SURV.2009.090109

Yu, R., Zhang, Y., Gjessing, S., Yuen, C., Xie, S., & Guizani, M. (2011). Cognitive radio based hierarchical communications infrastructure for smart grid. *IEEE Network*, *25*(5), 6–14. doi:10.1109/MNET.2011.6033030

Zanella, A. (2014). Internet of things for smart cities. *IEEE Internet of Things Journal, 1*(1), 22-32.

Zeng, Y., & Liang, Y. C. (2007, September). Maximum-minimum eigenvalue detection for cognitive radio. In *Personal, Indoor and Mobile Radio Communications, 2007. PIMRC 2007. IEEE 18th International Symposium on* (pp. 1-5). IEEE.

Zeng, Y., Liang, Y. C., & Chia, M. W. (2011, May). Edge based wideband sensing for cognitive radio: Algorithm and performance evaluation. In *New Frontiers in Dynamic Spectrum Access Networks (DySPAN), 2011 IEEE Symposium on* (pp. 538-544). IEEE.

Zeng, Y., Koh, C. L., & Liang, Y.-C. (2008). Maximum Eigenvalue Detection: Theory and Application. In *Proceedings of IEEE International Conference on Communications* (pp. 4160-4164). Beijing, China: IEEE.

Zeng, Y., & Liang, Y. C. (2009). Eigenvalue-based spectrum sensing algorithms for cognitive radio. *IEEE Transactions on Communications*, *57*(6).

Zeng, Y., & Liang, Y. C. (2009). Spectrum-sensing algorithms for cognitive radio based on statistical covariances. *IEEE Transactions on Vehicular Technology*, *58*(4), 1804–1815.

Zeng, Y., & Liang, Y.-C. (2009). Eigenvalue–Based Spectrum Sensing Algorithms for Cognitive Radio. *IEEE Transactions on Communications*, *57*(6), 1784–1793. doi:10.1109/TCOMM.2009.06.070402

Zeng, Y., & Liang, Y.-C. (2009). Spectrum-Sensing Algorithms for Cognitive Radio Based on Statistical Covariances. *IEEE Transactions on Vehicular Technology*, *58*(4), 1804–1815. doi:10.1109/TVT.2008.2005267

Zhang & Lysaker. Dynamic Spectrum Access in Cognitive Radio Wireless Networks. *Proceeding of IEEE International Conference on Communications,* 4927 – 4932.

Zhang, B., Shen, H., Yin, B., Lu, L., Chen, D., Wang, T., … Kishiyama, Y. (2016). A 5G Trial of Polar Code. IEEE Globecom Workshops, 1-6.

Zhang, K., & Huang, X. (2009). A Low-Complexity Rate-Compatible LDPC Decoder. In *Conference Record of the Forty-Third Asilomar Conference on Signals, Systems and Computers* (pp.749-753). Pacific Grove, CA: Academic Press.

Zhang, P., & Qiu, R. (2013). GLRT-Based Spectrum Sensing with Blindly Learned Feature under Rank-1 Assumption. *IEEE Transactions on Communications*, *61*(1), 87–96. doi:10.1109/TCOMM.2012.100912.120162

Zhang, W., & Letaief, K. B. (2008). Cooperative spectrum sensing with transmit and relay diversity in cognitive radio networks. *IEEE Transactions on Wireless Communications*, *7*(12), 4761–4766. doi:10.1109/T-WC.2008.060857

Zhang, Y., Zheng, J., & Chen, H. H. (Eds.). (2016). *Cognitive radio networks: architectures, protocols, and standards.* CRC Press.

Zhang, Z., & Xie, X. (2007). Intelligent cognitive radio: Research on learning and evaluation of CR based on neural network. *Proc. ITI 5th ICICT*, 33–37. 10.1109/ITICT.2007.4475612

Zhao, Y., Morales, L., Gaeddert, J., Bae, K. K., Um, J. S., & Reed, J. H. (2007, April). Applying radio environment maps to cognitive wireless regional area networks. In *New Frontiers in Dynamic Spectrum Access Networks, 2007. DySPAN 2007. 2nd IEEE International Symposium on* (pp. 115-118). IEEE. 10.1109/DYSPAN.2007.22

Zhou, L., Huang, W. C., Zhao, R., & He, Y. C. (2017). A Non-Greedy Puncturing Method for Rate-Compatible LDPC Codes. *Journal of Communications and Networks (Seoul)*, *19*(1), 32–40.

About the Contributors

Ashish Bagwari is Head of Electronics and Communication Engineering Department, Women's Institute of Technology (WIT) (Institute of State Government) Dehradun, Constituent College of Uttarakhand Technical University (State Government Technical University), Dehradun, India. He has more than 8.0 years' experience in industry, academic, and research. Dr. Bagwari is received various awards as the Best WIT Faculty award in 2013, 2015, Best Project Guide Award in 2015, Corps of Electrical and Mechanical Engineers Prize from the Institution of Engineers, India (IEI) in December 2015. Dr. Bagwari has published more than 65 research papers in various international journals (including ISI/SCI indexed) and IEEE international conferences. He is a member of IEEE-USA, MIR Laboratories India, IETE, ACM, and IAENG. He has also been an editor, an advisor, and a reviewer for several well-known international journals published by IEEE, Elsevier, Springer, etc. Dr. Bagwari organized several conferences and has been invited as a key speaker in various workshops/conferences. Dr. Bagwari has filed one patent. He has written two books and a few book chapters in CRC Press/other reputed publications.

Jyotshana Bagwari received the Bachelor of Technology degree (Hons.) and Master of Technology degree in Computer Science & Engineering from Uttarakhand Technical University, Dehradun. She has published several research papers in various reputed international SCI/ISI indexed journals and conferences including IEEE international conferences as well.

Geetam Singh Tomar has more than 30 years of working experience and has published good number of patents, papers and books. Currently working as Head of the Institute at Uttarakhand Government Institute.

* * *

Mina Abd-el-Malek is Professor of Engineering Mathematics. PhD in Applied Mathematics, Windsor University, Ontario, Canada. Fellow University of Liverpool, UK since 1986.

Karm Veer Arya did his Masters in Electrical Engineering from Indian Institute of Science (IISc) Bangalore, India and earned PhD in Computer Science & Engineering from Indian Institute of Technology Kanpur (IITK), Kanpur, India. His research area includes Image Processing, Biometrics, Information Security, and Wireless Adhoc Networks. Currently, he is Associate Professor at ABV-Indian Institute of Information Technology and Management, Gwalior, Madhya Pradesh, India. He has published more than 150 papers in international journals and peer reviewed conferences around the world. He has guided 08 Ph. D. and 82 Master's Thesis. He is on the Editorial board of many International journals. He is the member of organizing committee and programme committee for various national and international conferences and workshops. He has edited the Proceedings of International Conference on Industrial & Information Systems (ICIIS2014) held on 15th-17th, Dec. 2014 at Gwalior, India. Dr. Arya has co-authored a book titled "Routing in Wireless Sensor Networks: A Clustering based Approach" (LAP LAMBART, Germany). He is Senior Member of IEEE (USA), Fellow of IETE (INDIA), Member of ACM and life member of ISTE (INDIA).

Reda Benkhouya was invited as a visiting scholar at Toulouse institute of computer science research (IRIT), France, in 2004. Since 2006 he has been with Etisalat where he is currently an IT operations engineer. He is now pursuing the Ph.D. degree at Ibn Tofail University. His research interest includes channel coding and cognitive radio.

Vimal Bhatia received his Ph.D. from Institute for Digital Communications at The University of Edinburgh (UoE), UK in 2005. He is currently an Associate Professor with IIT Indore. His research interests are in the algorithms and solutions for future communication and optical systems.

Abhijit Bhowmick received his B.E (Hons) degree in Electronics and Telecomm. Engineering in 2002 from Burdwan University, West Bengal, India and M.Tech. degree in Telecommunication Engineering in 2009 from NIT Durgapur and PhD in Wireless Communication Engineering in 2016 from NIT, Durgapur, respectively. He worked for Cubix Control System Pvt. Ltd. from 2004 to 2006. After that he joined the Department of Electronics and Comm. Engineering, Bengal College of Engg. and Tech., Durgapur as a Lecturer in 2006. He joined VIT University, Vellore, TN, India in the School of Electronics Engineering (SENSE) as an Assistant Professor (SG) in June, 2016 and now he is serving as an Associate Professor there. As of today, he has published more than thirty five (35) research papers in various journals and conferences. He is a reviewer of several IEEE, Springer, Wiley and Elsevier journals.

Abhijeet Bishnu received his B.E. degree in Electronics and Communication Engineering from Technocrat Institute of Technology, Bhopal, India in 2010. He also received his M.E. degree in Electronics and Telecommunication Engineering from S.G.S.I.T.S. Indore, India in 2013. He is currently pursuing PhD from Indian Institute of Technology Indore, India. His research interests include OFDM system and signal processing for wireless communications.

Rajib Biswas completed his B. Sc from Dibrugarh University and attained his Masters from the same university. He pursued his Doctoral work from Council of Scientific and Industrial Research lab of NEIST in the area of instrumentation. From 2010, he is serving as Faculty member in Department of Physics, Tezpur University, India and actively engaged in Academics and Research.

Idriss Chana received the Ph.D. degree from Mohamed V University, Morocco in 2013. He is currently a full professor at Moulay Ismail University, Morocco. He is member of IEEE. His research interests include information and communication technologies. A large part of his research projects are related to Error correction codes and Turbo codes. Idriss Chana published more than 20 papers in major journals and conferences in Information Theory.

Sanjay Dhar Roy is an Assistant Professor at NIT Durgapur. He is working in the domain of wireless communication. He has published more than 100 papers in various reputed journals and conferences.

Said E. El-Khamy received the Ph.D. degree from the University of Massachusetts, Amherst, USA, in 1971. He joined the teaching staff of the Department of Electrical Engineering, Faculty of Engineering, Alexandria University, Alexandria, Egypt, since 1972 and he is currently an Emeritus Professor. He published more than three hundred scientific papers in national and international conferences and journals and took part in the organization of many local and international conferences. His Current research areas of interest include Wireless Communications, Wave Propagation in different media, Smart Antenna Arrays, Space-Time Coding, Modern Signal Processing Techniques. Image Processing and Watermarking, and Cognitive Radios. Prof. El-Khamy is a Fellow member of the IEEE since 1999. He has earned many national and international research awards and was awarded the IEEE Region 8 Volunteer award for 2010. He is a Fellow of the Electromagnetics Academy. Prof. El-Khamy is the president of Egypt's URSI National Radio Science Committee (URSI-NRSC).

Rajasi Gore did her Masters in Advance Networks from ABV-IIITM Gwalior, India and has a bachelor degree in Computer Science and Engineering. Presently pursuing her doctoral degree in Computer Science and Engineering at MNNIT Allahabad, India.

Shashank Gupta received his B.Tech. in Computer Science and Engineering from Jaypee Institute of Information Technology, Noida, India in 2017. He is pursuing his M.Tech. from the same college and is expected in 2018. His research interests include Machine Learning, Artificial Intelligence, Nature Inspired Computing, and Cognitive Radio Networks.

Youssef Hadi received his B. Sc. degree in Computer Sciences on June 2001 from the Faculty of Sciences and Techniques of Moulay Ismail University, Errachidia, Morocco, and his Master Sc. degree in Software Engineering and Advanced Technologies of Information on December 2003 at the Mohammedia School of Engineering (EMI) of Mohammed V University, Rabat, Morocco. In June 2008, he received his Ph.D. degree in Computer Sciences and Telecommunications from the Faculty of Sciences of the Mohammed V University, Rabat, Morocco. From 2009 to 2015, he was an Assistant Professor at Ibn Tofail University, Faculty of Sciences with LARIT laboratory. Currently he is an Associate Professor at the same university and he is the Vice Director of MISC Laboratory (Modeling of Information and Communication Systems). Also, he is a member of IEEE and he is Student Branch Member of Ibn Tofail University. Youssef Hadi research interests include Software Engineering, Model Driven Engineering, Cloud Computing and Signal Processing.

Pooja Joshi received the B.E degree from BRCM College(MDUniversity,Rohtak),M.Tech degree in computer science from Dehradun Institute of Technology (UTU, Dehradun) in 2008 and 2010 respectively. She is currently working as Assistant Professor in computer science Engineering Department at Women's Institute of Technology (WIT) (Institute of State Government) Dehradun, Constituent College of Uttarakhand Technical University (State Government Technical University), Dehradun, India. She has more than 6 years' experience in academic. Her research interest in Cognitive Radio and Spectrum Sensing Techniques in WRAN.

Sara Kamel is an Assistant Lecturer at the department of Engineering Mathematics, Faculty of Engineering, Alexandria University, Egypt. She received the B.Sc. (Hons) degree in Communications and Electronics Engineering in 2010 and the M.Sc. degree in Engineering Mathematics in 2014 and is currently pursuing a Ph.D degree in Engineering Mathematics at the same university. She received the Young Scientist Award at the URSI General Assembly held in Beijing, China in 2014. She is a member of IEEE, IEEE Women in Engineering and IEEE Young Professionals.

Rashmi Kittali completed her M.Tech from Baseveshwar Engineering College, Bagalkot-587102, Karantaka, India. Her area of interest includes Wireless Networks, Satellite communication, Wireless communication, MEMS.

Sumit Kundu received his B.E. (Hons.) degree in Electronics and Communication Engineering in 1991 from NIT, Durgapur, India and M.Tech. degree in Telecommunication Systems Engineering and Ph.D. in Wireless Communication Engineering from IIT, Kharagpur, India, respectively. He has been a faculty in the Department of ECE, National Institute of Technology, Durgapur since 1995 and is currently a Full time Professor there. His research interests include Wireless Ad Hoc and sensor networks, Cognitive Radio Networks, Cooperative Communication, Energy harvesting and Physical layer security in wireless networks. As of today, he has published more than one hundred fifty (150) research papers in various journals and conferences. He is a senior member of IEEE (Communication Society) and is a reviewer of several IEEE and Elsevier journals.

Ashish Negi is an alumni of Madan Mohan Malviya Engineering College (Now Madan Mohan Malaviya University of Technology, Gorakhpur, http://www.mmmut.ac.in/), Gorakhpur, Banaras Hindu University (http://www.bhu.ac.in/), Varanasi and Gurukula Kangri University (http://gkv.ac.in/), Haridwar he has 24 years of Academics and Research experience in the field of Computer Applications. His research area includes Computer Graphics, Image Compression. He has been actively involved in the research related to Fractal Geometry and its Applications in Computer Science for the past 10 years and has presented several papers in various national and international conferences. He is expertise in various fields of Programming Languages, Operating Systems and Analysis and Design of Algorithms. Presently working as Professor & Head, Department of Computer Science & Applications G.B. Pant Engineering College, Pauri Garhwal Uttarakhand.

Ala Eldin Omer has received the B.Sc. (magna cum laude), from University of Khartoum, Sudan, in 2014, in Electrical and Electronics Engineering with specialization in Communication Engineering. He has achieved his M.Sc. (summa cum laude) in Electrical Engineering from the American University of Sharjah (AUS). In addition, since 2014, he has been with the Department of Electrical Engineering in AUS as a Teaching and Research Assistant. He has been awarded the prestigious Singapore International Pre-Graduate Award (SIPGA). His research focus is on pattern classification, communication systems, wireless networks, video streaming and cognitive radio networks.

Bhuvaneswari P. T. V. is currently working as Assistant Professor (Senior Grade) in the Department of Electronics Engineering, Madras Institute of Technology, Anna University, Chennai, Tamil Nadu. She completed her undergraduate degree in Engineering from Madras University, Chennai, Post Graduate degree in Applied Electronics from Bharathiyar University, Coimbatore and PhD from Anna University,

Chennai. She possess more than 17 years of experience in teaching and research. She has completed several projects and has more than 60 publications to her credit.

Andallah Rhattoy obtained a Ph.D. in 1997 from national high school of electrotechnics, electronics, computer science, hydraulics and telecommunications (ENSEEIHT) of national polytechnic institute (INP), Toulouse, France. He has been teaching for 19 years. He is currently a Professor and the Head of department at Higher School of technology (EST), Moulay Ismail University, Meknes, Morocco. He conducts research and teaches information systems and communication theory and applications, and multimedia communication signal processing.

Shikha Singhal is an aspiring teacher, and completed her M.TECH in Computer Science from Jaypee Institute of Information Technology, Noida. She is also assisting teachers with Teacher's assistantship program at JIIT, gaining on-ground experience with the students of B.Tech. She loves teaching and doing research work, with a special affinity towards wireless communication and Cognitive Science. She excelled her GATE Score and has done several projects practicing the core principles of computing. She is constantly looking for the opportunities of learning and contributing to the society.

Adwitiya Sinha has received Doctor of Philosophy from Jawaharlal Nehru University (JNU), New Delhi, India in 2015. She has obtained Master of Technology in Computer Science and Technology in 2010 from JNU, New Delhi, India. She has also completed Master and Bachelor of Computer Applications in 2008 and 2006, respectively. She is a recipient of Senior Research Fellowship from Council of Scientific & Industrial Research, New Delhi, India and also awarded with UGC Research Scholarship for her research in data analytics across social and wireless domains. Presently, she is working as an Assistant Professor in Jaypee Institute of Information Technology (JIIT), Noida, India. Also, she is serving as NAAC Coordinator for Centre for Performance Analysis of Computing Systems; and Member of Doctoral Program Monitoring and Advisory Committee in JIIT, Noida, India. She has delivered MOOC on Networks & Graphs organized by CEC, UGC, New Delhi, India. Her publications mostly includes articles as book chapters and papers in international journals & conferences. Her application-based research is mainly focused on performance analysis of energy efficient wireless sensor & adhoc networks, cognitive networks, data aggregation/prediction techniques & confluence of sensor-based applications with social networking (social sensing).

Bino Stephen is currently a research scholar in the Department of Electronics Engineering, Madras Institute of Technology, Anna University, Chennai, Tamil Nadu. He has completed his Under Graduate degree in Electronics and Communication Engineering and Post Graduate degree in Computer and Communication from Anna University, Chennai. His area of interest are Wireless Networks, Software Defined Radio and Spectrum Sensing in Cognitive Radio Networks.

Ashok Sutagundar is working as Associate professor, Department of Baseveshwar Engineering College, Bagalkot-587102, Karantaka, India, He is the research scholar working in the area of Information management in wireless sensor networks, His area of interest includes agent technology, Multimedia Networks, Wireless Networks, Image Processing, Image Processing.

Index

F

G

I

L

M

N

O

P

R

S

T

U

W

www.ingramcontent.com/pod-product-compliance
Lightning Source LLC
LaVergne TN
LVHW081251011026
849162LV00006BA/1407